IN SEARCH OF THE RAIN FOREST

**NEW ECOLOGIES FOR
THE TWENTY-FIRST CENTURY**

SERIES EDITORS: ARTURO ESCOBAR,
UNIVERSITY OF NORTH CAROLINA, CHAPEL HILL:
DIANNE ROCHELEAU, CLARK UNIVERSITY

IN SEARCH OF THE RAIN FOREST

EDITED BY CANDACE SLATER

DUKE UNIVERSITY PRESS   DURHAM AND LONDON 2003

© 2003 Duke University Press
All rights reserved
Printed in the United States of
America on acid-free paper ∞
Designed by Rebecca Giménez
Typeset in Adobe Minion
by Keystone Typesetting, Inc.
Library of Congress Cataloging-
in-Publication Data appear on the
last printed page of this book.

CONTENTS

ABOUT THE SERIES

There is widespread agreement about the existence of a generalized ecological crisis in today's world. There is also a growing realization that the existing disciplines are not well equipped to account for this crisis, let alone furnish workable solutions; a broad consensus exists on the need for new models of thought, including more constructive engagement among the natural, social, and humanistic perspectives. At the same time, the proliferation of social movements that articulate their knowledge claims in cultural and ecological terms has become an undeniable social fact. This series is situated at the intersection of these two trends. We seek to join critical conversations in academic fields about nature, globalization, and culture, with intellectual and political conversations in social movements and among other popular and expert groups about environment, place, and alternative socio-natural orders. Our objective is to construct bridges among these theoretical and political developments in the disciplines and in non-academic arenas and to create synergies for thinking anew about the real promise of emergent ecologies. We are interested in those works that enable us to envision instances of ecological viability as well as more lasting and just ways of being-in-place and being-in-networks with a diversity of humans and other living and nonliving beings and artifacts.

New Ecologies for the Twenty-First Century aims at promoting a dialogue among those engaged in transforming our understanding and practice of the relation between nature and culture. This includes revisiting new fields (such as environmental history, historical ecology, ecological economics, or political ecology), tendencies (such as the application of theories of complexity to rethinking a range of questions, from evolution to ecosystems), and epistemological concerns (e.g., constructivists' sensitivity toward scientific analyses and scientists' openness to considering the immersion of material life in meaning-giving practices). We find this situation hopeful for a real dialogue among the natural, social, and human sciences. Similarly, the knowledge produced by social movements in their struggles is becoming essential for envisioning sustainability and conservation. We hope that these trends will become a point of convergence for forward-looking theory, policy, and practical action. We seek to provide a forum for authors and readers to widen—and perhaps reconstitute—the fields of theoretical inquiry, professional practice, and social struggles that characterize the environmental arena at present.

ACKNOWLEDGMENTS

The members of the Rain Forest Seminar are grateful to the University of California's nine-campus Humanities Research Institute (HRI) at Irvine for funding six months of collaborative research on rain forest images and their practical effects. We thank HRI directors Patricia O'Brien, Steven Mailloux, and David Theo Goldberg for their help before, during, and after our residence at the institute during spring 2000. We are similarly indebted to the HRI staff as well as our graduate research assistants Lyman Hong and Jacqueline Scoones for their energy and patience.

Though brief, our journey to the not-quite-rain forest of the Yucatán played a decisive role in shaping both this book and the larger seminar experience that underlies it. For this reason, we are particularly grateful to the colleagues who helped us in our preparations. We thank Professors Jaime Rodríguez, Rob Patch, and Steven Topik for sharing their expertise with us. We are deeply grateful to Mike Baker and Teresa Martelon for opening their impressive Yucatán vacation home to us. We were equally fortunate to enjoy the friendship and professional guidance of Gonzalo Merediz and Marcos Lazcano Barrera of the Sian Ka'an Biosphere Reserve. We also thank Cancún environmental activist Araceli Domínguez along with our guides at the Punta Laguna spider monkey reserve (Ricardo and Eulogio Canul), Cancún archaeological museum (Guillermo Ahuja O.), and Caste War Museum at Tihosuco for greatly enriching our field experience.

The visitors who helped us think through the essays that began emerging once we returned to Irvine also deserve special mention. Francis E. Putz and Claudia Romero, conservation biologists committed to biological preservation and social justice, allowed us to revisit our own assumptions from the vantage point of the natural sciences. As a science correspondent for National Public Radio, David Baron was able to provide a different, but equally valuable perspective on our aims and writing methods. Jenny Price helped us to begin thinking about writerly issues early on in this book's conception, and we thank her for her ongoing assistance to us as individuals as well as a group.

We owe much to the colleagues who filled in for us on our home campuses during the time we were at Irvine. Our debts to our families are even greater, and we thank them warmly for their patience, good humor, and support. We also extend special thanks to the people in the forest areas where we have done extended research. Each and every chapter in this book reflects their generous help.

This book's location within Duke's New Ecologies for the Twenty-First Century series is a source of satisfaction. We thank the series editors, Arturo Escobar and Dianne Rocheleau, as well as our initial manuscript reviewers for their many useful suggestions. We also thank our editors, Valerie Millholland and Pam Morrison, for their help throughout the publication process.

Finally, we are grateful to each other for a deeply exciting exchange of ideas and experiences—an academic ideal far too seldom realized in practice. Scott Fedick deserves special thanks for making possible our trip to the Yucatán, but each of the seminar participants made unique intellectual and personal contributions to the collective effort. "Your group gets on so well!" people often told us. "You really seem to enjoy each other's company." We did enjoy each other, and we learned much from our time together. Our hope is that the reader will find similar profit and excitement in this book.

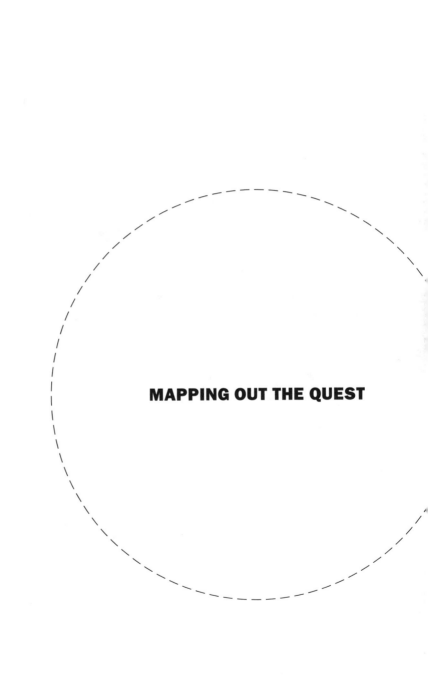

MAPPING OUT THE QUEST

CANDACE SLATER

IN SEARCH OF THE RAIN FOREST

"Save the Rain Forest!" Who hasn't seen these words emblazoned on a glossy poster of tall trees studded with monkeys and parrots and wreathed with majestic vines? The emerald allure of such calls to action is hard to resist. And yet, at the same time that these sorts of powerful slogans have fueled laudable rescue attempts, they have often done as much harm as good. A genuine source of inspiration, the idea of a single fragile forest has nonetheless tended to conceal the full biological variety of rain forests as well as the diversity of their human inhabitants. These same sorts of universalizing slogans have also obscured the specific local, global, and intranational interactions that are as much a part of today's rain forests as are their dazzling wealth of plants and animals.

The following pages lay out the search for a Rain Forest that is both a natural entity and a social history, an inhabited place and a shifting set of ideas (the capital letters indicate this overarching entity).[1] Our goal is to contribute to the growing field of environmental discourse studies—how nature gets talked about by whom and to what ends—by focusing on particular portrayals of rain forests and their consequences for forest inhabitants as well as outsiders.[2] *In Search of the Rain Forest* underscores the rich and varied—if often contentious—human presence that includes not just the easy-to-romanticize native populations who have shaped the land over millennia (solitary Indians, doe-eyed hunter-gatherers) but also all of the varied groups—miners, missionaries, tour operators, agro-industrialists, and small-time farmers—who live and work in forest areas today.[3]

While our humanities and social science backgrounds led the seven authors of this book to stress the crucial role of rain forest representations in such apparently straightforward problems as biopiracy and the defense of endangered species, our firsthand work in separate forests has made us well aware of their distinct, material features. The notion that these forests could be totally imagined spaces strikes us as no more productive than the idea that "real" forests could be wholly divorced from human conceptions of what such forests ought to be.

In one sense, our quest for the rain forest is nothing new. Humid, canopied forests rich in diverse life-forms have existed for millennia, and outsiders' concern for and fascination with a tropical nature variously described as "jungle," "wilderness," or simply "forest" goes back for centuries.[4] However, today's "rain forest" has different connotations from those surrounding forests of the past. A translation of the German word *Regenwald,* "rain forest" began to find its way into botanical treatises at the end of the nineteenth century. It then entered popular usage with the emergence of the global environmental movement in the late 1960s and early 1970s.[5] Bountiful, biodiverse, and yet supremely fragile, this rain forest of the popular imagination is light-years from the hostile jungle of news reports and adventure movies with which it continues to coexist.[6] Increasingly tropical, it is at once antithesis and direct extension of the temperate zone that is so concerned with its preservation.[7]

To the extent that our title echoes a long line of travel adventure books and romances (*In Search of the Lost Lagoon, In Search of Captain Zero, In Search of Dot.Calm,* and so forth), it is decidedly tongue-in-cheek.[8] Today, even schoolchildren can locate the green expanse of the Amazon, Pacific Northwest, or equatorial Africa on a map of the world. And yet, even though no one has to look too hard for present-day rain forests, the conceptual roots along with the symbolic and political dimensions of the threatened forest that inspires impassioned calls to action are far harder to pinpoint. For this reason, they demand a new sort of pursuit.[9]

In focusing on the role and suggesting the practical effects of specific rain forest depictions in these interactions, the book raises a number of essential questions. When, for instance, we join in the cry to "Save the Rain Forest," what exactly are we trying to save? Why do we want to save it? From whom? And for whom? Because overly simplistic images mask the rich variety of real rain forests and the competing interests inevitably at play within them, our quest for answers to these fundamental questions begins in every case with these portrayals.

MAPPING OUT THE QUEST

The author of *In Search of the Rain Forest* are seven academics from six different U.S. institutions who came together at the University of California's Humanities Research Institute in Irvine during the first half of 2000 for six months of collaborative writing and reflection on rain forest images and their consequences.[10] We represent a half-dozen different disciplines: literature (Candace Slater), cultural anthropology (Suzana Sawyer and Alex Greene), archaeology (Scott Fedick), history (Paul Greenough), sociology (Nancy Lee Peluso), and law (Charles Zerner).[11]

Conservation biologists Jack Putz and Claudia Romero joined us for an intensive three-week discussion of present perspectives on rain forests within the natural sciences. Because conservation biology and systematics, landscape ecology, ethnobiology, and biogeography still dominate the knowledge and policy field when it comes to rain forests, we were particularly interested in their views on current perceptions of rain forests and scientists' shifting ideas about their own role as researchers within specific forest settings.[12]

David Baron, National Public Radio science correspondent, also spent a week with us in extended conversations about how best to present our ideas to a public made up of both specialists and more general readers including graduate and undergraduate students.[13] We had already made the decision to make the essays as accessible as possible, but David pushed us harder. His insistence that we could draw in more general readers without sacrificing scholarly rigor encouraged us to try to write in unaccustomed ways.

Although all seven of us do fieldwork in tropical forest areas, our research sites range from the Amazon to the Yucatán Peninsula to India and Borneo.

Because each of us came to Irvine already engaged in a separate writing project, our immediate subject matter—which includes tiger reserves in India, virus-producing jungles in Africa, hybrid healing traditions in Belize, and the figure of the Borneo headhunter—is notably diverse.

These real differences in interests and methods made our conversations an ongoing challenge. The struggle to find a common language with which to address the mingled practical, material, and symbolic dimensions of rain forests not only permitted but regularly forced us to question assumptions that we had come to take for granted. How could one explain the diverse trajectories of the word "jungle" in Africa, India, and the Americas? Why were depictions of forest peoples different in Borneo and Brazil? Likewise, we did not always agree on whether the iconic forest hurt or harmed rain forest peoples, in part because different rain forest peoples had been affected by, and had themselves employed, these icons in diverse ways. We also had different ideas about the present status of particular rain forest representations as well as their relation to specific forest settings.

At the same time, our colleagues' firsthand knowledge of different rain forests helped make these places immediate to us in ways that they had not been before. Recordings of Dayak storytellers in Borneo or snapshots of Maya ruins amid tall trees in Belize and Mexico made those of us who worked in other places reflect in new ways on forests we had known exclusively through books. Similarly, one person's familiarity with, say, the present plight of tigers in India would push us to rethink the place of jaguars or alligators in the Amazon—or cougars in a distinctly nontropical American West.

A series of informal exchanges extended and enriched the more formal conversations that took place around the seminar table at HRI all day every Tuesday. Our visits to state parks, botanical gardens, movie theaters, and the local Rain Forest Café also served as preparation for a longer exploratory mission. Instead of bringing speakers to Irvine with the research monies available to us, the seven of us decided early on to visit and interrogate an actual tropical forest.

We had no illusions about the limited amount we could learn on a short trip to an unfamiliar place. Nonetheless, we hoped that the encounter with a forest both like and unlike our own research settings would help us better see the latter's particularities. We also thought that the shared experience of being specialists momentarily thrust into the role of (admittedly self-conscious) tourists would give us much to reflect on.[14] Our forest of choice turned out to be the central Yucatán Peninsula—an area that while clearly not dense-canopied rain forest, is increasingly billed as such by both the international tourist industry and a number of environmental groups.[15] As the site of an intense spatial, political, and cultural reconfiguration brought on by tourism, ecotourism, and wide-scale development, the Yucatán struck us as a prime example of a twenty-first-century ecology.

Only archaeologist Scott Fedick focuses his essay on the Yucatán, where he has been working for ten years. The rest of our pieces make only passing reference to the trip. However, as we had hoped, the months of planning for the intensive ten days that we spent together in Mexico, and our many subsequent discussions of the experience, had an impact on our thinking about our own areas of expertise. By making vividly apparent the larger questions that had initially brought each of us to Irvine, the trip furnished a number of the examples that appear in this introduction and provided our individual chapters with a unifying thread. Though our voices and approaches differ, we all speak of the uses and transformations of key symbols by particular groups. For this reason, *In Search of the Rain Forest* is both a series of essays on diverse forests and the record of a hard-won, genuinely collective quest.

ICONS AND SPECTACLES

Among the themes that reappear throughout this book is that of the rain forest as a series of "icons" and "spectacles." By no means original to us, these terms nonetheless acquired new and often quite particular meanings in our discussions of different rain forests. Above all, they helped us to describe and compare larger struggles for control over forest peoples and resources.[16]

Both terms with long histories reflecting different scholarly traditions, "icon" and "spectacle" have multiple, sometimes competing meanings.[17] The word "icon" has had a special resonance for art historians, scholars of linguistics, and communications theorists.[18] It is especially important to semioticians, who continue to look to writers such as Roland Barthes and Umberto Eco for cues.[19] Philosophers in particular have been quick to affirm icons' primacy or "firstness" in the communication of otherwise inexpressible ideas.[20]

Logicians' concern for the relationship between readily identifiable icons and the more abstract conceptions that they embody contrasts with social theorists' interest in icons as key elements in the struggle for political power. Not infrequently, these theorists have treated iconic entities as embodiments of the contradictions of contemporary consumer societies.[21]

Some theorists see iconic entities as fueling spectacles that function as pseudoworlds or negative inversions of lived experience (the impassioned celebration of untrammeled nature by apartment dwellers, for instance).[22] Others speak of the "image events" (spray painting baby seals to destroy their fur's commercial value) that environmentalist groups use to sway public opinion.[23] A growing number of political ecologists see battles for material resources as symbolic tugs-of-war in which one side's ability to appear as nature's true defender is often as vital as access to physical force.[24]

"Icons" in the sense in which we use this term are, above all, vivid simplifications that stand in for a far more complex set of places and people, and the ways they interact. These icons inevitably have unspoken implications or "iconic shadows." (If green people are the best in the world, then orange, gray, or purple people are necessarily inferior.)

"Spectacles," in turn, are icons in motion. Multilayered performances that surround attempts to create and control particular representations of nature, spectacles employ simultaneously enacted, often competing narratives that we call "bioscripts." While the meanings of all these terms will become clearer as the book progresses, their usefulness in approaching a wide range of rain forest situations makes it appropriate to introduce them here.

RAIN FOREST ICONS

Although iconic simplification can occur in many different contexts, the icons that concern us here are all stylizations of particular aspects of tropical nature, including forest peoples. The jade green forest full of monkeys and parrots that crops up on calendars, ice cream cartons, and the glossy covers of reports on corporate rain forest operations is one of the most common. Among the various other familiar depictions that appear in the following pages are the *virgin forest* that demands protection, the foreboding *jungle forest* that is home to wild beasts and dread diseases, the *wild forest* that alternately shrinks from or actively embraces violent humans, and the *forest library* rich in both commodities and encoded knowledge.

Simplifications such as the jade green forest resemble computer icons in their concrete quality (such as the hard-to-mistake postal carrier symbol for e-mail). Even though the forest suggests a whole range of larger and more abstract ideas—the purity of nonhuman nature or moral ambiguity of nontemperate places and peoples, for instance—it gives the appearance of something immediate and real. Deeply grounded in particular places with unique histories, these ideas of purity and moral ambiguity find expression in readily recognizable, easy-to-replicate images of exotic animals and lush vegetation. This grounded quality sets apart the icons and ideas we discuss here from more diffuse concepts such as truth, knowledge, or even biodiversity.[25]

Iconic simplifications also resemble computer icons in their ability to create a conceptual link among potentially disparate entities by representing them as members of a common domain ("My Documents" or "Business Letters"). Much as the stylized file folder provides access to electronic documents that are apt to be diverse in length, style, and content, so the iconic forest serves as a convenient umbrella for tropical expanses that under scrutiny, reveal a myriad of dissimilarities. In both cases, the icon permits entry to a larger universe, which it also helps define.

And yet, if the iconic simplifications of which we speak resemble the shortcuts on today's computers, they also include something of the force of a gilt-encrusted religious icon. The rain forest is not just an abbreviation; it is a concentrated form of a nature in which very different peoples have long perceived religious meanings and which often acquires the force of the sacred in an increasingly secular world.[26]

Every icon presupposes the existence of an iconic shadow—the logical extension or necessary flip side of ideas and qualities on which the icon dwells.[27] I have already offered the example of how proclamations of one people's superiority contains within it an implicit judgment of their peers. In much the same way, the iconic shadow of the idyllic forest crammed with exotic plants and animals is the all-too-familiar urban jungle, whose polluted character presents a striking contrast to the purity of nonhuman nature. Conversely, the logical extension of the dark, wild, and disease-ridden forest is the light-filled world of civilization that it threatens to eclipse. While the iconic shadow of the ecologically harmonious past is an environmentally destructive present, the bright future promised by development conjures up a far dimmer past.

In contrast to icons, whose outlines are clear and easily articulated, iconic shadows may remain largely invisible or unspoken. While icons mobilize and inspire through their direct emotional appeal, iconic shadows are more apt to function through insinuation and halftones. The icon of the cathedral-like rain forest that demands rescue from desecration is easy to identify and expand on. However, the logical corollary that the world outside the forest is less sacred, and therefore far less deserving of salvation, is rarely spelled out. Precisely because they do not encourage direct scrutiny or frank discussion, these conceptual shadows may have a power that extends beyond the more immediately vivid icons they accompany.

RAIN FOREST SPECTACLES

As shorthand for larger entities and ideas, icons are potential calls to action. Spectacles, in turn, are the concrete ways in which these calls are acted out. As such, they encompass all of the varied processes through which different groups and individuals create, transform, and infuse particular images with competing meanings. If, for instance, the threatened, biodiverse rain forest is the icon, then the fight to preserve, transform, or destroy it is the spectacle.

A noun derived from the Latin *spectare* meaning "to look or regard," spectacle emphasizes the public, performative character of these processes. As in theatrical productions, extralocal spectacles are always staged by particular actors for a designated audience. Unlike the play whose aim may be little more than lighthearted entertainment, however, the dramas we describe have serious, often directly material effects. The staging of a forest as a chaotic jungle inhabited by fearsome people, for instance, is all but certain to spur official attempts to control and transform both.

The spectacles that appear in the following pages are inevitably "extralocal"—a term that includes, even while it transcends, the near at hand. The struggle to preserve rain forest flora and fauna, for instance, involves a series of simultaneous dramas in which local, regional, national, and transnational elements regularly collide and intertwine. While local populations may be interested above all in preserving access to lands on which they have long farmed, gathered nuts, or extracted rubber, their allies on the national and international levels are likely to paint them as innate preservationists.[28] Other players on the regional or national level may decry these same populations' inefficient use of lands that multinational corporations promise to render more productive—meaning more profitable to themselves and their investors.

As this example suggests, even while spectacles reveal a high degree of improvisation, their stagers regularly fall back on formulaic, predefined roles (victim, villain, developer, defender, and so on). The actions that spectacles encompass are inevitably elicited and then condemned or justified through rhetorical strategies that we call bioscripts.

Not all of these bioscripts employ the same vocabulary or rhetoric. All, however, make reference to nature or environmental issues. While some, for instance, focus on wilderness as a primary value, others exploit the tension between wild animals and violent people. Still other bioscripts revolve around the possibilities or limits of local healing knowledge, or the need to protect a fragile, virgin nature that may appear as a capricious femme fatale in other times and places.

Different groups and individuals regularly rework what may look like the same bioscript for their own purposes.[29] A multinational corporation is apt to stress the preservationist dimensions of its "green" technologies when seeking access to mineral-rich lands. Yet a native group demanding collective title to this land may place a similar emphasis on preservation in its court appeals.

Seemingly identical scripts often take on different emphases and new inflections in different circumstances. For instance, a company's presentations to international environmental groups are apt to highlight quite different aspects of a project than are its presentations to shareholders. Even though both may dwell on the theme of corporate stewardship of nature, these differences in emphasis alter what is perceived and said. Moreover, even while the presentations utilize a common vocabulary of environmental preservation, their real concern is less the environment than the effective marketing of a product and a particular image of the corporation.

Somewhat abstract in theory, icon and spectacle are easier to define in context. In the two examples from our trip to the Yucatán that follow, it is quite clear who is using which icons for which audience and toward what ends. Although the mega–theme park of Xcaret and much smaller, far less carefully packaged spider monkey reserve of Punta Laguna both dwell on an idyllic, primal nature, the motives for this focus and the ways in which their representatives ultimately define it are by no means the same.[30]

Often described as "the Mayan Disneyland," Xcaret (pronounced "Shkah RET") is by far the single most popular day trip undertaken by the Mexican and international tourists who flock to Cancún's long strip of high-rise, largely high-end resort hotels. Of the approximately 700,000 tourists who visit this lush 250-acre "eco-archaeological park" every year, a full 90 percent are from outside Mexico.[31] Early every morning, a fleet of plush, air-conditioned buses zooms down the approximately thirty miles of super-highway that link Xcaret to Cancún. At the end of the day, the buses redeliver passengers to their accommodations in the city or at the nearby Playa del Carmen resort.

The ostensibly primal paradise that is the buses' destination was constructed in the early 1990s by a group of Mexican developers with close family ties to the governor of the state of nearby Tabasco.[32] Human use of the site is nonetheless much older: archaeological evidence suggests that Xcaret was almost certainly founded in the Mayan Early Classic period (A.D. 200–600). An important trading center and port for the island of Cozumel during Late Postclassic times (A.D. 1250 until European contact in 1520), the verdant expanse was once one of a number of sprawling and densely settled urbanized communities stretching along the east coast of Quintana Roo.[33]

The name Xcaret means "little inlet" in Maya, and the theme park's natural freshwater wells or cenotes likely served as purificatory baths for travelers headed for the shrine to the goddess of fertility, Ixchel, on Cozumel. Xcaret, however, expands on the theme of regenerative moisture to include an artificial lagoon, pools, and fountains that underscore the idea of an oasis in a land with little surface water.

The park's largely constructed character does not keep its promoters from describing it over and over again as "Nature's Sacred Paradise" and "a natural park carved in the lowland jungle."[34] This paradise finds visual expression in the lush rain forest that appears on any number of promotional posters and pamphlets distributed both directly by the park and by various hotels and tourist agencies. The same marvelous collection of flora and fauna also crops up in descriptions of the park on the Internet. A toucan nestles in the "c" of "Xcaret," and a flower springs up between the "e" and "t" in one ad. On another Internet site, three butterflies flit beguilingly across the screen. "Paradise is just a click away!" the text for the ad exclaims.[35]

A day in Eden does not come cheap, however. Adult admission to the park (dollars, pesos, and major credit cards all gladly accepted) is a hefty $40. In turn, "Xuberant, Xciting, Xotic" Xcaret does its best to give visitors something for their hard-earned cash. Its prime attractions are all iconic representations of a mysterious, exotic, and sometimes tantalizingly dangerous (if always carefully controlled) nature. They include a bat cave, butterfly pavilion, a man-made island ringed by an artificial moat from which real alligators stare up at real jaguars, a coral reef, an orchid nursery, a mushroom garden, a turtle farm, and a dolphinarium. "These experiences," explains the text on one website, "allow visitors to take home with them a new sensation for nature that gives meaning to the real values of the world we belong to."[36]

Although the gem-bright forest crammed with exotic plants and animals remains the park's primary logo, Xcaret is also "the place where you'll discover the mysterious and marvelous past of the Maya close to the modern Cancún of today."[37] An eco-park that celebrates a "nature-loving" civilization, its secondary icon is thus an ancient people whose environmental sensitivities happen to coincide almost exactly with those of twenty-first-century tourists.

Xcaret is indeed home to several dozen bona fide Maya ruins.[38] However, the two structures that appear time and again in ads for the park are from the 1990s. One of these is the "traditional Maya village" whose rustic houses rise above a newly drilled cenote ringed by baskets of polyester squash and rubber tomatoes. The other is a carefully built replica of an ancient Maya ball court with two sloping walls of limestone bricks and a sideways, doughnutlike, vertical stone "basket" jutting out from either side.

The ancient past finds a complement in a third icon: the more contemporary regional dances performed by the Mexican Folklore Ballet as part of a regularly scheduled evening entertainment program called *Xcaret at Night*.[39] The finale of this program involves the release of white doves over the darkened arena. The image of these birds poised over a group of brightly costumed dancers on tourist brochures suggests the close relationship between nature and a regional and national folklore that an international public can easily comprehend and enjoy.

All three of the icons that serve as shorthand for the park—the abundant forest, harmonious past, and folkloric present—have distinct iconic shadows. The iconic shadow of the primal rain forest is the booming "Maya Riviera" whose creators also financed the construction of Xcaret. ("I didn't realize there'd be quite so many big hotels in Cancún," one young French woman photographing jaguars confides with a sigh. "The ocean is beautiful, but I really like being here in the rain forest far away from all those tourists on the beach.")[40]

The iconic shadow of the traditional village is the often considerably less picturesque urban slums where many contemporary Maya live. Although the park makes a point of selling contemporary "native crafts" in its gift store, it is entirely possible for visitors to come away from Xcaret with the impression that the Maya existed only in the distant past.[41] ("Too bad all those people went extinct, Mom!" one teenage tourist exclaims sadly.) The Maya are, of course, decidedly present in the form of low-paid employees who tend the quaint, small houses that purport to mirror their homes.[42]

Likewise, the iconic shadow of the regional folklore that finds expression in the dove-encircled dancers is an indigenous popular culture that resists easy definition and absorption. The wide array of dances significantly does not include a single number from the Yucatán—a part of Mexico where bitter memories of the nineteenth-century Caste Wars tend to live on among the poorer, heavily Maya segments of the population.[43]

The many-layered spectacle of Xcaret posits connections between individual visitors and a larger, comfortably universal nature, which a series of hands-on activities allow visitors to apprehend. For an added fee, tourists can scuba dive, snorkel around the coral reef, take a horseback ride along the ocean front, or swim with one of a troop of resident dolphins.[44] They are also free to sniff the over eighty varieties of indigenous trees and plants in the botanical garden, run a cautious finger across the mottled surface of a manta ray or starfish, and get close (though not too close) to the spider monkeys, jaguars, and assorted birds. "In Xcaret," effuses one public relations writer, "one can watch, in real life, how a quail is born; how a caterpillar turns into a butterfly; explore the fossilized corals while swimming in the underground river; or simply stroll through the exuberant vegetation native to the area."[45]

The ball game staged at the end of every afternoon provides a similar sort of face-to-face encounter with the Maya past. Following the gyrations of a scantily clad female dancer, two groups of male players arrayed in body paint and feathers file out onto the court. Although the unfamiliar sight of players bouncing a hard leather ball off their hips and eventually up into a stone ring intensifies the air of exoticism, the game's competitive intensity makes it seem less alien. Even while the players' movements, like the accompanying music, remain strange and unfamiliar, their skill and palpable excitement gives the performance—and with it, the ancient past—a decided air of the here and now.

And yet, even while the ball game temporarily becomes a part of the present, the extraordinary character of the spectacle remains clear. Although the rupture between past and present is bridged anew every afternoon within the park's leafy confines, the momentary character of this fusion ultimately reaffirms the depth of the divide. Without "the magic of Xcaret," the ball game would remain part of a past that rarely obtrudes on the modern world.

In much the same way that the ball game temporarily links the past with the present, so the nightly "folklore program" brings together nature and the nation in another sort of symbolic performance. A ceremony in which each visitor is directed to light a small white candle in honor of the day that he or she has just spent in the presence of a supposedly sacred nature prefaces the dances. Once the candles have been blown out, women in long skirts and mantillas reminiscent of Mexico's colonial past swish the air with fans as men in white suits stamp and strut across the stage. Horses prance and snort as mariachi musicians fill the air with joyous sound and yet more dancers crowd the stage. The release of a dozen chalk-white doves trailing white confetti transforms the finale into a sort of secular benediction.

An explicit concluding declaration of nature's restorative and unifying powers reminds one that in Xcaret, "Earth, Life and modern technology fuse in an environmental frame with tremendous natural and cultural wealth."[46] The sea of light created as one visitor passes on the flame to his or her neighbor becomes a graphic illustration of nature's power to transcend individual and national differences. Likewise, the ease with which the cries of "¡Viva Mexico!" that greet the doves' release are taken up by many non-Spanish-speaking members of the public suggests the ability of a particular sort of popular culture to transcend the nation it all too easily defines. The doves themselves, with their strong religious connotations (Noah's dove, the Christian Holy Spirit), are clearly meant to reaffirm the sacredness of nature.[47]

ICON AND SPECTACLE IN PUNTA LAGUNA

Located directly inland from Xcaret in a swatch of high tropical forest on the border of the Mexican states of Yucatán and Quintana Roo not far from the imposing Maya ruins at Cobá, the Punta Laguna (literally "Lagoon Point") spider monkey reserve initially appears to share little with "Nature's Sacred Paradise." No multimillion-dollar tourist destination, the reserve is instead a small Maya community or *ejido* that small, if growing numbers of eco-visitors seek out. In contrast to Xcaret's heavily reinvented landscape, Punta Laguna, with its Maya ruins scattered amid the tall trees through which the monkeys scamper, bears witness to the ongoing interactions between humans and nonhuman nature.[48] However, Punta Laguna's growing identification as a rain forest (the reserve is "the closest tropical rain forest habitat for anyone living in the continental United States and Canada," declares one writer) invites a series of comparisons and contrasts between it and the giant eco-park.[49]

Arriving in Punta Laguna is not at all like arriving in Xcaret. The bumpy road that leads to the reserve bears scant resemblance to the Xcaret super-highway, and the clearing in the woods that serves as a makeshift parking lot could never handle a fleet of tour buses. A modest thatched-roof shelter in the Maya style takes the place of Xcaret's large and luxurious visitors' center. Here, a trail down which visitors shuffle single file behind a local guide in blue jeans replaces the smooth paths tended by uniformed employees.

Although some of the Maya ruins along the trail date back a thousand years, itinerant chicle gatherers drawn to the high, dense forest established the present-day community of Punta Laguna roughly four decades ago.[50] The hundred or so residents depend for a living on maize cultivation and forest products (fruit, honey, medicinal plants, and lumber for house construction).[51] Presently seeking recognition as an official eco-reserve from the

Mexican government, Punta Laguna has received small, but critically important amounts of financial support and technical assistance over the last decade. The donors have been not big developers but rather various national and international environmental organizations, including Pronatura de Yucatán, the Mesoamerica Foundation, and Community Conservation.[52] These organizations have helped community members to design the small-scale ecotours that now provide an additional source of income.

Like Xcaret, Punta Laguna brings these visitors face-to-face with something that they (though not necessarily the residents) find easy to identify as a harmonious nature. In contrast to the dense swatch of greenery that serves as shorthand for Xcaret, however, the primary icon for Punta Laguna is the spider monkey.[53] Although these intelligent, large-bodied, but extremely agile animals represent only one of several hundred species of birds, turtles, and other wild animals (including howler monkeys) found within the forest, they have become all but synonymous with Punta Laguna over the last ten years.[54]

According to Ricardo, the teenage guide who led our group through the forest the day after our visit to Xcaret, the close association between Punta Laguna and the spider monkeys dates back to the present community's founder—Ricardo's grandfather.[55] On discovering that the adults were being killed or carried off by hunters, Don Nacho Canul sought to protect their offspring. The ensuing close association between Punta Laguna and the spider monkeys also has much to do with the presence of foreign and Mexican biologists who began to study its monkey population in the early 1990s.

This concentration on the spider monkey is one of the factors that makes the eco-spectacle unfolding at Punta Laguna different from the process played out at Xcaret. While the giant theme park loses no opportunity to assert the necessity of environmental preservation, the exuberant nature within its borders makes it appear to be a refuge in which the visitor can momentarily forget about the environmental problems that abound outside its gates.[56]

In contrast, nature's vulnerability is a large part of the spectacle of appreciative discovery in which Punta Laguna invites visitors to participate. The spider monkey's status as both an indicator species (one found only in areas containing a high degree of biodiversity) and umbrella species (one whose protection implies the protection of this same biodiversity) makes it a convenient symbol for all of threatened nature. The habitat destruction that has caused a precipitous drop in the spider monkey population in other parts of the Yucatán makes Punta Laguna an oasis in a rising ocean of destruction that the guides here, unlike those in Xcaret, make no attempt to downplay.[57] To the contrary, ongoing threats to the forest and its creatures are a recurring theme in a way that they could not be in a faux rain forest where the aim is to make people feel good.

What does get swept under the forest rug at Punta Laguna, at least initially, are more immediate forms of social conflict and political dispute. Thus, at the beginning of our tour, Ricardo kept the conversation tightly focused on the monkeys who clambered above us in the trees. "That one you see there is Monica," he announced as we squinted up into the thick and lacy branches. "Pancho is just above her, and that's Benito high up there in the branches of that ceiba to the right. You don't have to worry about frightening them. No one here has ever hurt them, and so, they don't run away from people. When they get thirsty, they come to drink water out of the plants here on the forest floor."[58]

Although the charismatic monkeys and tall trees piqued our interest, there was clearly more to Punta Laguna than the tranquil scene before our eyes. The iconic shadow of an idyllic nature became quickly apparent when we asked Ricardo who had toppled one particularly tall tree still sprawled out across the mossy floor. "Those other ejidos used to want to take our lumber and now they want to claim our monkeys for themselves," Ricardo told us with a frown. "But the Mexicans won't let them do this because those are our monkeys and our trees."[59]

In short, at the same time that the eco-guides of Punta Laguna work hard to encourage a Xcaret-like sense of a nature that predates humans, they clearly see the forest as communal property to which they have a right. For Ricardo and his family, the real spectacle of present-day Punta Laguna lies in the transformation of a nature on which they have always depended for survival into a new form of livelihood. Previously a supplier of products destined for local consumption (herbs, nuts, fruits, and game) as well as regional export (chicle and hardwoods), this forest has become a source of a far less tangible harmony with nature that attracts a growing national and international clientele.[60]

Much as the jaguars and brightly colored birds who were once forest fixtures have become charming hand-carved souvenirs available for purchase, so the monkeys who were previously just one more part of daily life have become the object of a modest international pilgrimage.[61] Moreover, in the sense that most visitors come to the reserve not just to see the animals but to observe in hushed fascination a "traditional" people at harmony with their surroundings, the Maya themselves have become living commodities.

However, even while the forest has become a new sort of living icon, it has retained other meanings for the community. Although to elaborate all of these meanings would require a detailed ethnography, some are immediately apparent. Ricardo's uncle, for instance, is a *h'men* or shamanic priest and healer who is frequently called on to perform cures and rites such as the rain ceremony in which the children chirp like frogs in order to bring on a downpour. The rock on which a previous group of tourists had left a pale pink seashell is a shrine on which he and his ancestors have long left their own, quite different offerings.

The bioscripts in evidence at Xcaret and Punta Laguna bring home the differences in their use of Sacred Nature. While the first script is steeped in the vocabulary of international ecotourism, the second reflects a play between several languages: that dialectic of English one might label "Environmentalese" or "Eco-speak," the Spanish community members learn in school, and the Yucatec Maya most still speak among themselves.

The growing significance of outsiders' speech is obvious in the community's adoption of the names bestowed on the spider monkeys a few years ago by a visiting University of Pennsylvania graduate student in biology. These names—Monica (with its echoes of Monica Lewinsky), Pancho (the Mexican rebel leader Pancho Villa), and Benito (Benito Juárez)—signal larger changes in the way that the residents of Punta Laguna talk about their world among themselves and to outsiders.[62]

Then too, while Xcaret employs the language of communion into a new sort of global society that bills itself as different from the past, Punta Laguna favors the language of initiation of outsiders into an enduring harmony. The universal and transcendent Nature that Xcaret celebrates is ultimately different from the intensely local nature that visitors encounter in Punta Laguna. Since the visitor who embarked on a solitary trek through the forest would be unlikely to catch the slightest glimpse of the spider monkeys, this nature demands translation by local interpreters. Only native guides can render the concealed apparent; only their willingness to share the forest's secrets can bring the monkeys into view. As a result, while Xcaret seeks at every turn to invoke and capitalize on a past markedly different from which it purports to offer respite, Punta Laguna emphasizes the seamless connections between the past and a present in which the Maya continue to live in harmony with their surroundings.

Accordingly, if Xcaret relies on language that denies or effaces the local in favor of an easily intelligible global language ("¡Viva Mexico! ¡Viva la naturaleza!"), Punta Laguna presents local stewardship as the key to preserving a nature that outsiders want and need for their own survival. In so doing, the residents of the ejido manage to redefine what these same outsiders might, in another context, consider to be backward, alien, and potentially threatening into something that is appealingly traditional to outsiders (Maya language, Maya rain gods, Maya-style clothing).

Through a trilingual spectacle in which they emerge as long-standing guardians of a unique planetary heritage, in short, the residents of Punta Laguna use the icon of sacred nature to carve out a place for themselves and their children in a globalizing world where cultural difference may be either a sign of backwardness or a guarantee of environmental wisdom. In so doing, they offer a striking contrast to Xcaret, in which the same iconic nature is both big business and a universal good.

THE RAIN FOREST AT THE NEW MILLENNIUM:
A PRELIMINARY GUIDE

New groups of people with differing aims regularly transform the bioscripts that they have inherited or that they encounter. At the same time that Punta Laguna and Xcaret underscore the strategic uses of the rain forest as "sacred nature," they also highlight a number of features common in portrayals of tropical forests at the beginning of the twenty-first century.[63] Often intensifications or reworkings of much older elements, these features are apt to find a place in novel constellations that recall previous depictions even while they stand apart from them. Although the variety and fluidity of rain forest imagery make any set list impossible, the examples of Xcaret and Punta

Laguna suggest a handful of larger themes that reappear in a number of the chapters in this book.

Perhaps the most salient characteristic of the "new" rain forest is the increasing lack of a single unified portrayal capable of persuading very different publics to participate in a common rescue operation. In the early 1970s, the perceived fragility of tropical nature prompted urgent calls for coordinated international efforts to avert a planetary crisis. The rain forest quickly became a verdant poster child as well as an enormous green umbrella under which a host of diverse groups found common cause. In contrast, the conceptually as well as physically fragmented rain forest at the beginning of a new millennium is far more likely to invite sparring among competing entities eager to assign blame for its ongoing destruction. ("If only X had done Y, the forest would not be in such bad shape today!") This finger pointing implies, when it does not make explicit, that a particular rain forest would be better off in other hands. ("Certainly, *we* would not pollute rivers or allow those age-old trees to burn.")

These more divisive uses of fragility are readily evident in Punta Laguna, where the endangered status of the spider monkeys has allowed the local population to make an effective case for itself as a collective steward in an ongoing spectacle of environmental preservation. By insisting that irresponsible outsiders pose threats to the monkeys and their forest home, the community has succeeded in enlisting support from a national and international audience concerned about preserving rain forest plants and animals.[64]

Punta Laguna, however, represents a happy exception. All too often, the fragility of the rain forest becomes an arm directed against, instead of by, local populations. The idea of the rain forest as a paradise under siege from irresponsible insiders is examined in my essay "The Road to El Dorado," where I discuss mainstream journalistic coverage of a series of particularly severe fires that swept the Brazilian Amazon in 1998.

Over and over, reports in major U.S. dailies such as the *Washington Post* and the *New York Times* portrayed "slash-and-burn subsistence farmers" as ravaging the virgin forest. These "ragtag" and "uneducated" newcomers caused grave harm not just to trees and plants but to defenseless "Stone Age tribes." Both their incendiary habits and the failure of the Brazilian government to adequately address the ensuing conflagrations prompted repeated calls for action by more responsible outsiders. The icon of the threatened forest thus inspired a spectacle of (often highly justified) demands for environmental rescue.

Interestingly enough, a similar rash of fires in Florida during the late 1990s prompted a very different set of stories in which disaster descended on unsuspecting humans. While these differences reflect, in part, real dissimilarities in fire patterns, public policies, and ecosystems, they are also reflections of longstanding images of a distant tropical nature that has little resemblance to the nature just outside our door.[65] If the icon in the Amazonian case is a fragile Virgin Forest, in Florida it is a capricious Mother Nature who unleashes a spectacle of destruction.

A second, related difference between today's rain forest and various earlier incarnations lies in a growing return to the nonhuman attributes of tropical nature. While depictions of the rain forest in the early 1970s focused on a green world full of trees, vines, and parrots, the grassroots political movements of the 1980s made it hard to ignore the human presence. Yet ongoing threats to nonhuman nature and the desire to defend a shrinking natural world have led to a reaffirmation of today's rain forests as precious realms of flora and fauna.[66] A desire to mute the increasingly vocal demands of local peoples, whose vision of—and for—the forest may be quite different from that of outsiders, also fuels some of this renewed insistence on an idyllic, nonhuman nature.

Xcaret's celebration of flora and fauna, for instance, tends to drown out the demands of people who cannot afford the price of admission to the Sacred Nature Show. A similar muting process is clear in the third essay, where Suzana Sawyer discusses multinational petroleum companies' portrayals of the oil-rich Ecuadorian rain forest as a picture-perfect Eden. In dwelling on how state-of-the-art technologies such as the so-called invisible pipeline have minimized potential damage to the region's marvelous flora and fauna, the oil giant ARCO (now BP-Amoco) has effectively downplayed the presence of a variety of native groups who do not want a pipeline of any sort in their backyard. The bioscript of high-tech conservation thus facilitates erasure of a human presence.

The depiction of the Amazonian rain forest as a glittering realm of nature that the corporation has bent over backward to protect is clearly meant to impress consumers back home. ("But why are these local people so upset if ARCO is taking such pains to protect the rain forest?" a reader in Kansas or Kyoto is apt to demand.) The heavy stress on an ostensibly primordial nature casts an iconic shadow in which the disgruntled natives look like late arrivals, if not outright intruders, to a preexisting, as yet unsullied, world.

The case of the pipeline cloaked in vegetation suggests not just a return to nonhuman nature but also the "new" rain forest's tendency to appear as a source of overtly symbolic commodities. Although more traditional rain forest products—gold, jute, teak, sarsaparilla—do possess a symbolic dimension, their obvious materiality contrasts with the considerably more diffuse harmony and purity that enterprises such as Xcaret openly market. Moreover, there is no way that a made-to-order eco-park could masquerade as a "real" rain forest if this symbolic element were not so important to one set of present-day consumers. ("Just being here with nature makes me feel energized," one of our fellow visitors to the butterfly pavilion, a German businessman, confided.) Likewise, although visitors travel to Punta Laguna to observe real monkeys, the promise of a renewed sense of connection to the earth is a big part of the reserve's appeal.

The increasingly overt symbolic value of today's rain forest is one of the primary themes in the following essay, where Alex Greene reflects on the rain forest as an iconic living library in today's Belize. He shows how major multinational pharmaceutical companies have effectively cultivated outsiders' perceptions of the country's forests as an encyclopedic treasure trove of remedies not just for individual headaches and upset stomachs but for ailments around the globe.

Culled directly from what these companies portray as an Edenic forest-garden, the cures are also steeped in the sort of "ancient Maya" tradition that permeates Xcaret. Despite the distinctly hybrid quality of healing practices in a country where indigenous groups, black slaves, and English and Chinese immigrants have long rubbed shoulders, promotional campaigns for these remedies emphasize their uniquely Maya roots. This iconic simplification is in part a pragmatic response to the complicated legal issues associated with ethno-botanical property rights. At the same time, the Xcaret-like insistence on the "mystery and magic" of the Maya enhances the perceived therapeutic value of their herbal products by evoking ties to an ostensibly healthier past.

The iconic shadow of this insistence on the Maya results in the exclusion of other segments of the population, which receive no intellectual or economic recognition for their part in the cures.

The "mystery and magic" of prime concern to Alex Greene resurfaces in Scott Fedick's discussion of three icons of the Maya forest. Poised between three essays on rain forest icons and three on spectacles of wildness, Fedick's essay takes on a pivotal position in the book. While the first set of depictions of the Maya forest he discusses are clearly variations on the tropical Eden that dominates the first trio of essays, his second set of images foregrounds the wild, harder to control jungle that provides the focus for the chapters that follow. The last icon he presents—the "managed mosaic"—stands apart from both rain forest and jungle in its stress on human interactions with the land.

Like many writers in this collection, Fedick gives special attention to popular representations (in this case, mass-market publications such as *National Geographic* and the adventure movie *El Dorado*). He also emphasizes the porous character of the dividing line between academic visions of the Maya forest and their mass-market counterparts. Present in different forms in several essays, these sorts of specialist/nonspecialist connections are particularly evident in his discussion of oscillating visions of the Maya. Although *National Geographic* continues to favor presentations of the Maya as children of a bountiful nature, the magazine has also been quick to translate into more popular language scholars' portrayals of the Maya as environmental illiterates whose disrespect for their surroundings assures their own demise.

Both the paradisiacal rain forest that satisfies the Mayas' every need in the first case and the chaotic jungle into which these same Maya convert a once-lush nature in the second strike Fedick as inadequate descriptions of the forest and its peoples. He therefore goes on to argue for what he sees as a third, more accurate vision of the Yucatán as a "managed mosaic," in which human beings both transform and are transformed by their surroundings.

By contrasting first the rain forest and then the jungle with the more nuanced mosaic, Fedick suggests links between these initial two, apparently opposing icons. Even though the fragile if abundant forest would be hard to confuse with the tangled jungle full of beasts, maladies, and fearsome people, both accord equally narrow roles to human beings. In this sense, even the most radiant forest and the darkest jungle are far more like each other than might be apparent.

In the first three essays of this book, images of the rain forest happen to coincide with New World forests, while in the last three, the jungle is associated with Africa and Asia. Fedick's examination of the Maya forest as both rain forest and jungle makes clear that this division is by no means fixed. Not only can rain forests and jungles exist side by side, but a place described as a rain forest in the opening sentence of a newspaper report may morph into a jungle just a few lines later.[67]

Moreover, yesterday's jungle may be today's rain forest. Such transformations over time are particularly clear in the triad of chapters that analyze present-day spectacles surrounding wildness. Because wildness itself is differently defined in each of the three examples (wild tigers, savage head hunters, a terrifying virus), the jungle-like forests that appear in this section are as different from each other as they are from the dazzling forest in the first part of the book.

In the past, wild humans were often an embodiment of, and primary icon for, an equally wild nature. The savage, if alluring, Amazons, for instance, were direct extensions of an equally savage, if alluring, New World full of an immense, unfamiliar vegetation and wondrous, fearsome beasts. Today, however, environmental loss and degradation have imbued wild nature with newly positive connotations. At the same time, they have encouraged the projection of traditionally more negative aspects of wildness onto human beings.

This transfer is clear in Nancy Lee Peluso's examination of how the Indonesian military government of the 1960s appropriated older colonial images of the fearsome Borneo Headhunter to justify its own brutal suppression of the largely forest-dwelling Dayak people during this period. At the same time, it used these images to incite the Dayaks to bloody actions against the Chinese immigrants, whom it saw as agents of communism and therefore wanted to dislodge.

The ensuing spectacle of violence allowed the military to portray the razing of Borneo's rain forests (quickly replaced with oil palm and other export crop plantations) as a regrettable necessity in the fight against native barbarism. The bioscript that found its way into newspaper headlines regularly coupled bloody warfare with environmental destruction. This script resurfaces today in international press reports that describe the Dayaks as "savage tribesmen." Instead of rain forests that demand careful preservation, these "modern-day indigenous headhunters" inhabit "lush jungles" that appear to cry out for transformation and control.[68]

The jungle exercises a similar ambiguous presence in Paul Greenough's analysis of tiger reserves in India. While the "bio-ironies" that he describes are by no means limited to jungles (the "rain forest" of Xcaret is every bit as rich in contradictions), they are particularly striking in the case of the national tiger reserves. Greenough shows how the tiger—a standard icon for the sort of dark, dangerous animal kingdom that appears in the nineteenth-century *Jungle Books* of Rudyard Kipling—has become the symbol of a diminished nature that nonetheless succeeds in threatening humans in new and unexpected ways.[69]

Created in response to international pressure, India's tiger reserves resulted in the forced evacuation of peaceful peasants from these areas. As the preservation program proceeded to augment the once dwindling tiger population, the animals began spilling out of their forest confines to attack neighboring peasant communities. Unsurprisingly, the frightened residents retaliated by killing the very tigers that the reserves had been intended to protect. The iconic shadow of tiger preservation in Greenough's presentation is thus the destruction wrought by, as well as that visited upon, particular human groups.

Changes over time in the identifying contours of the jungle are every bit as obvious in Charles Zerner's discussion of popular portrayals of Ebola and a number of similar scourges as products of the present-day African rain forest. Although this forest-jungle has unmistakable iconic roots in older visions of Africa such as Joseph Conrad's *Heart of Darkness*, these later representations of widespread contamination suggest a crucial twist.[70]

The colonialist adventurers in Conrad's classic novel push their way ever deeper into the miasmic jungle, itself a close cousin of the Amazonian Green Inferno pictured in Colombian writer José Eustacio Rivera's aptly titled *The Vortex (La vorágine)*.[71] The viral forest, in contrast, does not spiral inward but rather erupts outward onto a globalizing world marked by the dissolution of boundaries.[72] In so doing, the spectacle of uncontainable infection reinterprets the old bioscript of a dangerous forest that retaliates upon trespassers unable or unwilling to resist its lure.

The new-style, all-too-readily-transportable jungle that Zerner locates in mass-market movies such as *Outbreak* is the dark, foreboding flip side of Xcaret's fantastically permeable Nature. The same freedom from national borders that permits the Japanese or German tourist to shout "Viva Mexico!" in the middle of an at once regional and universal folk dance helps catapult Ebola onto a planetary stage. That the hero of the film should be a sympathetic male military doctor has other implications that the essay explores.

THE IMPORTANCE OF THE SEARCH

Representations of species extinction, deforestation, and pollution that create a sense that there are no alternatives for rain forests or the planet fuel a desire for quick and radical solutions.[73] Increasingly vocal calls for fencing off, and for exiling human populations from, the dwindling remnants of "wild nature" are among the most obvious responses to this sense of crisis and frustration.[74] Although these calls are by no means uniform, they mark a backlash in some quarters against the increasingly widespread idea that local populations are fundamental to forest preservation.

The essays in this book offer compelling evidence that these sorts of newly conservative approaches won't work. While their simplicity is appealing, they are also often patently unjust. Iconic representations of the ancient Maya forest as either a once and always Garden of Eden or a ravaged jungle obscure a more nuanced vision of it as a shifting mosaic within which different regional subgroups interact with nature in different ways. The radical simplification of Belize's marvelously varied healing heritage into a single Maya tradition falsifies its richness even as it denies non-Maya ethnic groups their fair share of the profits from its commercialization. The spectacle of "ragtag" Amazonians who blithely burn down age-old forests makes convenient scapegoats of people whose economic circumstances curtail their sense of choice.

Likewise, the use of green technologies to drown out native voices invites sabotage of oil operations in Ecuador's rain forests. The forced evacuation of tiger reserves drives out peaceful peasants, who are often replaced not just by tigers but also by bandits, rebels, drug traffickers, and poachers. The iconic equation of locally significant headhunting with state-managed war in Borneo elicits genuinely violent reactions from a people who feel pushed out of a forest that the fighting further despoils. The staging of the African rain forest as a breeding ground for dreaded diseases averts attention from the factors that foster, if not ensure, their spread within the "civilized" world.

If only because individual rain forests and different peoples' claims upon these are constantly changing, the search for the rain forest that we describe here is clearly destined to go on. The real surprises in this book are not the open-ended nature of the search, but the often unexpected differences in the uses of superficially similar icons as well as in the similarities that may emerge from seemingly disparate spectacles. These unexpected couplings and collisions hold out hope that we can come to see the iconic forest, and with it, individual forests and forest peoples, in new ways that will permit equally new perspectives and solutions. By recasting this totalizing forest as a series of distinctive, constantly evolving forests with different social and natural histories, we can make the quest to comprehend and to protect them more likely to succeed.

When our group first arrived in the Yucatán, those of us who work in high-canopy rain forests were struck by the absence of surface rivers. "But where's all the water?" I remember whispering to my fellow Amazonianist, Suzana Sawyer. Scott Fedick had to remind us that the sparkling cenotes that appeared as isolated openings in the limestone surface were actually gateways into an immense system of underground rivers, which, if stretched out into a single thread, would be longer than the Amazon.

Just as in the case of these initially separate-looking cenotes, the connections among and between the following essays emerge gradually. Difficult to weigh or measure, the icons and spectacles on which we focus are nonetheless as revealing as the most finely calibrated readings of soil erosion or the sharpest Landsat photos. Every bit as real as charred trees, dying tigers, or hungry children, they have direct consequences for us and for the forests we would sally forth to save.

NOTES

1. For a now classic discussion of Nature as intertwining material reality and symbol, see Raymond Williams, "Ideas of Nature," in *Ecology: The Shaping Enquiry*, ed. Jonathan Benthall (London: Longman, 1972), and Williams's discussion of Nature in his *Keywords: A Vocabulary of Culture and Society* (London: Fontana, 1976). Some of the themes that Williams lays out are

taken up from varying perspectives in the essays in William Cronon, ed., *Uncommon Ground: Rethinking the Human Place in Nature* (New York: W. W. Norton, 1996).

2. We are thinking here of a variety of studies, including Cronon, *Uncommon Ground*; A. Maarten Hajer, *The Politics of Environmental Discourse* (Clarendon: Oxford University Press, 1995); E. Melanie Dupuis and Peter Vandergeest, *Creating the Countryside: The Politics of Rural and Environmental Discourse* (Philadelphia: Temple University Press, 1995); Kay Milton, *Environmentalism and Cultural Theory: Exploring the Role of Anthropology in Environmental Discourse* (London: Routledge, 1996); John S. Dryzek, *The Politics of the Earth: Environmental Discourses* (New York: Oxford University Press, 1997); Ron Harre, Peter Muhlhausler, and Jens Brockmeier, *Greenspeak: A Study of Environmental Discourse* (London: Sage Publications, 1998); Eric Darier, *Discourses of the Environment* (Oxford: Blackwell Publishers, 1999); Lisa Benton and John Rennie Short, *Environmental Discourse and Practice* (Oxford: Blackwell Publishers, 1999), and *Environmental Discourses: A Reader* (Oxford: Blackwell Publishers, 2000); and Nancy W. Coppola and Bill Karis, *Technical Communication, Deliberative Rhetoric, and Environmental Discourse: Connections and Directions* (Stamford, Conn.: Ablex Publishing, 2000).

Like many of these other works, our volume makes a strong case for considering representations as powerful forces informing policy, strategy, and visions. However, while a number of the authors cited above focus on specifically environmental discourses involving indigenous land battles, forest preservation, or political organization, we are interested as well in places and processes reflecting more implicit cultural biases and ideological concerns. Although portrayals of the Borneo headhunter, Indian tiger, or Ecuadorian "invisible pipeline" may not appear, at first glance, to have as direct consequences as a Greenpeace campaign to save the whales, they nonetheless offer a vivid demonstration of how environmental knowledge and policies are inevitably shaped in part by politics, representational histories, and culture.

3. For a useful introduction to the question of "natural" versus "unnatural" people and the allure of the exotic native, see Alcida Rita Ramos, *Indigenism: Ethnic Politics in Brazil* (Madison: University of Wisconsin Press, 1998).

4. For a discussion of these terms, see Candace Slater, "Amazonia as Edenic Narrative," in Cronon, *Uncommon Ground*. For an interesting parallel in Spanish, see Ileana Rodríguez, "Naturaleza/nación: Lo salvaje-civil escribiendo Amazonia," *Revista de Critica Literaria Latinoamericana* 23, no. 45 (1997): 26–42. Rain forests, to be sure, are temperate as well as tropical. Our focus in this book, however, is on the tropical variety.

5. Various writers (the nineteenth-century British naturalist Alfred Russel Wallace, for example) used the term before it became standard botanical usage.

There is a large and growing scholarly literature on the environmental movement(s) of the latter part of the twentieth century. For an introduction, see J. Peter Brosius, "Analysis and Interventions: Anthropological Engagements with Environmentalism," *Current Anthropology* 40, no. 3 (1999): 277–310; and Anna Tsing, "Transitions as Translations," in *Transitions, Environments, Translations: Feminisms in International Politics*, ed. Joan W. Scott, Cora Kaplan, and Debra Keates (New York: Routledge, 1997), 253–272. See also "Ecologies for Tomorrow: Reading Rappaport Today," ed. Aletta Biersack, the special issue in *American Anthropologist* 101 (1999).

6. Here we are thinking of mass-market films such as the *Jaws*-like *Anaconda* and news reports that center on warlike indigenous peoples or lethal "jungle" diseases.

7. For a discussion of ideas of tropicality, see the section on the invention of tropicality in David Arnold, *The Problem of Nature: Environment, Culture, and European Expansion. New Perspectives on the Past* (Oxford: Blackwell Publishers, 1996). See also the special issue in the *Singapore Journal of Tropical Geography* titled "Constructing the Tropics," 21, no. 1 (March 2000). For a vivid pictorial account of how rain forest areas have been conceived over time, see E. Francis Putz and N. Michele Holbrook, "Tropical Rain-Forest Images," in *People of the Tropical Rain Forest*, ed. Christine Padoch and Julie Sloan Denslow (Berkeley: University of California Press and Washington: Smithsonian Institution Traveling Exhibition Service, 1988), 37–52; and Nancy Leys Stepan, *Picturing Tropical Nature* (Ithaca: Cornell University Press, 2001).

8. Perhaps the best known of these "in search of" books is P. D. Ouspensky, *In Search of the Miraculous: Fragments of an Unknown Teaching* (New York: Harcourt, Brace and World, 1949). A search on Amazon.com for the title words "in search of" brought up a whopping 3,351 matches of books in print. A search of the University of California general catalog indicated an excess of 10,000 titles. A sizable number of the more recent titles on Amazon.com are specifically about rain forests or rain forest creatures. See, for instance, Joyce Ann Powzky, *In Search of Lemurs: My Days and Nights in a Madagascar Rain Forest* (Washington: National Geographic Society, 1998); Don Stap, *A Parrot without a Name: The Search for the Last Unknown Birds on Earth* (Austin: University of Texas Press, 1991); and Martha L. Crump, *In Search of the Golden Frog* (Chicago: University of Chicago Press, 2000), which one Amazon.com reviewer enthusiastically recommends as a survey of "exotic amphibians of faraway tropical rain forests."

9. We by no means claim to be the only ones embarked on this sort of search for the rain forest or new ways of understanding natural entities. Many scholars in various disciplines are similarly engaged. For just a few examples, see the work of a number of the authors represented in Noel Castree and Bruce Willems-Braun, eds., *Remaking Reality: Nature at the Millennium* (London: Routledge, 1998); and Laura Rival, ed., *The Social Life of Trees* (Oxford: Berg, 1998). See to the essays in Christine Padoch and Julie Sloan Denslow, eds., *People of the Tropical Rain Forest* (Berkeley: University of California Press and Washington: Smithsonian Institution Traveling Exhibition Service, 1988).

10. The initial idea for the seminar emerged in a brief conversation between me and Anna Tsing at HRI when I made some comment about images of the Amazonian rain forest and Anna replied, "But in Borneo, it's not that way at all." I would like here to acknowledge also my debts to a previous HRI seminar, Reinventing Nature, directed by William Cronon, in which I was fortunate enough to participate for five months during 1994.

11. Fuller descriptions of the contributors and their publications appear at the back of this book.

12. Scientists are not the only actors in the policy arena. Certainly, economists, communications experts, anthropologists, and many others also participate in conservation policy making. Also, given the increasingly vocal role of political economists in debates about forest development and preservation, scientists are not the only commentators on the "material" aspects of forests. However conservation biologists have retained a position of prominence in these debates, as witnessed by E. O. Wilson's still largely unchallenged authority in the biodiversity field.

13. Here we also thank Jenny Price for an intensive session with the group on "writerly" concerns. We also acknowledge our debts to her study of the cultural uses of birds in *Flight Maps: Adventures with Nature in Modern America* (New York: Basic Books, 1999).

14. We were, to be sure, particularly self-conscious tourists with an atypical interest in larger representations of tropical nature. For an introduction to the ample bibliography on tourists and tourism, see Dean MacCannell, *Empty Meeting Grounds: The Tourist Papers* (New York: Routledge, 1992), and *The Tourist: A New Theory of the Leisure Class* (New York: Schocken, 1976).

15. Definitions of what constitutes a rain forest vary. Although general-use dictionaries tend to identify rain forests in terms of rainfall (more than a hundred inches per year) and type of vegetation (closed-canopy, nondeciduous trees), scholars have their own criteria. See the standard classification scheme by L. R. Holdridge, *Forest Environments in Tropical Life Zones* (Oxford: Pergamon Press, 1971). By this classification, true rain forest (wet, tropical, perennial forest) in the Maya region is restricted to the southern Petén of Guatemala, southwestern portion of Belize that borders the Petén, Lacandón forest area of Chiapas (Mexico), and southeast Tabasco, also in Mexico. However, there are other areas of the Yucatán that while technically not rain forest, are nonetheless covered by lush, high forest.

16. Both "icon" and "spectacle" are used, for instance, by MacCannell in his chapter on spectacles in *Empty Meeting Grounds*, 230–254. MacCannell draws heavily on the work of Charles Peirce and Guy Debord. The meanings he gives these words, however, are somewhat different than those we employ here. A "spectacle," for instance, retains for him a more narrowly theatrical identity, while we see it as a performance in the broadest sense.

17. See the multiple entries for "icon" and "spectacle" in the *Oxford English Dictionary*. In both cases, these entries date back to the early thirteenth century.

18. See, for instance, Herbert Edward Read, *Icon and Idea: The Function of Art in the Development of Human Consciousness* (Cambridge: Harvard University Press, 1955); Troels Degn Johansson, Martin Skov, and Berit Brogaard, eds., *Iconicity: A Fundamental Problem in Semiotics* (Arhus, Denmark: NSU Press, 1999); *Iconicity: Essays on the Nature of Culture*, Paul Bouissac, Michael Herzfeld, Roland Posner, eds. (Tübingen: Stauffenburg Verlag, 1986); George Bornstein and Theresa Tinkle, ed., *The Iconic Page in Manuscript, Print, and Digital Culture* (Ann Arbor: University of Michigan Press, 1999); and Raffaele Simone, ed., *Iconicity in Language* (Amsterdam: J. Benjamins, 1994).

19. For a good sense of the semiotics of cultural analysis, including the mythicization of scenery, see Roland Barthes, *Mythologies* (New York: Hill and Wang, 1972); and the essays in Umberto Eco, *Travels in Hyper-Reality* (San Diego: Harcourt, Brace and Company, 1986).

20. See the section on "The Icon, Index, and Symbol" in Charles Sanders Peirce, *Elements of Logic*, in *The Collected Papers of Charles Sanders Peirce* (Cambridge: Harvard University Press, 1960), 155–161.

21. In one particularly dramatic example, the Mall of America—the largest themed retail and entertainment complex in the United States—emerges as a kind of collective dreamhouse where fantasies of authentic life become displaced onto commodities such as the exuberant, if largely synthetic flora and fauna that adorn the Rainforest Café. See Jon Goss, "Once-upon-a-Time in

the Commodity World: An Unofficial Guide to Mall of America," *Annals of the Association of American Geographers* 89, no. 1 (1999): 45–75.

22. See, in particular, Guy Debord, *The Society of the Spectacle*, trans. Donald Nicholson-Smith (New York: Zone Books, 1994.) For a very different example, see Susan G. Davis, *Spectacular Nature: The Sea World Experience* (Berkeley: University of California Press, 1997).

23. For a fuller discussion of this and other examples, see Kevin Michael DeLuca, *Image Politics: The New Rhetoric of Environmental Activism* (New York: Guilford Press, 1999). DeLuca provides a helpful discussion (165) of the ways in which his idea of "image events" resembles and departs from related concepts, including Debord's spectacle, Daniel Boorstin's "pseudo-events," and John Fiske's "media events."

24. See, for example, Roderick P. Neumann's discussion of the intermeshing of symbol and political maneuvering in *Imposing Wilderness: Struggles over Livelihood and Nature Preservation in Africa* (Berkeley: University of California Press, 1998).

25. When people define biodiversity, they almost always rely on examples of actual places. Rain forests were *the* prime example of biodiversity in the early 1980s when conservation biologist E. O. Wilson introduced the term, but today scientists are apt to give a range of illustrations.

26. For a discussion of the transference of religious sentiment onto Nature from the age of European expansionism onward, see Richard Grove, *Green Imperialism: Colonial Expansion, Tropical Island Edens, and the Origins of Environmentalism, 1600–1860* (Cambridge: Cambridge University Press, 1996).

27. Here, we are indebted to ideas summed up in the title of Robert Pogue Harrison's *Forests: The Shadow of Civilization* (Chicago: University of Chicago Press, 1992).

28. This process is particularly clear in the case of the Brazilian rubber tappers. See Margaret E. Keck, "Social Equity and Environmental Politics in Brazil: Lessons from the Rubber Tappers of Acre," *Comparative Politics* 27 (1995): 409–24. For a discussion of how the idea of the Noble Savage has served different groups, including foreign anthropologists, see Kent H. Redford, "The Ecologically Noble Savage," *Orion Nature Quarterly* 9, no. 3 (1990): 25–29; and Ter Ellington, *The Myth of the Noble Savage* (Berkeley: University of California Press, 2001). Kay Milton has also examined the "myth of primitive environmental wisdom" in convincing detail in her *Environmentalism and Cultural Theory: Exploring the Role of Anthropology in Cultural Discourse* (London: Routledge, 1996).

29. These different uses are crystal clear in Bruce Willems-Braun, "Buried Epistemologies: The Politics of Nature in (Post)colonial British Columbia," *Annals of the Association of American Geographers* 87, no. 1 (1997): 3–31.

30. For a study of one particular theme park that provides an introduction to larger questions surrounding theme parks, see Davis, *Spectacular Nature*.

31. The numbers are from http://laplayadelcarmen.com/english/xcaret.html. A grand total of over two million Americans and Canadians alone are estimated to visit the greater Cancún area every year.

32. The developers were Miguel Quintana-Pali and the brothers Carlos, Marcos, and Oscar Constandse Madrazo. Oscar Constandse Madrazo is the cousin of the official then governing Tabasco.

The construction of Xcaret is described with scathing sarcasm in a piece on a previously existing website (http://laplayadelcarmen.com/english/xcaret.html), in which the unidentified writer begins by lauding Quintana-Pali's ingenuity. "Actually," the writer continues, "it took much more ingenuity to build 80% of the Xcaret park without permits from the local government and, in fact, Quintana Pali originally said that he was buying the property for the park in order to build a small house on only 5 hectares of land. Today that 'small house' is the largest ecotourism destination in the Yucatán if not in the entire world. Xcaret truly is a product of 'Ecoarchaeological magic!' "

33. The archaeological site of Xcaret covers an estimated ten square kilometers, extending along the coast for about five kilometers and inland for another two. The site is full of small clusters of ruined residential structures set within house lots marked by low stone walls. These lots probably contained both homes and home gardens not unlike those found in traditional Maya communities today. For an introduction to the archaeology of Xcaret, see E. Wyllys Andrews IV and Anthony P. Andrews, *A Preliminary Study of the Ruins of Xcaret, Quintana Roo, Mexico, with Notes on Other Archaeological Remains on the Central East Coast of the Yucatán Peninsula*, no. 40 (New Orleans: Middle American Research Institute, Tulane University, 1975). See also Marís José Con, "Trabajos Recientes en Xcaret, Quintana Roo," *Estudios de Cultura Maya* 18 (1991): 65–129; and Guillermo Antonio Goñi Motilla, *Solares Prehispanicos en la Península de Yucatán*, thesis for licenciado en arqueología, Escuela Nacional de Antropología e Historia, Mexico, 1993.

34. Paris Permenter and John Bigley, "Xcaret, Secret of the Mayas," Fine Travel series, http://www.finetravel.com/mexico/xcaret.html. Note the choice of the word "jungle" in place of "rain forest."

35. See "The Most Beautiful Entertainment Center . . . Nature's Sacred Paradise Xcaret Cancún," http://www.xcaretcancun.com.html.

36. See "Behind the Scenes in Xcaret," http://www.xcaretcancun.com/welcom.html.

37. Http://www.cancunsunrise.com/xcaret.html.

38. Most of these structures date back to the Mid and Late Postclassic Period (1200–1520 A.D.) when Xcaret was also known as "Polé."

39. The group, known as the Ballet Folklórico de México in Spanish, does numerous international tours as well.

40. Cancún presently has approximately twenty-three thousand hotel rooms, the great majority of which are part of large chain resort hotels.

41. The promoters of Xcaret are hardly alone in creating this impression. In an article titled "Looting a Lost Civilization" (*San Francisco Chronicle*, 7 June 2001, A13), Jeremy McDermott makes the astonishing assertion that "the Maya disappeared in the 11th century and scholars are unsure why, offering a range of theories, including drought, war, and disease."

42. Donna Haraway would have a good deal to say about the reconstitution of the Maya for capital and capitalized nature. Not only are they cheap labor for tourist complexes but they are also spokespeople for a commodified pure village very different from any place where they would or could live.

43. According to environmental activist Araceli Domínguez (interview by HRI group, Cancún,

6 March 2000), approximately 70 percent of the population of Cancún lives in *palitos* or tar paper houses. Although a number of these are immigrants from other parts of Mexico, a significant portion are Maya.

44. Consorting with captured dolphins does not come cheap. A dive with a dolphin is $130, a swim is $80, a thirty-minute "interactive experience" is $55, and a fifteen-minute educational program is $30 (see http://cancunsouth.com/cit-xcaret.html). For disquieting information on the apparently questionable manner in which the dolphins were obtained and their high mortality rate, see http://laplayadelCarmen.com/English/xcaret.html.

45. http://www.xcaretcancun.com/welcom.html.

46. "Everything in the park supports that modern development, conservation, and fun are not exclusive from one another" (http://xcaretcancun.com/welcom.html).

47. Ibid.

48. In contrast to Xcaret, whose core of monumental and administrative architecture has been extensively mapped, Punta Laguna has been the site of relatively little archaeological research. Although the full extent of the site is not known, preliminary studies suggest that it covers at least 1.3 square kilometers. Most of the architecture is characteristic of the Late Postclassic period (approximately 1250–1520 A.D.). Nonetheless, the evidence indicates a longer occupation stretching back into the Preclassic period, and even earlier than 100 B.C. For an introduction, see Antonio Benavides Castillo and Renée Lorelei Zapata Peraza, "Punta Laguna: Un Sítio Prehispánico de Quintana Roo," *Estudios de Cultura Maya* 18 (1991): 23–53, which includes a good bibliography as well as some recent history of the community. See also Jason H. Curtis, David A. Hodell, and Mark Brenner, "Climate Variability on the Yucatán Peninsula (Mexico) during the Past 3500 Years and the Implications for Maya Cultural Evolution," *Quaternary Research* 46 (1996): 37–47.

49. The quote is from http://www.stevensunpress.com/PunLag.html. Though subject to frequent droughts that can keep rainfall as low as one thousand millimeters annually, the forest can receive over 2,000 millimeters of rain in wetter years. The one-thousand-millimeters measurement contrasts with an average thirty-five hundred millimeters in the Lacandón rain forest of southern Mexico. (The rainfall figure generally associated with rain forests is one hundred inches of rain, though not all scholars would accept this figure.) Not just subject to droughts, Punta Laguna is also located in a hurricane corridor and is the site of periodic torrential storms. For more on the climate of this region, see Curtis, Hodell, and Brenner, "Climate Variability on the Yucatán Peninsula."

50. Estimates of its founding vary from between thirty and fifty years. The founder, Don Nacho Canul, came from the small settlement of Chemax in search of chicle and fertile land. For an introduction to the community, see María Teresa Puig de Silveira, "Los mayas de Punta Laguna," *Pronatura* 7 (1999): 49–53.

51. According to our guides, the community of Punta Laguna consists of about twenty families, with several related by marriage. The ejido is larger, consisting of about seven hundred persons.

52. Pronatura underwrote a comprehensive bird census in 1992. The first spider monkey studies were done by Laura G. Vick and David M. Taub in 1994. Vick is associated with Community

Conservation, a nonprofit organization originally formed in 1989 to support experimental grassroots conservation efforts in Belize. Official government recognition of the reserve was first proposed in 1997.

53. Of seventy-seven species of neotropical primates, spider monkeys (*Ateles geoffroyi*) are the most common. They are found in the area between the Yucatán Peninsula and northern Argentina. Social groups are generally small, ranging from ten to twenty-five adults of both sexes. Female spider monkeys give birth every three to four years in most areas. For a general scientific description, see M. G. M. Van Roosmalen and L. L. Klein, "The Spider Monkeys, Genus *Ateles*," *Ecology and Behavior of Neotropical Primates* 7, ed. R. A. Mittermeier and A. B. Rylands (Washington, D.C.: World Wildlife Fund, 1987); and A. Estrada and R. Coates Estrada, "Tropical Rain Forest Conversion and Perspectives in the Conservation of Wild Primates (*Alouatta* and *Ateles*) in Mexico," *American Journal of Primatology* 14 (1988): 315–27. For information specifically on the spider monkeys of Punta Laguna, see L. G. Vick and D. M. Taub, "Ecology and Behavior of Spider Monkeys (*Ateles geoffroyi*) at Punta Laguna, Mexico," *American Journal of Primatology* 36, no. 2 (1995): 160; and Gabriel Ramos-Fernández, David M. Taub, and Laura G. Vick, "The Spider Monkey Project at Punta Laguna," http:www.sas.upenn.edu/-ramosfer/spmkpl.html. See too Gabriel Ramos-Fernández, "Patterns of Association, Feeding Competition, and Vocal Communication in Spider Monkeys," Ph.D. diss., University of Pennsylvania, 2001; and G. Ramos-Fernández and B. Ayala-Orozco, "Case Study: Population Size, Habitat Use, and Social Organization in Spider Monkeys in a Fragment of Forest in Yucatán, Mexico," in *Primates in Fragments: Ecology and Conservation*, ed. Laura K. Marsh (New York: Kluwer Academic/Plenum Publishers, 2003).

54. According to Ramos-Fernández, Taub, and Vick ("The Spider Monkey Project at Punta Laguna," 2), there are 87 kinds of snakes, 161 species of birds (114 resident, 44 migratory), and 93 types of mammals, including white-tailed deer, *tigrillo*, and jaguar. There are also 8 species of fishes in Punta Laguna's cenotes.

55. We thank Ricardo Canul Aban, age nineteen, and his older brother Eulogio for their generous introduction to the Punta Laguna reserve.

56. The irony here is that Xcaret necessitated considerable environmental devastation in order to construct this simulacrum of a rain forest. Moreover, the ongoing waste that tourist sites such as this theme park continue to generate has created significant degradation of the cenotes in particular.

57. The drop in the monkey population also reflects a thriving illegal pet trade. Spider monkeys are a common sight in hotel courtyards along much of the coast.

58. The plants from which the monkeys drink are bromeliads, which hold rainwater in their waxy, cup-shaped leaves.

59. As this comment makes clear, some Maya think of "Mexican" as a category that does not include them, even though they are incorporated into the state from an early age through government schools in which instruction is conducted in Spanish.

60. For a discussion of symbolic goods and an accompanying definition of symbolic capital, see Pierre Bourdieu, *Outline of a Theory of Practice* (Cambridge: Cambridge University Press, 1977), 176.

61. The actual jaguars and birds are still there. The jaguars eat deer and boars as well as the monkeys.

62. The student is Gabriel Ramos-Fernández, a Mexican citizen who received his Ph.D. from the University of Pennsylvania in 2001, and the names (based on "acquaintances, cartoon characters, and legends") were chosen in order to continue tracking monkeys who he and his team had reliably identified. I thank Dr. Ramos-Fernández for his e-mail answers (16 March 2001) to my questions regarding the spider monkey project that was the basis for his dissertation.

63. For a fuller discussion of some of these features, see the section titled "Roots of the Rain Forest" in Candace Slater, *Entangled Edens: Visions of the Amazon* (Berkeley: University of California Press, 2002).

64. For parallels in Amazonia involving rhetorical strategies through which native peoples both communicate among themselves and attempt to enlist outside aid, see Alcida Rita Ramos, "Indian Voices: Contact Experienced and Expressed," in *Rethinking History and Myth: Indigenous South American Perspectives on the Past,* ed. Jonathan D. Hill (Urbana: University of Illinois Press, 1988), 214–234; and Sylvia Caiuby Novaes, *The Play of Mirrors: The Representation of Self as Mirrored in the Other* (Austin: University of Texas Press, 1997).

65. In this insistence on the value of a distant nature, the rain forest is not unlike the wilderness. See William Cronon, "The Trouble with Wilderness; or, Getting Back to the Wrong Nature," in *Uncommon Ground,* 69–90.

66. The parameters of the debate are clear in the diverse articles included in the special issue of *Conservation Biology* devoted to rain forests (vol. 13, no. 5, October 2000). Obviously, not all conservation biologists take the same stance. Particularly in Latin America, biologists are increasingly sensitized to more political approaches. At the same time that local social movements are more apt to face repression today than they were a decade earlier, conservation approaches that completely bracket local people and movements are increasingly rare.

67. See, for just one of myriad examples, Anthony Faiola, "Welcome to My Jungle Nightmare: Four Hundred and Fifty Miles of Jungle at the Back of the Bus," *Washington Post,* 6 April 1998, 21, in which the Amazon is both rain forest and jungle, depending on how hot and miserable the author is at the time.

68. The rain forest quickly became a jungle or lush jungle in a number of newspaper articles on the Dayak confrontations with the Madurese in February 2001. See, for example, Richard C. Paddock, "One Hundred and Eighteen Ethnic Refugees in Borneo Massacred after Police Flee," *Los Angeles Times,* 28 February 2001, A1; and Calvin Sims, "Borneo Backwater's Clashes Draw Little Notice," *New York Times,* 28 February 2001, A3, in which the author describes Indonesian Borneo as "a backwater region of lush jungles and indigenous tribes." For an interesting counterpoint by a travel writer, see Daffyd Roderick, "Go Wild in the Heart of Borneo" in "Travel Watch," *Time Asia,* 22 November 1999, online at http://www.time.com/time/asia/travel/99/1122), which promises the reader a "green vacation." "If you think Borneo is all pith helmets, knee socks and headhunters, think again," asserts the writer.

69. Rudyard Kipling, *The Jungle Books,* ed. W. W. Robson (Oxford: Oxford University Press, 1998). The differences between the jungle that appears in Kipling and the one that appears in Disney's remake of the novel are striking and significant. For equally stark differences as these

relate to the African jungle, see Edward Rice Burrough's Tarzan novels and the various movie treatments of the Tarzan theme. See too Mariana Torgovnick's discussion of the allure of the ostensibly primeval in her *Gone Primitive: Savage Intellects, Modern Lives* (Chicago: University of Chicago Press, 1990).

70. Joseph Conrad, *Heart of Darkness* (New York: Penguin Books, 1999). The novel was originally published in 1902. Proof that the "heart of darkness" concept lives on can be found in Redmond O'Hanlon, *No Mercy: A Journey to the Heart of Congo* (New York: Alfred A. Knopf, 1997), a travel narrative that reinforces the idea of the African jungle as a deep, dark, dangerous place.

71. José Eustacio Rivera, *La vorágine*, 3rd ed. (Buenos Aires: Losada, 1953). The novel was originally published in Spanish in 1924. It appeared in English translation as *The Vortex.*

72. "Globalization" has been defined in many different ways and is increasingly taken to include a variety of often contradictory processes. Many writers, however, stress the new porousness of national boundaries as one of its primary attributes. For a fuller discussion, see Arjun Appadurai, *Modernity at Large: Cultural Dimensions of Globalization* (Minneapolis: University of Minnesota Press, 1996); and Michael Burawoy et al., *Global Ethnography: Forces, Connections, and Imaginations in a Postmodern World* (Berkeley: University of California Press, 2000).

73. The problems that rain forests face—such as mining, the erosion of soils and local practices, and gene hunting—are amply documented. See, for instance, Roger D. Stone and Claudia D'Andrea, *Tropical Forests and the Human Spirit: Journeys to the Brink of Hope* (Berkeley: University of California Press, 2002). What these documentations often fail to point out is the degree to which particular depictions of tropical rain forests (either as the green hell demanding cultivation or fragile rain forest that defies any sort of transformation) are a large part of these real problems.

74. For examples of such calls, see John Terborgh, *Requiem for Nature* (Washington: Island Press, 1999); and Thomas T. Struhsaker, *Ecology of an African Rain Forest: Logging in Kibale and the Conflict between Conservation and Exploitation* (Gainesville: University Press of Florida, 1997).

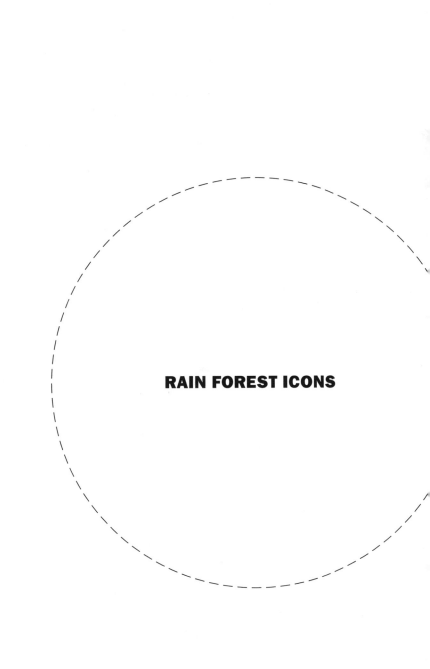

RAIN FOREST ICONS

CANDACE SLATER

FIRE IN EL DORADO, OR

IMAGES OF TROPICAL NATURE

AND THEIR PRACTICAL EFFECTS

During the week that the HRI group was busy exploring the Yucatán, the DreamWorks movie *The Road to El Dorado* opened in theaters in Orange County.[1] Our curiosity piqued by the distinctly Maya-looking setting that had jumped out at us in the previews, we quickly mounted an expedition to the nearest multiplex cinema on our return.

True to the neon-colored posting on the DreamWorks website, the movie told the story of two sixteenth-century con artists who find their way to El Dorado—the fabled golden city that colonial writers were wont to locate in various parts of the New World including the Amazon, the Orinoco, the Andes, and Mesoamerica.[2] However, despite the abundance of Maya architectural elements that had lured us to the theater, the DreamWorks El Dorado turned out to be a purposefully imprecise location and largely generic treasure trove.[3]

The attempts of the basically good-hearted Tulio and Miguel to find, and then defend, the golden city are less a full-fledged story than a pretext for a series of kaleidoscopic visual effects. ("Pretty cool," our five-year-old guest critic, Clayton, was quick to opine.)[4] And yet, although *The Road to El Dorado* shed little light on either the Maya or one of the most peripatetic and enduring legends of the Americas, scenes from the movie continued to pop into my head long after the lights had come back on.

The more I thought about the DreamWorks El Dorado in the days that followed, the more clearly I saw links between it and the newspaper accounts of massive fires that had swept the Amazon in 1998. The reports, about which I had been thinking for some time, interested me because although the authors regularly began by citing the El Niño ocean current as one of the fires' primary causes, they inevitably went on to portray humans as the destroyers of a "terrestrial paradise." This insistence on humans' role in the fires contrasted with accounts of the El Niño–related fires during the same period in Florida, in which people appeared time and again as innocent victims of a capricious Mother Nature.[5]

The lush, biodiverse paradise wreathed in flames is thus the primary icon in the Amazon stories. This paradise-like forest connects the somber depictions of environmental catastrophe that appeared in the *New York Times* and *Wall Street Journal* with the high-spirited, high-tech adventure story that the group had watched over popcorn.

EL DORADO AND ASSUMPTIONS
ABOUT TROPICAL NATURE

Set in the early 1500s, *The Road to El Dorado* is an unabashed fiction whose goal is to entertain. And yet, to the extent that its sixteenth-century golden kingdom turns out to be an earlier incarnation of a contemporary rain forest, *The Road to El Dorado* highlights a whole series of expectations about tropical nature not limited to adventure films or descriptions of the colonial past.[6] In underscoring these thoroughly present-day assumptions, the film leaves no doubt about the power of seemingly innocuous images to influence perceptions, and hence, both individual actions and official policies.

The technicolor Eden into which Tulio and Miguel stumble is a genuine biological marvel. Human embellishments on the landscape simply confirm its identity as "Shangri-la," "the promised land," and "virgin mystery undefiled"—all terms that appear in the accompanying lyrics. "Paradise is close at hand," composer Elton John croons as Tulio and Miguel approach the gleaming city.

Beautiful in the extreme, this El Dorado is also a source of useful commodities. In contrast to the monotonous, shark-infested ocean that the adventurers must cross in order to reach the new world, El Dorado abounds in the brilliant birds, glistening trees, and dazzling waterfalls that lead Miguel to decide to make his home there even as Tulio prepares to head back to Spain in a boat crammed with gold. Along with the butterflies that resemble winged flowers, there are (uncharacteristically friendly) hedgehogs and brightly colored fish that make children in the audience giggle in delight when one bites one of the con artists on the seat of his pants.

As exotic as the butterflies and bright fish, the people of El Dorado live in a New Age–like harmony with their lush surroundings. The evil sorcerer—who exhibits a disturbing fondness for human sacrifice and a nasty habit of conjuring up monsters—serves as a reminder of the idyllic forest's savage underside. However, the great majority of the natives are thoroughly charming, childlike beings. Largely oblivious to the cash value of the wealth that surrounds them, they regularly dump large heaps of golden objects into the water's depths as a show of tribute to the gods. Not surprisingly, this practice causes the gold-hungry Tulio and Miguel (whom their hosts take for visiting deities) no little consternation.

These same natives display an almost total inability to defend themselves and their glittering, but vulnerable surroundings against outside attack. Even though Tulio's new girlfriend, Chel, and the endearingly tubby chief are more worldly-wise than the other residents of El Dorado, it is Tulio and Miguel who must step in to save the day when the evil sorcerer teams up with the equally ruthless Hernán Cortés.[7] In good Hollywood fashion, the duo of adventurers succeeds in blocking the lake that had provided access to the city with just seconds to spare. Although this daring maneuver results in their exile from El Dorado, Tulio, Miguel, and Chel remain convinced of their ability to find their way back into paradise.

The idea of an abundant tropical nature that permeates the DreamWorks film is not without foundation. Today, almost anyone who watches television knows that rain forests are both threatened and biologically important. This particular icon of a fragile earthly paradise has definite implications, however. The iconic shadow in this case is the innocent, but inept native who is not up to the defense of the forest and its marvelous riches. Just so, even while the events that the Amazon articles describe are factually true (the fires *did* happen), their insistence on a precious paradise whose problems demand outside intervention bears a discomfiting similarity to the DreamWorks movie. The iconic connections between the film and the reports highlight how easily particular ideas about tropical nature cross the boundaries between fiction and nonfiction to influence many different areas of daily life.

THE AMAZON IN FLAMES

During the fall of 1997 and spring of 1998, the disruption of the ocean-atmosphere system in the tropical Pacific known as the El Niño Southern Oscillation brought heavier-than-usual rains to some parts of the world such as California, Arizona, and Peru.[8] At the same time, this shift imposed severe drought conditions on a number of other regions including Indonesia, Australia, parts of Africa and Mexico, Florida, and the normally extremely humid Amazon. In diverting rain from the dense, water-hungry vegetation that covers many of these areas, El Niño produced a vast and rapidly drying stock of potential tinder, dramatically increasing the likelihood of fire.

In the Amazon, the effects of the drought were especially noteworthy in regions where brushland or grassy savanna bordered on rain forest. From the fall of 1997 on into the spring of 1998, slow-moving, low-intensity ground fires surrounded by clouds of heavy smoke burned through vast areas of the Brazilian states of Mato Grosso, Tocantins, Rondônia, Pará, and Amazonas.[9] However, it was the France-sized state of Roraima in Brazil's extreme north that would garner the bulk of media attention. The high concentration of indigenous peoples within its borders (roughly one-tenth of the total population of 250,000) helped propel the state's name onto the front pages of newspapers around the world when its governor formally declared the blazes out of control in March 1998.

The fires themselves were nothing new. Massive blazes have periodically swept the Amazon area for at least the past ten thousand years, and Amazonian native peoples have long used fire as part of the slash-and-burn gardening on which they still depend, along with hunting and fishing, for food. What differentiated the 1998 fires from those of past centuries—and indeed, past decades—was the presence of an ever greater number of small farmers as well as large cattle ranchers eager to clear increasingly large areas of land.

New roads designed to link the interior to burgeoning cities served as conduits for the fires. At the same time, widening chinks in the forest canopy caused by logging and mining activities allowed high winds to whip up flames from fires initially set by humans and to knock over trees, which provided further fuel.[10]

The destructive capacities of the fires that eventually consumed an area as large as Belgium, Lebanon, or West Virginia, were greatly heightened by the regional, and then national, government's ineffective response. Although officials in Brasília eventually moved two thousand firefighters from all over the nation into Roraima, critics both within and outside Brazil denounced its actions as woefully inadequate.[11] Mounting pressure from both within and without Brazil forced an initially reluctant government to accept offers of international aid. Nevertheless, the delay allowed the fires to reach crisis proportions in many parts of a far-flung region, including patches of forest near the large Yanomami Indian reserve located on the border between Brazil and Venezuela.[12]

A TALE OF TWO FIRES

Environmental, demographic, and political factors played an important role in shaping the ensuing news reports. However, they are not the whole explanation for these reports' insistent vision of a paradise in flames.[13] The larger spectacle underlying reports of the Amazon fires becomes much clearer in contrast with reports of the rash of wildfires that swept Florida between May and July of the same year. I have therefore set out to compare the two sets of narratives with their in-some-ways similar, and yet ultimately different, bioscripts.

In each case, I have looked at news reports from major U.S. dailies such as the *New York Times*, *Washington Post*, and *Wall Street Journal* as well as from electronic postings by news services such as the Associated Press and CNN that appeared at the height of the crisis. In both instances, I collected a full hundred articles, many of which turned out to be highly repetitive in terms of content, authorship, and perspective.[14] Although I did look at alternative and academic publications, along with news releases originating in Europe and Latin America, I was primarily interested in the kind of mainstream journalism that reaches a general public in the United States. Moreover, while I found telling correspondences among the reports of fires in the Amazon and others in Indonesia during roughly the same time period, I only mention these in passing here.[15]

My choice of these reports as a point of comparison is partly a response to the two fires' prominence in U.S. newspapers at roughly the same point in time. Coverage of the Amazon fires peaked in March and April of 1998; coverage of the Florida fires was heaviest in June and July of the same year. This comparison also reflects my proximity to both places, as I had been doing research in the Amazon since 1988 and was scheduled to assume a visiting professorship at the University of Florida in Gainesville for the spring of 1999.[16] First and foremost, though, I was interested in the two sets of reports because they were at once so alike and so different. Reporters in both cases regularly and invariably identified El Niño as a major cause of the fires, and both sets of reports describe the ensuing destruction. However, while accounts of the Amazon stressed the role of humans in ravaging the Virgin Forest, accounts of Florida were far more likely to blame a capricious Mother Nature for the ensuing disaster.

Because Amazonia and Florida are clearly different ecosystems, my comparison of reports about the fires by no means implies a direct equation of the two in environmental terms.[17] While the Amazon contains vast stretches of high-canopy tropical rain forest, Florida is (or once was) a largely subtropical collection of longleaf pine–dominated savannas and sand pine forests, with scattered oaks and other hardwoods rooted in sandy limestone soils.

Moreover, while the Amazon is presently undergoing intensive deforestation, Florida is in the midst of an intense reforestation effort. The state is home to growing numbers of extensive, highly flammable slash pine plantations that have largely replaced the native pine forests and savannas that previously covered much of the southeastern United States.[18] As a result, while the two resemble each other more than either does, say, a northern conifer forest, they reveal diverse and even inverse environmental processes (largely chaotic deforestation versus planned forest growth).

These differences in the evolution of both geography and economics ensure a varying role for the fires in each case. In the Amazon, where naturally occurring fires have historically been far less frequent than in Florida, the number of fires set deliberately by humans is growing dramatically. This increase is occurring in a region where fire is apt to burn away, rather than replenish the soil. In Florida, by contrast, the number of wildfires has plummeted as the owners of pulpwood plantations and houses in expensive new residential developments bordering wooded areas attempt to protect their investments.[19] As a result, the frequent low-intensity fires that used to regularly clean out the undergrowth of grass straw, palmetto, and gallberry have decreased. When, however, fires do break out, the increased store of fire-friendly vegetation makes them far harder to control.

The elevated number of Amazon fires in 1998 and the vastness of the region they affected set them apart from the Florida blazes. At the beginning of April 1998, the United Nations Disaster Assessment Coordination declared that in Roraima alone, the fires had burned a full 33,000 square kilometers (12,741 square miles), or over thirteen times the area affected by fire in Florida, almost one-third of which had been covered by forest.[20] In contrast, although by July 1998 the U.S. government had proclaimed 40 of Florida's 67 counties federal disaster areas, the blazes affected only about a half a million acres, or some 770 square miles.[21] While these fires were often extremely destructive (a good portion of the 2,200 wildfires that broke out between Memorial Day and 13 July were high-intensity fires that burned all the way up into the forest crown), the net damage to forest areas was decidedly less extensive than it was in the Amazon.

Undeniably important, these ecological differences still fail to fully explain the intense dissimilarities in journalistic coverage of two sets of fires in which reporters assigned El Niño a similar starring role. It is the larger sorts of expectations about rain forests and their inhabitants that find open expression in *The Road to El Dorado* and the political expediencies with which these are associated, to which one again must look in order to comprehend peculiarities that go far beyond environmental fact.

NARRATIVE DISSIMILARITIES

Both major, long-running stories, the fires in Florida and the Amazon nonetheless occupied noticeably different niches in the daily news. In the United States, as in a number of other countries outside Latin America, reports about the Amazonian fires tended to appear either on the environmental or international pages. Less often, the fires also showed up in the editorial or science sections.[22] While the scope and duration of the Florida fires also assured them ample coverage in major newspapers throughout the United States, they unsurprisingly occupied a more prominent place in regional and region-based newspapers. In the *Miami Herald*, for instance, they remained front-page news for almost two months. In newspapers published outside Florida, the fires were often presented as general domestic news.

The disparate positioning within U.S. newspapers of these articles about Florida and Amazon fires suggests differing levels of familiarity in regard to the places and events described. In the case of the Amazon, reporters regularly felt obliged to spell out for their readers why the fires were important to people in New York City, Tuscaloosa, or Des Moines. "The destruction of trees wipes out nature's best way of absorbing noxious carbon gases from the atmosphere, while the burning releases more gases into the air. Carbon emissions not only pollute but are believed to speed up global warming, risking dangerous climactic changes within two generations," explains the author of one of many such accounts.[23]

Within the United States, in contrast, and especially within Florida, no one had to explain the wildfires' impact on people's daily lives. Often, the articles preceded or included a travel advisory hotline that readers could call to check on highway conditions in various parts of the state. The reports about the postponement of the popular Daytona Beach car races, prohibition of traditional Independence Day firework displays, and temporary closing of the main road into Cape Canaveral with which the fires shared news space during the first week of July left no doubt about the practical consequences.

The Amazon and Florida reports also varied in their descriptions of the fires as physical phenomena. Pieces on the Amazon included relatively few firsthand depictions of the fires as such. Instead, time and again, the writers emphasize not the fires but their effects on the forest, its wildlife, and native rain forest peoples. The articles are crammed with horrific descriptions of turtles, fish, and lizards cooked to death in boiling rivers or fried alive in a gigantic barbecue. "Fire has turned 100-year-old plants to ash, threatened snakes, crocodiles, turtles, armadillos," asserts an article in the *Miami Herald*. "Those who survive may die of thirst, 50 species of mammals could die of hunger."[24] In a marked departure from stories of the Indonesian fires, which stress the effects on orangutans but rarely mention the sufferings of the local population, the writers also describe Yanomami Indian grand-

1. "Amazon Threatened by Fires." Idalino Cordeiro de Souza, thirty-four, cuts a tree with a chain saw near where he lives in the Amazon forest, north of the city of Manaus, Brazil. Photo by AP/World Wide Photos. Reprinted by permission of the Associated Press.

fathers and small children choking on ashes or seeking admission to make-shift hospital wards set up to deal with fire-related respiratory ailments.[25]

The relative absence of fire as a palpable, firsthand presence in most of these articles almost certainly has something to do with the difficulties of access by foreign journalists to the actual areas consumed by flames. "Welcome to My Jungle Nightmare: Four Hundred and Fifty Miles of Jungle at the Back of the Bus," reads the headline on an article by one reporter who decided to visit the scene.[26] At the same time, however, the secondary, largely offstage character of the fires suggests that the real story resides in the more generalized destruction of an Amazon rain forest whose riches bring to mind the DreamWorks El Dorado. The fires themselves thus become one more part of a drama that is nothing less than a struggle for the survival of what more than one reporter labels "the most significant ecosystem in the world" (figures 1 and 2).[27]

The Florida articles, for their part, abound in close-ups of the blazes. No mere backdrop, the fire becomes a full-fledged actor in the events at hand. "Columns of smoke towered skyward," declares the author of a report steeped in a macabre lyricism. "The earth lay cloaked in a gray shadow, as

2. "Firefighter Awed by Blazes' Rage." A backfire, set by U.S. Forest Service officials to stop another advancing fire, rages on Friday, 3 July 1998, in the Indian Trail neighborhood of Palm Coast, Florida. Photo by AP/World Wide Photos. Reprinted by permission of the Associated Press.

if the sun were permanently eclipsed." He then goes on to compare how the first raindrops fell like tears of joy from a sky the color of "dirty cotton."[28] Another article describes "the mud-colored heavens" crammed with "foot-long fragments of ash and an eerie blood-red sun, its light filtered by smoke."[29]

Not simply a bit player, the fires become the actual protagonist of a number of the Florida reports. In striking contrast to the Amazon stories, where personification occurs in a mere four of the one hundred articles that I examined, fire regularly assumes either human or animal characteristics in an impressive twenty-nine of the corresponding reports on the Florida fires. "Swarms" of wildfires sweep through the landscape, "devour" scores of homes, and "attack" the forest. Time and again, a headstrong "Mother Nature" goes on the rampage, as firestorms "march" or "race" from one county to another, or "rebellious" blazes taunt bone-weary firefighters, refusing to be "tamed."

At least one reporter suggests that the Florida fires have become a near-human presence for people accustomed to living in their midst. "The fires have been here so long, they have assumed almost human qualities in the minds of people here in central Florida," she declares. "They 'slumber' at night, they 'wake' with the sun, they 'take over' roads. And above all, they 'just won't go away.' "[30] Longtime residents in this case are people who have lived in a place "almost all their lives" and, in some cases, had parents and "even grandparents" who have lived in the same town, if not the same house, before them.

The idea that fires have long constituted an equally essential part of the Amazonian environment is largely absent from the international reports. So is any sense that the native peoples whose roots in Amazonia may go back not just centuries but whole millennia have used fire to actively transform the land.[31] As a result, the fires not only play separate roles within the two groups of news stories but they also appear within a human and environmental context that reveals dramatic differences.

Over and over again, reporters for the major dailies describe the Amazon as virgin forest in which human transformation of any sort represents a pernicious aberration from a millennial norm. "It doesn't seem possible, this talk of fire in the tropical Amazon rain forest—the largest forest on earth, an emerald-green biological wonder with 20 percent of the world's fresh water and 15,000 kinds of animals, including one-third of all known fish species," declares one reporter for the Knight Ridder news service.[32] Laments the author of another article, titled "Fire Ravaging Large Section of Amazon Rain Forest": "Even pristine rainforest has become vulnerable due to uncontrolled exploitation, global warming, and El Niño." And as a writer for the Associated Press observes,"This year's burning season in Brazil's Amazon rain forest is so bad that even a lake is on fire."[33]

The bioscript of paradise destroyed finds a complement in references to hell and brimstone as the blazes threaten "the extinction of thousands of animals and plant species, many of them unstudied for their potential clues to medical cures and advances in agricultural knowledge." "For environmentalists," explains the author of an article subtitled "Blazes Ravaging Precious Brazilian Rain Forest," the conflagrations offer "a shocking apocalyptic vision of the Amazon's future as, weakened by drought and the encroachment of civilization, it becomes increasingly susceptible to fire."[34] This writer joins a number of other journalists in alluding to a specifically "Yanomami" Armageddon. "The tribe's oral tradition associates great quantities of smoke with the coming of hell—and flying over their territory by plane through gusts of heated smoke, hell is just how parts of their jungle paradise appear," he declares.[35]

Invocations of cataclysm, with its connotations of a violent, often cosmic upheaval of great proportions, are far less common in accounts of the Florida blazes. While acknowledging the severity of the fires that managed to defy a massive fire-fighting effort for weeks on end, the majority of reports speak in terms of an environmental disaster or calamity rather than an outright cataclysm.[36] Moreover, even those writers who describe the blazes as an "inferno" rarely talk about the destruction of a second Eden.

The notably smaller scope (if often higher intensity) of the Florida fires in relation to the massive ground fires in the Amazon explains in part why journalists covering the former would be slower to speak in terms of heaven versus hell. So, to be sure, does the fact that Florida has burned many times before. Recurring references to the state's "fire-accustomed environment" or "fire-dependent ecosystem"—which offers a genuine contrast to the Amazonian case—also makes it harder to invoke a supposedly primeval paradise. Time and again, reporters point out that the land has weathered periodic burns. "Florida's Flat Land and Shrubs a Haven for Fires," reads the headline of one article.[37] They are also quick to call attention to the blazes' beneficial consequences. "Don't Cry for Bambi," proclaims the headline of one story, which goes on to describe how the fires recycle nutrients. "Florida's wildlife have evolved in woodlands and grasslands that are regularly set ablaze by lightning," the author notes. "Believe it or not, the fires set a bountiful table for birds, deer and other animals."[38]

By and large, the Florida reports shy away from the vocabulary of irreparable loss so common in the Amazonian case, where reporters routinely talk about "the worst environmental catastrophe of the century" and even "an environmental disaster without precedent on this planet."[39] Most writers stress the excellent chances for recovery in areas affected by the blazes. "I assure you most of these areas will be back with lush growth by next summer," says Nick Wiley, chief of the state's Bureau of Wildlife Management, in one characteristically upbeat article.[40] While it is again true that Florida has long been the site of recurrent fires, their iconic "naturalness" (and the downplaying of human responsibility for the blazes) is as much, or more, narrative decision as environmental fact.

The routine minimization of destructive human action that provides the true spectacle in reports of the Florida fires finds expression in an array of details. The reporters covering these fires tend to slip in touches of humor that lighten and undercut what in stories about the Amazon, remain overwhelmingly doleful recitations of loss. References to a sign in a parish church ringed by thick black clouds that proclaims the building to be a "smoke-free environment" make it hard to see the Florida fires as part of a cosmic drama. So do descriptions of the baleful pet iguana forced to evacuate its home along with all the members of its human family. Pictures and verbal portraits of dogs lapping contentedly at dixie cups full of ice cubes do little to summon up visions of Armageddon.[41] Unlike the Amazon, which regularly appears as the last bastion of untouched nature, Florida comes across in these articles above all as a place where people much like the reader live.

The fact that there is considerably more rain forest available for saving than there is longleaf pine savanna helps explain the sense of urgency pervading descriptions of the Amazon. However, this urgency only reaffirms the emphasis on a supposedly pristine nature in the Amazonian case.[42] What is, in Florida, an ongoing local or regional drama becomes a novel planetary spectacle in the Amazon.

The reporters who present the Amazon as primeval nature dutifully acknowledge the impact of El Niño. Yet it is inevitably humans who allow what otherwise would be isolated outbreaks to spiral out of control. While the idea that Floridians should abandon their homes out of environmental considerations would evoke howls of laughter if not angry protest, the notion that the great majority of Amazonians are trespassers on a land that ought not to have people crops up again and again in these reports and, often, accompanying editorials.[43]

A number of the Amazonian articles provide close-ups of the small farmers who have set fire to their land to prepare it for cultivation, only to see everything they own disappear in a cloud of smoke. "It was already so dry," Maurício Pereira, forty-four, a farmer in far northern Boca da Mata, explains sadly to a reporter. "We just used one match [to set a whole expanse of forest on fire]."[44] Often openly sympathetic to the disaster-stricken individuals, the reports nonetheless place the blame for the fires on the shoulders of "slash-and-burn agriculturists" as a group. If Maurício and all those hundreds of his ragtag cousins had not set fire to their land, journalists imply when they do not state it directly, the rain forest would not be in the predicament that it is today.

Not just an insistence on human agency but a ready division of the human beings in question into victims and villains is obvious in a good number of the Amazonian reports. If the small farmers are unwitting perpetrators of destruction, the Yanomami, in contrast, regularly appear in international news reports as hapless victims who demand the active protection of the international community. Like the natives in the DreamWorks El Dorado, they quickly prove unable to defend themselves. "Where are they supposed to go? Florida?" demands a spokesperson for one Brazilian environmentalist organization with a greater than intended dose of irony.[45]

More than just another motive for concern, the Yanomami's inability to protect themselves against an army of ignorant intruders becomes a rallying cry for active intervention by outsiders in a number of these reports. Much as Tulio and Miguel must come up with a plan to ward off El Dorado's would-be invaders in the face of the natives' inability to defend this dazzling realm of nature, so international agencies devoted to protecting rain forests and rain forest people must devise a strategy to save the Indians, forest, and planet.

This division of the Amazon into victims and villains is also characteristic of at least some mainline journalistic accounts within Brazil. However, the identity of the villains is often somewhat different. This change is particularly clear in publications such as the widely circulated newsweekly *Veja*— the Brazilian equivalent of *Time* or *Newsweek*.[46] Unlike the international newspapers that almost always paint the Yanomami as innocent (and thoroughly ingenuous) targets of environmental injustice, *Veja* excoriates the Indians, as well as the small-scale colonists, for the "capillary and quotidian devastation that occurs, simultaneously in various locations, and without the slightest control."[47]

Writers for *Veja* do acknowledge the role of the federal government, multinational corporations, and large-scale logging enterprises in promoting deforestation and related environmental transformations that foster fire. They nonetheless insist that "a good part of the destruction results from the agricultural activity of Indians and small-time settlers in invasions sponsored by the Sem-Terra (Landless Rural Workers') Movement."[48] As a result, the authors of one article make a point of taking issue with a widely circulated photo of an abandoned Yanomami hut surrounded by fallen trees that foreign journalists have used to illustrate the fire's dire effects on long-suffering native peoples.

For the reporters in question, this same photo is graphic proof of the Indians' role in the Amazon's destruction.[49] "The abandoned home of the Yanomami Indians surrounded by recently-cut trees: Wrongly-identified victims," reads the caption on this photo. And yet, at the same time that their anger at Amazonian Indians distinguishes them from the great majority of reporters for international publications, *Veja*'s writers corroborate a more general spectacle in which the fires become the product of human action.

The idea of the fires as the result of human intervention is not entirely absent from mainstream accounts of the Florida fires.[50] From time to time, a writer for the *Los Angeles Times* or *Boston Globe* may point out humans' role in fostering the fires.[51] Some refer openly to home buyers' growing taste for luxury houses in the midst of fire-dependent ecosystems where spontaneous lightning-triggered fires previously thinned out vegetation that had become dangerously dense. " 'The greatest challenge is trying to save these homes,' argued Phil Theiler, a Wisconsin forest ranger taking a break Sunday morning after 29 hours on the line. 'It's just a dangerous situation. You have all of these houses built right within this fuel.' "[52]

Moreover, while few mainstream journalists make any mention of the timber industry's routine alteration of the landscape, at least a handful do point out the ironies of public opposition to controlled burns. Asserts a reporter for CNN, "Some people who returned to find their homes still standing were angry that the lush greenery of palmettos and pines that had been a big reason for their moving to Flagler County was gone. Some also found that their neighborhoods had been plowed up by firefighters using bulldozers to create fire breaks."[53] However, the writers rarely drive home the essential "environy" or bio-irony that underlies the larger story—namely, that the same people who don't want controlled fires to impinge on their high-priced landscape are incensed when an uncontrollable blaze sweeps the vegetation that has become a tinderbox. Consequently, the quotes' critical import remains largely unrealized.

By and large, reports in mainstream publications and on their Web versions treat the fires as something that nature has arbitrarily inflicted on unsuspecting human beings. Although the writers do acknowledge the role of human carelessness or arson in some of the conflagrations, most are quick to point out that the experts pin the blame for some 90 percent of the blazes on natural causes such as lightning.[54] "It has been almost a week since a horrific windshift from the west turned dozens of difficult blazes into firestorms that raced across Volusia, Brevard, and Flagler counties," one typical report maintains.[55] According to this bioscript, humans become bystanders and, often, innocent victims. "It just breaks your heart," states one relief worker. "It's the saddest thing I've ever seen."[56]

The human relationship to nature is necessarily different in the two cases because nature itself tends to be differently defined. Much like the sparkling rain forest that appears in *The Road to El Dorado*, the Amazon of the fire reports is a beautiful, but fragile maiden who demands protection from outside. The adjectives "pristine," "precious," and "fragile" appear over and over again in these reports. "Fires ravaged delicate rain forests" reads the caption beneath a photo of burned trees in one typical report.[57] Under assault from blazes set by ignorant small farmers, the lush "wilderness" is similarly "violated" by loggers and miners interested only in short-term profits.[58]

If "nature" regularly equals "virgin forest" in the case of the Amazon, the Florida articles equate this same nature with the fire that attacks people's homes. Both the reporters and the people who they interview frequently decry a capricious and powerful woman's fury. Thus, one report quotes Florida Agriculture Commissioner Bob Crawford as declaring, "There's just nothing you can do until Mother Nature gives us a break."[59] Normally beneficent, nature nonetheless has spells in which she supposedly goes on the rampage and works her mischief by lashing out against humans.

Not only do these two largely unrelated "women" enter into qualitatively different relationships with human beings but people themselves come to occupy dissimilar roles in the two instances. Even when the small-scale Amazonian farmers described by reporters are sympathetic characters, their poverty and often outright ignorance in regard to complex environmental phenomena such as El Niño create an all-but-unbridgeable gulf between them and readers half a world away. "Raemundo Firmino has never heard of El Niño," one reporter declares. "He knows nothing of the world's fascination with the rain forest in his backyard and even less about the scale of the environmental disaster beneath its smoky skies."[60]

This sense of Amazonians as fundamentally different from U.S. newspaper readers is even more accentuated when it comes to the Yanomami (and less often, Macuxi, Wapixana, and Taurepang) Indians. "These are people whose immune systems are wholly different than ours," says one of the reporters interviewed on the *Jim Lehrer NewsHour*.[61] This assertion has a certain truth. As recent debates about a 1960s' measles epidemic among the Yanomami have made clear, Amazonian natives *are* susceptible to diseases to which most *NewsHour* viewers are habituated. And yet, even though not false, this stress on the Indians' total otherness (their immune systems are "*wholly* different" from ours) nonetheless underscores a larger sense of these "Stone Age" natives as members of an appealingly alien world.[62] A full three-quarters of the reports on the Amazon fires that I examined include at least one reference to the Yanomami, and about a third offer a detailed account of their plight. For a number of reporters, the real story is not simply the devastation of the Amazon's rain forests but the accompanying destruction of an allegedly Stone Age culture capable of transporting weary newspaper readers back into an ostensibly more genuine, more harmonious past.[63] The icon of the fiery forest thus, once more, contains within it another icon of a peaceful Golden Age.

Descriptions of Yanomami villages that serve as home to women clad in bright red loincloths and men armed with bows and arrows who sport "simple strings of jungle vines around their waists" reinforce the air of exoticism. In Demini, "about 200 miles from Boa Vista, the capital of the northern state of Roraima," explains one reporter, "villagers live in a *maloca*, a large, round communal hut with an open-air center and thatched sides where the Yanomami sleep at night in skillfully woven hammocks. On a recent visit to the village during the Amazon fires, a lone wild jungle fowl of luminescent purple wandered around freely, nipping at fallen bits of manioc."[64] Accounts of intriguing "ancient customs" such as the rainmaking ritual that is followed by heavy showers the next day similarly emphasize the Yanomami's distance from the modern world. References to legends that predict the dire consequences of human meddling with the earth complement and render poignant the grim facts and figures about ongoing environmental devastation in the Amazon.[65]

Sometimes endearingly homespun, but rarely, if ever exotic in the manner of these vine-swathed Amazonian natives, the people in the articles about the Florida fires are apt to emerge as plucky folks "much like you or me." Intentionally folksy details—such as the "Froot Loops" that weary firefighters gobble down along with peanut butter on toast, sausage, eggs, and fresh-brewed coffee for breakfast; the birdbath that stands guard over a vanished garden; and the bright green tricycle that remains intact on the doorway of a charred house—invite newspaper readers to project themselves into the disaster scene.[66]

Though they too are often cast as victims, the Floridians in these reports respond quite differently to misfortune. While the Yanomami emerge as a kind of endangered species whose plight demands concerted action by outsiders, the Floridians of these reports appear decidedly more able to fend for themselves. Driven from their homes by a towering wall of flames, they fight back, if only by turning on the garden hose full force before they stage their own escape in the family car. Quick to demand help from state authorities, they sally forth to rebuild charred homes and businesses while the Yanomami sit around in stunned silence in makeshift relief camps set up for them by international agencies.

The portrayal of Amazonians as largely mute and frequently passive or disoriented victims whose dazed inertia contrasts with the resilient energy of their Florida counterparts has a good deal to do with the relative lack of official government response to the Roraima fires. Time and again, reporters from both within and outside Brazil cite a myriad of depressing statistics regarding the government's almost complete lack of preparation for the massive blazes. They point out, for instance, that Brazil's total fire-fighting budget was only around U.S.$562,000 in 1998. Due in part to a more general fiscal crisis that prompted cuts in a wide variety of government programs, this budget represented a mere sixth of the allocation for 1996, when there were far fewer fires.[67]

The critics in the Brazilian case also describe official reluctance in the face of pressure from the military. Fearful about damage to national security in a region that outsiders have long been seen to covet, some high-ranking officers lobbied hard to keep out international aid as the fires worsened—a policy that *Veja* sourly describes as "a retrograde nationalism."[68] In some instances, the lack of official action emerges as the single greatest cause of the severe and far-flung destruction initiated by the wildfires. Usually, however, it is Amazonians who bear the lion's share of the blame for the neglect of a nature regularly billed as a panhuman heritage.

Conversely, the great majority of newspaper reports portray the Florida state government as reacting in massive, speedy, and largely exemplary fashion to a natural calamity. They quote then-governor Lawton Chiles's assertions that Florida has done everything humanly possible to combat the fires. For instance, "the governor assured people that every state resource was deployed but he would widen the search," according to one article.[69] There were, the stories make clear, at least 7,000 firefighters from 45 states in Florida at the height of the fires there, or almost four times as many as there were on a 400-kilometer-long (or 248.5-mile) front in Roraima, where the blazes went largely unchecked for almost seven months.

The governor and other Florida officials do not escape blame entirely in mainstream newspaper accounts; "Finger pointing starts in Florida," reads one Associated Press dispatch.[70] However, there is no comparison between the scattered complaints that find expression in mainstream publications with the flood of recriminations that greet Brazilian president Fernando Henrique Cardoso's spokesperson, Sérgio Amaral, when he suggests, "There are limits to what we can do."[71] Although the two governments—one state, one national—acted differently, the gulf in coverage again reflects far more than the actual events. Even if the Brazilian government had responded more adequately to the fires in Roraima, it is unlikely that foreign reporters would have portrayed nature as the chaotic force that makes a routine appearance in articles about the Florida fires.

The differences outlined here add up to two superficially similar stories played out in separate ways. Although both revolve about the image of a burning forest, this image is deployed in ways that prompt dissimilar reactions in the reader. The Florida reports tell a thoroughly North American story of individual fearlessness and the rightful victory of human beings over nature. The Amazon pieces, in contrast, are more apt to portray people either as hapless victims or vandals and environmental illiterates.

These differences are summed up in the role of prayers within the two sets of narratives. In both cases, a number of accounts either begin or end with references to the fire victims' calls for help on the part of a divine being or beings. These prayers, however, reveal important differences in content and purpose within the larger narratives.

People in the Amazon inevitably pray for rain because they cannot rely on other human beings for solutions. Bereft of official protection, the Yanomami are thrown back on what the reports describe as picturesque rainmaking rituals. Likewise, the small farmers who find their fields scorched by fires meant to ensure a fertile harvest are portrayed as having little other recourse than to place their destiny in the hands of plaster saints.

Prayer occupies a similarly central place in some of the Florida articles. Nevertheless, rather than a response to human inaction or official incompetence, entreaties to a divine being appear as confirmation that humans have done everything possible to control a temporarily chaotic nature. Now it is God's turn to reward their efforts by restoring order in the cosmos. "Florida Prays for Rain as Firefighters Gain Upper Hand" affirms the headline of one CNN online post.[72]

Not infrequently, the larger spectacle enacted in the Florida reports is one of unity and healing. "God is calling people together to help us here," Dave Cooper, education minister of the First Baptist Church, explains. "Out of all of this, I sense that God wants . . . to heal our land and our people."[73] Sometimes, the fires appear as a reminder to the members of an openly consumerist society that material goods are less important than is compassion and solidarity among human beings momentarily (but only momentarily) stripped of their possessions.

The underlying moral is different in the Amazonian stories, where prayer may shade off into environmental protest. The Yanomami's insistence on the catastrophic consequences of humans' meddling with nature that finds indirect expression in some of these reports, often becomes a pointed denunciation of the miners, loggers, and other intruders (many from outside Amazonia) who continue to cast a greedy eye on Yanomami lands. The Yanomami's specific protests, however, are often plugged into a larger, more general story about deforestation, which has already stripped an area of the Amazon as large as California and threatens to destroy what writers regularly describe as "the earth's last remaining paradise."

These concerns about the fate of paradise bring us back to *The Road to El Dorado*. Political concerns help explain a number of the differences between accounts of the El Niño fires in Florida and the Amazon. The Yanomami, for example, unquestionably conduct their rainmaking ritual partially for the benefit of news-hungry journalists who they see as vehicles in getting their own message out into the world. Yet, these contemporary, practical objectives are routinely couched in a much older symbolic language. Reporters covering the "rain forest fires" repeatedly draw on a familiar, El Doradian vocabulary to encourage a particular set of preservationist actions.

The ensuing bioscript (great, if elusive wealth that demands transformation by outsiders) may be utilized for what most readers would consider laudable ends. Reporters may invoke the idea of the Amazon as a terrestrial paradise in order to pressure the Brazilian government to call in international rescue missions. They may make use of these same ideas to effect changes in laws that had previously allowed settlers to burn up to eight acres of forested land without any sort of official authorization.

At the same time and far less fortunately, the use of El Doradian elements often encourages scapegoating of Amazonians of mixed blood. The iconic simplification of a vast and intricate Amazon into a homogeneous paradise reduces a complex issue of social and environmental causes and effects. In the ensuing spectacle, local populations are forced either to play the villain (pyromaniacal small farmers) or appear as victims (dazed, long-suffering Indians). Laced with vocal calls for outside intervention, the ensuing bioscript lends legitimacy to programs that would buy up and fence in large tracts of tropical nature with little or no regard for the people who consider it their home.[74]

Tulio and Miguel save the day in El Dorado. They defeat the evil sorcerer, trick the greedy Cortés into leaving, and ride off into a glittering forest that they have helped to preserve for themselves as well as the natives. No wonder a discriminating five-year-old like Clayton would pronounce their technicolor exploits "pretty cool."

Unfortunately, in the real world, Cortés didn't go home. Instead, he and others like him stuck around to plunder much of Mexico. El Dorado could not be lost because its pursuers never found it. However, the conquistadores effectively laid waste to any number of lands and peoples in their quest for the mythic city's golden treasures. The outsiders' magic that saves a distant dream of paradise in a Hollywood movie is unlikely to help any of the real-life natives whose descendants are apt to find themselves providing cheap labor for the kinds of massive development projects that are making a tourists' paradise of Cancún and the rest of the so-called Maya Riviera.

And yet, if the DreamWorks movie clearly cannot extinguish the fires that continue to spring up in Amazonia, it can help us to confront our own images of tropical nature in relation to the nature we find in our backyard. In making visible the customarily transparent icons and the larger assumptions that underlie them, the city on the silver screen allows us to rethink the domestic and international policies in which our own actions (or inactions) have a role. The Amazon and Florida may be very different places, but they are part of the same planet and, ultimately, the same symbolic universe. "Florida's struggle for balance between civilization and nature likely will continue," opines the state's fire management administrator.[75] He could just as well have been referring to the Amazon.

NOTES

1. *The Road to El Dorado* was directed by Don Hall. The principal voices were those of Kevin Kline (Tulio) and Kenneth Branagh (Miguel).
2. For an introduction to colonial accounts of El Dorado, see Juan Gustavo Cobo Borda, ed., *Fábulas y leyendas de El Dorado* (Barcelona: Tusquets Editores, 1987); and John Hemming, *The Search for El Dorado* (London: Joseph, 1978).

3. As Scott Fedick points out in his discussion of shifting visions of the Maya, these elements represent a mélange drawn from different monuments in different parts of the far-flung Maya homeland. There are also Olmec and Aztec borrowings that suggest ties to the more northern part of Mexico.

4. Most of the film critics for major dailies were less enthusiastic. The film got tepid reviews at best from the critics, who noted the similarities between the plot and that of *The Man Who Would Be King*, a 1975 screen success starring Sean Connery. The DreamWorks film, which had attempted to challenge the Disney monopoly on animated movies, grossed a total of slightly over $50 million in the United States.

5. This sort of sharp division is less evident in the reports on the massive wildfires in Colorado and Arizona that broke out in June 2002, just as this book entered production. These later reports on the western fires acknowledge human agency more than the Florida reports do. However, their tendency to place the blame on a single human (a white, female, National Parks Service employee in Colorado; a twenty-nine-year-old Apache firefighter in Arizona) diverts attention from the other sorts of human caused problems (namely, home construction in wooded areas) that they acknowledge but rarely develop.

6. For a discussion of the concept of tropicality, see David Arnold's chapter on "The Invention of Tropicality," in his *The Problem of Nature: Environment, Culture, and European Expansion* (Oxford: Blackwell Publishers, 1996), 141–168. See also Candace Slater, *Entangled Edens: Visions of the Amazon* (Berkeley: University of California Press, 2001).

7. These caricatures of native peoples have not been lost on a number of critics—above all, those writing for a public with a good number of Hispanic readers. See, for example, Olin Tezcatlipoca, "*The Road to El Dorado* has no Respect for History," *Los Angeles Times*, 10 April 2000, F3; and Peter Ritter, "*The Road to El Dorado* Turns the Rape of the Americas into a Ride through the Magic Kingdom," *City Pages* (online version) 21, no. 1,008, article 8,548, 28 March 2000.

8. During an El Niño year, the trade winds relax in the central and western Pacific leading to the eastward displacement of the atmospheric heat overlaying the warmest water. This displacement results in large changes in the global atmospheric circulation, which in turn forces changes in weather in regions far removed from the tropical Pacific. El Niño ("The Boy" in Spanish) is a reference to the Christ Child. It was originally recognized by fishermen off the coast of South America as the appearance of unusually warm water in the Pacific Ocean around Christmastime. For an introduction, see S. G. H. Philander, *El Niño, La Niña, and the Southern Oscillation* (San Diego: Academic Press, 1990).

9. In 1997, fully half of the fires registered were in the state of Mato Grosso. For a series of useful statistics and ongoing reports on Amazon fires, see the statistics available from the Environmental Defense Fund, Woods Hole Research Institute, and Institute of Environmental Research in the Amazon. Data based on satellite images of the fires from the National Institute for Space Research are available on the Internet at http://condor.dsa.inpre.br.mapas. For a helpful overview of fires in the Amazon, see Daniel C. Nepstad, Adriana G. Moreira, and Ane A. Alencar, *Flames in the Rain Forest: Origins, Impacts, and Alternatives to Amazonian Fire* (Brasília: Pilot

Program to Conserve the Brazilian Rain Forest, 1999). The book includes a useful, primarily English-language bibliography (141–150).

10. There is a vast bibliography on deforestation within the Amazon. For a summary of some of the latest work, see Charles H. Wood and Roberto Porró, eds., *Patterns and Processes of Land Use and Deforestation in the Amazon* (Gainesville: University Press of Florida, 2002).

11. The budget figures vary, depending on the source. One report cites U.S.$3 million in emergency aid to indigenous communities, and $17 million to stimulate the economy. Another cites $6.6 million spent on fire control. What is certain is that the two thousand firefighters who moved into the area cannot compare with the seven thousand who battled the fires in Florida.

12. The reserve, which is 68,331 square miles, or approximately the size of Missouri, spans the Brazilian-Venezuelan border. The Brazilian portion was decreed by then-president Fernando Collor in 1991.

13. For an introduction to fire as a natural phenomenon that is also a potent symbol and cultural presence, see the writings of Stephen J. Pyne, including *World Fire: The Culture of Fire on Earth* (New York: Henry Holt and Company, 1995), and *Fire in America: A Cultural History of Wildland and Rural Fire* (Princeton: Princeton University Press, 1982). The first book contains an article specifically on the Amazon ("Queimada Para Limpeza," 60–75). It is worth comparing the treatment of the Indonesian fires during this same period. Although many of the causes and effects are similar, reports of the Amazon offer much heavier coverage of native peoples.

14. This repetitiveness is hardly surprising given many smaller newspapers' dependence on major news services such as the Associated Press.

15. There are also significant similarities in coverage of the Florida fires and those that swept much of the American West in 1999 and 2000.

16. I am grateful to the Center for Latin American Studies at the University of Florida for the opportunity to spend a semester there as the Bacardi Family Eminent Scholar during spring 1999.

17. There is a large bibliography on both Amazonian and Floridian ecosystems. For starters in the Amazon case, see Nigel J. H. Smith, *The Amazon River Forest: A Natural History of Plants, Animals, and People* (New York: Oxford University Press, 1999). For Florida, see William J. Platt, "Southeastern Pine Savannas," in *Savannas, Barrens, and Rock Outcrop Plant Communities of North America*, ed. Roger C. Anderson, James S. Fralish, and Jerry M. Baskin (Cambridge: Cambridge University Press, 1998), 23–51. The article includes an ample bibliography (43–51).

18. Originally grown for timber on a fifteen-year cycle, these trees are now harvested for pulp after a mere three years. I am grateful to Professor Francis E. Putz and Gary Peterson for helping me better to understand the dynamics of the Florida fires.

19. "There are an awful lot of subdivisions being built in a way in which they try to maintain as much of the natural landscape as possible," notes Steve Lindeman, resources manager at Tall Timbers, a nonprofit ecological research station in Tallahassee. "The landscape around the house would have burned naturally within 3 to 15 years, and now those subdivisions are 15 years old." Cited in Mike Oliver, "A Preventable Tragedy," *Orlando Sentinel*, 7 July 1998.

20. The state government declared a state of emergency in January 1998. In March, the federal government assumed control.

21. The statistics vary somewhat with the source. According to a *New York Times* report of 6 July 1998, the Federal Emergency Management Agency said that 1,946 fires had burned 458,288 acres since 1 June, causing an estimated $276 million in damage and making the fires the worst natural disaster since Hurricane Andrew. See Rick Bragg, "In Florida, Stoicism and Thankfulness for What's Left," *New York Times* online, 6 July 1998. The fires got hot enough to burn through the top organic layer of soil in only 10 to 20 percent of this area.

22. Articles in these sections tended to be more critical than others labeled as straight news. They are nonetheless in the minority.

23. Katherine Ellison, "Fire Ravaging Large Section of Amazon Rain Forest," *San Luis Obispo County Times-Tribune*, 28 March 1998, C9. The article is a dispatch from the Knight Ridder news service that was picked up by many local newspapers.

24. "New Fires Add to Trouble in Drought-Parched Amazon," *Miami Herald*, 24 March 1998.

25. This difference may be less meaningful than it first appears since the narrative treatment of the Yanomami approximates that of U.S. endangered species. See, by way of background, Candace Slater, "Amazonia as Edenic Narrative," in *Uncommon Ground: Rethinking the Human Place in Nature*, ed. William Cronon (New York: W. W. Norton, 1996), 119–121.

26. Anthony Faiola, "Welcome to My Jungle Nightmare: 450 Miles of Jungle at the Back of the Bus," *Washington Post*, 6 April 1998, 21.

27. Anthony Faiola, "Amazon Going Up in Flames: Blazes Ravaging Precious Brazilian Rain Forest," *Washington Post*, 27 March 1998, A01. Here, my point is not that this description of the Amazon as the most significant ecosystem is necessarily untrue but that the privileging of nonhuman nature in the Amazon has significant practical effects in terms of how events that occur there are understood by outsiders.

28. Michael Browning, "Winds, Rain Bring Relief; Flagler County Evacuated," *Miami Herald*, 3 July 1998.

29. Lori Rozsa, Gary Long, and Martin Merzer, "Swarms of Wildfires Sweep through Urban Areas," *Miami Herald*, 2 July 1998.

30. Heather Mahar, http://www.ABCNEW.com, 24 June 1998.

31. For an introduction to native peoples' handling of the land, see Darrel Posey and William Balée, ed., *Resource Management in Amazonia: Indigenous and Folk Strategies* (New York: New York Botanical Garden, 1989). See also the pioneering work of Anna Roosevelt, Hugh Raffles, and Emilio Moran.

32. Monica Yant, "As Rain Forest Burns, Brazil's Farmers Pray for Divine Intervention," *San Luis Obispo County Times-Tribune*, 11 April 1998, E-5.

33. Ellison, "Fire Ravaging Large Section of Amazon Rain Forest"; Michael Astor, "Amazon Threatened by Fires," *San Francisco Chronicle* (AP news service, online version), 29 October 1997.

34. Faiola, "Amazon Going up in Flames."

35. Anthony Faiola, "Flames Impart Life to Indian Legend: Slash-and-Burn Amazon Clearing Triggers Yanomami 'Armageddon,'" *Washington Post*, 8 April 1998, A01.

36. Synonyms for "disaster" include "mischance," "misfortune," "mishap," "accident," and "reverse." The term "cataclysm," in contrast, comes from the Greek word for flood, and con-

notes "a sudden and violent upheaval." See the entries for both words in *The Random House Dictionary of the English Language*, 2d ed. (New York: Random House, 1987).

37. Matthew L. Wald, "Florida's Flat Land and Shrubs a Haven for Fires," *New York Times*, 8 July 1998, A14.

38. Cyril T. Zaneski, "Don't Cry for Bambi," *Miami Herald*, 8 July 1998.

39. "Rains Quench Fires," Associated Press, 1 April 1998. The speaker is Carlos Pereira Monteiro, head of a UN team of fire fighting experts.

40. Cited on ABCNEWS online, 8 July 1998.

41. All of these details appear in Michael Browning, "An Eerie Landscape of Flame, Smoke, Ash," *Miami Herald*, 2 July 1998.

42. Approximately 98 percent of southeastern longleaf pine savanna has disappeared. Although estimates vary on how much of the Amazon's rain forest still survives, the percentage is many times higher and explains, in part, the note of urgency in calls to "Save the Rain Forest" of the sort described in the introduction to this book.

43. See, for instance, the minutes of 8 May 1998 for the European Working Group on Amazonia, in which Johan Bosman, the member representing a Flemish support group for indigenous peoples, states outright that land reform has no place in the Amazon. Instead, "people migrating into this region should be stopped while those [non-Indians] already inhabiting the Amazon should be encouraged to leave." The full text of this lengthy document is available at http://www.amazonia/org.br/ewga.htm.

44. Cited in Yant, "As Rain Forest Burns, Brazil's Farmers pray for Divine Intervention." The name is almost certainly Maurício in Portuguese.

45. The speaker is Claudia Andujar, director of the Pro-Yanomami Commission. See Ellison, "Fire Ravaging Large Section of Amazon Rain Forest."

46. *Veja* began publication in the early 1970s, during the height of censorship and repression on the part of the military government that ruled Brazil between 1964 and 1984. Reflecting the interests of the middle and upper classes in the country's industrial south, it is presently on the center Right of the Brazilian political spectrum.

47. "Gato por lebre," *Veja*, 22 April 1998, 48. The title literally means "A Cat for a Hare" and refers to a case of mistaken identity. In Portuguese, the quote reads: "A devastação capilar e cotidiana, que ocorre, em vários pontos simultâneamente e sem nenhum controle." The hostility toward Amazonian Indians evident in this article is part of a much larger phenomenon. For an introduction, see Alcida Ramos, *Indigenism: Ethnic Politics in Brazil* (Madison: University of Wisconsin Press, 1998).

48. "Gato por lebre." The original sentence is: "Boa parte da destruição resulta da atividade agrícola de índios e colonos de invasões promovidas pelos sem-terra." For an introduction to the important Sem-Terra movement, see Christa Berger, *Campos em confronto: A terra e o texto* (Porto Alegre: Editora da Universidade Federal do Rio Grande do Sul, 1998). A similar logic to that pervading the *Veja* reports can be found in Indonesia, where the government blamed the slash-and-burn technique of the peasants for the fires that turned out to be tightly linked to large-scale agricultural enterprises, above all oil palm plantations.

49. The photo appears, among other places, in "UN Helps Fight Amazon Fires," BBC News online, 25 March 1998.

50. This idea is certainly present in alternative accounts. While many of these are by academics, there are also a number of more popular reports. See, for example, Ron Matus, "Clearing the Path," *Gainesville Sun*, 8 June 1998, 1A, 6A.

51. For example, see James Gerstenzang, "Florida Fires Show Dangers When Man Tries to Tame Nature," *San Francisco Chronicle*, reprinted from the *Los Angeles Times*, 25 July 1998, A3; and Mike Oliver, "A Preventable Tragedy," *Orlando Sentinel*, 7 July 1998, A1.

52. Meg James and Richard Browning, "Relief and Struggle," *Miami Herald*, 6 July 1998.

53. "Time to Head Home; Now What?" CNN online, 7 July 1998.

54. "Florida's wildlife have evolved in woodlands and grasslands that are regularly set ablaze by lightning, the scientists say" (Zaneski, "Don't Cry for Bambi"). A CNN posting on the Web on 22 June 1998 affirms that "an average of 80 new fires begin every day, 90 percent of them caused by lightning."

55. Phil Long, "Light at the End of a Smoky Tunnel," *Miami Herald*, 8 July 1998.

56. Cited in Rozsa, Long, and Merzer, "Swarms of Wildfires Sweep through Urban Areas."

57. "Rain Douses Brazil's Amazon Fires," BBC news, 2 April 1998.

58. The rape of the forest is a theme not limited to the Amazon. For a useful comparison, see reports about the Indonesian rain forest, which suffered disastrous blazes somewhat earlier.

59. Cited in Rozsa, Long, and Merzer, "Swarms of Wildfires Sweep through Urban Areas."

60. Yant, "As Rain Forest Burns, Brazil's Farmers Pray." Raemundo is almost certainly a misspelling of Raimundo.

61. "Ecological Disaster," *Jim Lehrer NewsHour*, 14 April 1998.

62. Much has been made of the Yanomami's lack of immunity from common diseases such as measles. See Patrick Tierney, *Darkness in El Dorado* (New York: W. W. Norton, 2000); and the various critiques of the book's allegations regarding the measles epidemic that killed hundreds of Amazonian Indians in 1967–1968.

63. See, for example, Daniela Hart, "Brazilian Wildfires Threaten Indians," *Washington Post*, 20 March 1998, A29. The Yanomami are regularly described as "primitive peoples." Also, although the existence of this largest remaining group of Amazonian Indians has been documented at various points in the twentieth century, they often appear as "recently discovered hunters and gatherers." Their discovery by the international media is, in fact, relatively recent. For a discussion of the differences in journalistic treatment of the Yanomami and Kayapó, see Slater, "Amazonia as Edenic Narrative," 114–131.

64. Faiola, "Flames Impart Life to Indian Legend."

65. For a discussion of a similar presentation of the Yanomami as beings not quite of this world, see Slater, "Amazonia as Edenic Narrative," 114–131.

66. The Froot Loops appear in James and Browning, "Relief and Struggle." Class difference emerges here as much or more than regional difference.

67. Ronaldo de Oliveira, "Mais Queimadas, Menos Verbas," *Correio Braziliense*, 5 September 1998. "The fires are so vast and the operation to contain it so small that it is like using needle

and thread to sew up a gash in the Hindenburg," one U.S. reporter says with an almost audible sigh (Faiola, "Amazon Going Up in Flames").

68. Klester Cavalanti and Vladimir Neto, "Fogo, Omissão e Bravatas," *Veja*, 1 April 1998.

69. Rozsa, Long, and Merzer, "Swarms of Wildfires Sweep through Urban Areas."

70. John Pacenti, "Finger Pointing Starts in Florida," Associated Press. The article documents homeowners' complaints that red tape contributed to the devastation and that the lack of controlled burns by state officials added to the loss of 484,000 acres in six weeks, making the wildfires some of the worst recorded in the state during all of that century.

71. Cited on BBC online, 20 April 1998.

72. "Florida Prays for Rain as Firefighters Gain Upper Hand," CNN online, 28 June 1998.

73. Cited in James and Browning, "Relief and Struggle."

74. I am thinking here of the minority of conservation biologists who are arguing for the new bracketing off of local populations as a means of preserving forests. For two expressions of this position, see John Terborgh, *Requiem for Nature* (Washington: Island Press, 1999); and Thomas T. Struhsaker, *Ecology of an African Rain Forest: Logging in Kibale and the Conflict between Conservation and Exploitation* (Gainesville: University Press of Florida, 1997).

75. Jim Brenner, cited in Mike Oliver, "A Preventable Tragedy," *Orlando Sentinel*, 7 July 1998. "The public attitude is the biggest hurdle," Brenner says at another point in this article.

SUZANA SAWYER

SUBTERRANEAN TECHNIQUES:

CORPORATE ENVIRONMENTALISM,

OIL OPERATIONS, AND SOCIAL INJUSTICE

IN THE ECUADORIAN RAIN FOREST

In late 1999, ARCO published a coffee-table book commemorating its oil operations in the rain forests of the Upper Amazon.[1] Titled *The Villano Project: Preserving the Effort in Words and Pictures*, ARCO's book documents the technological wonders that ARCO Oriente, Inc. (ARCO's subsidiary in Ecuador) conceived in designing and constructing the infrastructure needed to discover, develop, and pump petroleum out of the Upper Amazon. "Ecuador's rainforests," the book notes, "rank among the world's most biologically diverse regions." And it was ARCO's concern for protecting this "incredibly rich environment" that spurred its geophysicists and engineers to harness their "technological ingenuity" and build oil operations that would leave only a "minimal footprint" in the rain forest.[2] As the caption to the book's cover assures, the "Villano well site sets the new standard for [oil] development in the green sea"—that is, the verdant, wet tropical forest.

The Villano Project is a photo gallery, and the more carefully I leafed through its pages, the more intrigued I became. It is packed with vibrant and compelling images that fill me with the sights, sounds, and smells of Pastaza—Ecuador's central Amazonian province where both ARCO and I have worked over the past ten years: ARCO as an oil company and myself as a cultural anthropologist. The numerous aerial photographs of a seemingly endless mantle of lush rain forest brought back for me the thrill of flying low over a densely forested landscape. Pictures of the forest elegantly shrouded in morning mist almost allowed me once more to feel the cold, dank dawn air seep into my bones. The photo spreads of tree frogs and Hercules beetles triggered an inward chuckle I have often felt when staring at the forest's somewhat fanciful life. Even the pictures of the rain-drenched and mud-splattered working conditions reminded me of the familiar loamy smell of Upper Amazonian soils after a storm.

Interspersed in ARCO's book among these pictures of biological wonders are photos of technological ones: the heavy equipment, sophisticated technology, elaborate infrastructure, and crucial labor power that was necessary to construct oil operations in the neotropics. *The Villano Project* chronicles the engineering acrobatics that ARCO performed in meeting the challenges that the Upper Amazon posed (with its mountainous terrain, intense rainfall, and fragile ecosystem), concentrating on the more unique aspects of its undertaking (such as the fact that the Villano well site was comparatively small with multiple oil wells emanating from this single location; that the company located its facilities for processing crude outside the rain forest; that it built the pipeline that travels through the rain forest without a road; and that all the equipment and labor needed to build ARCO's rain forest operations were transported by helicopter). Given the massive materials and machinery needed to stabilize a well site, build a drilling platform, drill ten two-mile-deep oil wells, and construct a pipeline in a region where the soils are unstable and the earth is riven with geologic faults, this was no easy feat. ARCO's message rings clear: oil development and environmental preservation are not antithetical. Together they form the core icon in ARCO's story: petroleum prowess and ecological conservation go hand in hand.

Yet ARCO's book fascinates me not simply because its pictures bring to life a place that I know. Nor does it draw me in simply because, being the granddaughter of a petroleum engineer and the daughter of a petroleum geologist, I can appreciate the sophisticated technologies that ARCO used in building its operations. I find *The Villano Project* fascinating because it also disturbs me. In looking at and reading ARCO's book, I am torn about its message. Why do I want to believe that ARCO's message is true (oil development and environmental conservation can go hand in hand), yet still have deep apprehensions? Although the images in ARCO's picture book do capture much that I recognized in Pastaza, there is also much that they leave out. Despite the fact that I can recall my thrill at flying over Pastaza's rain forest expanse, I can even more clearly recollect that of Marco, a young Indian leader. Marco's exhilaration, however, came more from pointing out the abandoned settlement of a fabled Indian warrior, the site of his next *purina* (distant agricultural and hunting zone), and his grandmother's *chacra* (forest fields of swidden agriculture) than it did from seeing the forest from above. Even though images of the forest steeped in mist bring back memories, I more fondly remember *Miquia* Magdalena chanting about her dream the previous night as I warmed myself by her early dawn hearth.[3] Similarly, though pictures of fantastic critters trigger wonder, I am more moved by my two-day forest trek with Carmen and her seven children to her purina and *purun* (managed fallow). And while the water-drenched working conditions bring back a familiar scent, I can remember more vividly the smell of over three hundred bodies crowded in an assembly hall to protest ARCO's activities during a torrential rain.

My point is not that ARCO's picture book should show all these other images. Like all picture books, ARCO's is made up of photographs carefully selected for a particular purpose—in this case, to exhibit the company's achievements. Rather, my point is that by portraying the forest as an exclusively biophysical realm, ARCO negates its deep social character. This bioscript—that the environment is a biological and geophysical entity—makes all the difference for how ARCO, as well as readers such as you and I, think about its operations. By depicting the rain forest as a peopleless place of biological nature, ARCO set up its challenge, solution, and the measure for evaluating its success along restricted technical lines. The challenge, for ARCO, was to extract crude oil from the rain forest without undermining the forest's ecological integrity. The solution was in developing the right technological expertise and equipment. And the degree to which ARCO achieved its goal could be assessed by the extent of environmental disruption. By contrast, a different bioscript—one of a peopled rain forest—would make it difficult to present ARCO's operations as an unequivocal success.[4]

Rather than assess the environmental soundness of ARCO's oil operations in the Ecuadorian rain forest, I examine here how a technological rationality—or way of thinking—comes to replace an ethical and political one.[5] That is, I am troubled by a growing trend within the hydrocarbon industry: how a scientific, managerial authority overrides and even silences questions that challenge the legitimacy of resource extraction. How, more specifically, did ARCO manage to lay claim to and control processes within the rain forest while simultaneously muffling the voices of others? How were ecologically sensitive petroleum techniques represented as near-noble feats in the service of our global industrial society? And how did a narrow focus on environmental soundness come to limit debate over petroleum operations and the fate of the people in the forest? While I focus here on ARCO and Ecuador, the processes I describe are present throughout Latin America and the Upper Amazon in particular—a kind of tropical Arabia that ships more oil to the United States than all the Persian Gulf countries combined.[6] I suggest here that ARCO's highly technical and rational focus on how to make a messy business clean and green effectively deflected, even buried, many social, political, and economic issues from public discussion. An overemphasis on the technological calculus for making oil operations ecologically correct depoliticized the profoundly unequal context in which transnational petroleum extraction occurred.

ARCO's sharp focus on the environment, I suggest, went hand in hand with its continual attempt to contain the presence of indigenous peoples and erase the role of the military in securing its oil operations. From the moment it began exploring for oil in 1988, the corporation publicly championed its corporate compassion for the environment while simultaneously concealing its part in setting local indigenous communities against each other, weakening indigenous political capacity, and using both state and private security forces to stifle local protests. Both the limited role of indigenous peoples and the use of force reflect the violence at the core of petroleum exploitation. ARCO used subterranean technologies in managing both the geophysical *and* social terrain of its oil operations. Not only were the company's cutting edge technologies for extracting crude oil "subterranean"; so too were the undermining practices that ARCO employed in preparing the social ground for its operations to proceed unhindered.

The first image appears on the dust jacket of ARCO's picture book (figure 1). It offers a grand, aerial view of a rain forest that sweeps around the book's bounded edge and spills across its back cover. The forest is dense and vibrant. It appears teeming with life as it undulates across the landscape as far as the eye can see. In the center of the photograph is the Villano well site—submerged in the forested "green sea," yet towering as a technological wonder. Below the well site, bold print announces: "The Villano Project"—a citadel of progress in the midst of pure nature. The well site itself is compact and contained, encompassing only four hectares (ten acres) and housing up to a hundred workers. Most impressively, this symbol of late industrial civilization has neither destroyed its luxuriant natural surroundings nor apparently interfered in ecological processes. ARCO's 210-foot oil rig stands alongside majestic trees.[7] The only humans in sight are the speck-size individuals at the base of the Villano oil rig. And the only artifact of a human presence is the well site itself. In the center of the book's back cover, a rainbow arches through the sky and plunges into the forest. Both the well site and rainbow point to riches: the black gold (oil) lodged deep below the rain forest and the new green gold (biodiversity) housed within it.[8] Technological progress and nature happily coexist.

The angle of this photograph offers a perspective one associates with grandeur and transparent vision—a vantage point that one scholar has called the "god-trick," that ability "to see everywhere from nowhere."[9] This is the view that ARCO invites the reader to share. Extracting crude oil from the rain forest is no easy task. Only one hundred miles further to the north, oil operations in the Upper Amazon have led to extensive and devastating industrial contamination.[10] Yet ARCO appears to have succeeded where others have not. Deus ex machina: like the god in a Greek drama who appears from above to unravel complexity and resolve a plot, ARCO has managed to extract petroleum while preserving the forest. In the struggle between the global desire for hydrocarbon fuels and the global desire to save biodiversity, ARCO has reconciled the paradox and, as if through divine intervention, created the Villano Project. Both literally and figuratively the well site dropped from the sky.

1. The Villano Project. ARCO 1999, courtesy of BP America Inc.

But what is the bioscript in the picture above that allows the rain forest to take on the aspect of pure nature and ARCO's operations to embody benevolent development?[11] I raise this question to encourage pause over the wave of techno-scientific environmental expertise that has flooded dominant understandings of the rain forest.[12] There is a politics, I suggest, to an exclusively biophysical understanding of nature; it sets the conditions that enable some groups (while disabling others) to define the rain forest in very particular ways. Few people today are against conservation in and of itself. The problem arises when a strict environmental bioscript that authorizes conservation-minded action ends up simultaneously normalizing the profoundly unequal social, political, and economic terrain on which such practices take place.

The book's origin story of the Villano Project is a good place to start unraveling how and why ARCO imagined the rain forest as a site of pure nature. In 1988, ARCO acquired rights to explore for and exploit hydrocarbons from the Block 10 oil concession. Inaccessible by road, the 200,000-hectare concession was located in what the company called the "primary rainforest," by which it meant an untouched, timeless forest. In 1989, ARCO's seismic exploration identified potential oil deposits, and in 1992, the company discovered an oil field that it deemed to have "excellent reservoir characteristics."[13] A second exploratory well in 1993 confirmed the discovery and fact that the Villano field could be a commercial success. Yet according to the company, it would not be easy to develop this newly found "asset." First, because the field was located in a sensitive physical environment, its development would require, in ARCO's words, "extraordinary measures." Second, a "number of incidents" (that is, demonstrations against oil operations) at various petroleum installations in Ecuador made ARCO "uneasy about developing the Villano discovery." And third, "protests at [ARCO's] headquarters in Los Angeles" in opposition to its operations in the Ecuadorian rain forest further compelled the company to think twice. ARCO was "concerned about potential demonstrations and bad publicity" if it decided "to proceed with the Block 10 development." U.S. executives would, however, agree to proceed with the project on the condition that they received "assurance that a development plan could be created [that would] protect the environment and [thus] shield the company from valid criticism by environmental and indigenous groups." In 1994, such a plan was developed, although its "implementation was delayed."[14] In 1997, ARCO signed an agreement with the Ecuadorian state to develop the oil field and the Villano Project chronicled in the picture book began.[15]

At first glance this story seems straightforward. ARCO is a corporation that has long prided itself on being environmentally responsible.[16] Since Block 10 was situated in such a sensitive environment, only after the right "technological ingenuity" was devised to protect the rain forest's ecology could ARCO proceed in good conscience. But two things needed to be assumed for the magic of technology to assuage all concerns: first, ARCO had to label all "incidents"—that is, protests and demonstrations—as "environmental," and second, it had to see the "environment" in strictly biophysical terms.

Since the late 1960s, debate around the oil industry has largely been framed in dichotomous terms: the industry versus the environment. Over the decades, environmental activists have focused a sorely needed spotlight on the deadly effects of petroleum contamination. Significantly, criticism from the environmental movement has critically reshaped the workings of the industry in First World ecologies, and slowly, environmental concerns are affecting how petroleum is produced in Third World ecologies too. Given the devastating effects that petroleum contamination has had in the Ecuadorian Amazon (as well as elsewhere), the importance of this cannot be overemphasized.[17]

Yet framing concerns about hydrocarbon extraction in dichotomous terms (industry versus environment) has also had contradictory effects. As visible in industry advertising and ARCO's book, oil companies have sought to prove that not only have they taken environmental critiques to heart but that with their newfound green consciousness, corporations are just as concerned about the environment as environmentalists.[18] An anecdote captured in ARCO's 1998 *Annual Report* under the heading "Telling Differences?" underscores this point. Allegedly, in a meeting among corporate executives in the early 1960s, the meeting chair said, "Let's have the environmentalists sit on this side of the table, and all the businessmen on the other side." To which the then-president of ARCO replied, "But Mr. Chairman, how do you tell the difference?"[19]

In many ways, the petroleum industry has surpassed the environmental lobby at its own game. Not only have a number of oil companies attempted to make their operations squeaky-clean but they have also roundly broadcast their support for an array of environmental conservation projects to strengthen their legitimacy all the more.[20] In the case of ARCO, it disarmingly turns the spotlight on its own practices and opens its operations up to public scrutiny.[21] We are asked to look at ARCO's operations and judge the company's accountability with our own eyes. The coffee-table book's images and text bear witness to ARCO's ultimate concern for the environment, exacting technical expertise for confronting environmental challenges, and unparalleled responsiveness to public concerns. Yet wittingly or not, a subtle sleight of hand is at work here. Without even being aware, the reader is included in a singular "we," a collectivity where everyone is presumed to share a common commitment to saving the rain forest. In an age of hyperconcern for the environment (especially for sensitive, fragile ecologies), ARCO's "relentless commitment to protecting the environment" becomes the measure against which to assess the validity and legitimacy of its operations.[22] What gets left out, however, are the voices not preoccupied with the "environment" per se, but very much concerned about the rain forest as a lived landscape.

The anecdote from ARCO's 1998 *Annual Report* appeared in a section titled "Putting Technology to Work: Innovative Approaches Save Money and the Environment." The article highlighted the company's "technological ingenuity" by juxtaposing its operations in two "fragile environments": the frozen tundra of Alaska's North Slope and the Ecuadorian rainforest.[23] Despite fierce environmental challenges, technology has allowed ARCO to exploit petroleum from both the North Slope and rain forest while saving money and the environment too. It is not surprising that ARCO would allude to saving money and the environment in the same breath; the corporation has long held that both make "good business sense." Indeed, both are forms of capital. Saving money generates fiscal capital: it boosts profits, increases corporate coffers, and pleases stockholders. Saving the environment generates cultural capital: it "enhances" ARCO's "reputation as a responsible global citizen," increases consumer confidence, and also pleases stockholders.[24] Saving money is a measure of profitability; saving nature is a measure of legitimacy. Yet there is something disturbing here. As an empty standard of value, money acts to obscure the social relations that produce it. By equating nature with money, the environment can serve to do the same.[25]

The "site map" (see Map 1) that ARCO presents in its picture book is instructive along these lines.[26] The map depicts Ecuador, a handful of cities, the TransAndean pipeline, Block 10 (ARCO's oil concession), and the pipeline that connects it to the company's processing facilities and the Trans-Andean pipeline. All maps, of course, are schematic; what is intriguing is *how* they portray a schema. ARCO's map is disconnected. There is no relief to indicate the country's geologic terrain; there is no geopolitical context to let a reader know that violence-torn Colombia lies to the north and Peru to the east and south, or that the Pacific Ocean forms Ecuador's western border; there are no sociopolitical divisions to locate oil operations in relation to regional and provincial authorities; and there are no delimitations of indigenous lands. Block 10 rests both literally and figuratively within a "gray" space devoid of any specificity. What we see are simply ARCO's operations. The oil concession itself is displaced from its sociocultural surroundings and its historical economic context. It is superimposed on an

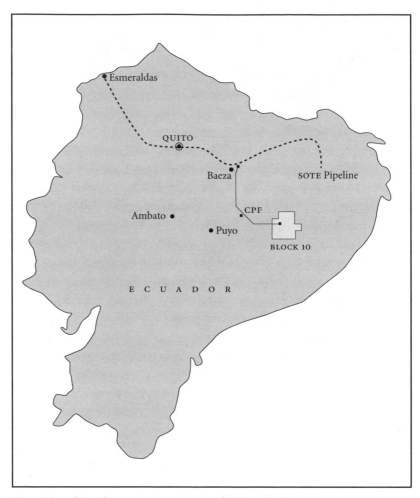

Map 1. Map of Ecuador. ARCO 1999, courtesy of BP America Inc.

unmarked and unclaimed space filled only with "pristine rainforest." In many ways, the area of ARCO's concession is simultaneously any fragile environment and no fragile environment. Like the Alaskan North Slope, Europe's North Sea, or the Indonesian coral reef (other places where ARCO drills for oil), the Ecuadorian rain forest is abstract, empty, and peopleless— a space of delicate nature predisposed as the ground for ARCO's rational, environmentally sensitive exploitation all for the benefit of you and me.

Extracting a place from its lived context is a crucial maneuver for claiming authority and authorship over that space; denying a locale its historical and cultural specificity was a central strategy in many colonial ventures.[27] In the case of the Villano Project, ARCO has removed nature from its Upper Amazonian context and repositioned it within the unmarked and unquestioned context of conservation biology, as well as the market and global society. The idea that Pastaza's rain forest expanse is "primary" or "pristine" grants the illusion that the forest is a place untouched by humans, and hence a space with no competing claims. Marco's claim over the land for his next purina, his grandmother's claim over the area she cultivates as her chacra, or those of the legendary warrior's descendants to his abandoned settlement have no place here. The rain-forest-equals-biodiversity thesis entitles ARCO to explore for and exploit hydrocarbon resources so long as it does so in an ecologically responsible way. Similarly, it deflects attention away from politically charged questions of rights, resources, and property regimes.

Yet the pristine rain forest of the Villano Project is anything but pristine. Today, approximately twenty thousand indigenous peoples (Quichua, Achuar, Shiwiar, and Zaparo) live in Pastaza province and approximately twelve to fifteen thousand live in the forested region.[28] Throughout history, people have played an integral role in creating what may appear to many as the untouched rain forest. On our trek to her purina, Carmen pointed out the plants she had planted along the path (the one that acted like soap, the one that helped stop bleeding, and the one that soothed diaper rash). The first night, we camped in a chonta palm grove she and her husband had planted five years before. On the second day, her two older children collected cinnamon bark from another grove. What lowland Quichua Indians call jatun sacha (the big forest) is very much a human environment where households living in multiple homes spread across the landscape shape the species content, distribution, and diversity of the forest through planting, transplanting, and selectively managing different vegetation.[29] Moreover, the past five centuries have witnessed multiple shifts and transformations among people connected to extractive activities in the Upper Amazon: gold from Spanish colonialism to the present, cinchona and sarsaparilla in the seventeenth and eighteenth centuries, rubber during the late nineteenth and early twentieth centuries, and petroleum exploration since the 1920s.[30] Royal

Dutch Shell, which worked in Pastaza Province from the 1930s to 1950, drilled the first Villano oil well in 1949. In short, humans have long been a part of this rain forest.

If one looks carefully, even the cover picture to ARCO's coffee-table book reveals this human presence. Along the perimeter of the Villano well site is the faint outline of a tall, chain-linked fence topped with razor wire. The concern here is not to keep out man-eating tigers but rather human beings. Among the "incidents" that made ARCO wary about developing the Villano oil field were specific acts of sabotage, demonstrations, and protests by indigenous peoples in Pastaza against ARCO's activities. Built to protect the corporation's multimillion dollar investment, the fence implies that there are local peoples opposed to ARCO's operations.

Given this, the project's name is curious. Villano is the name of the river near ARCO's well site.[31] Similarly, Villano has come to define a small, yet distinct region ever since multinational capitalism and the Ecuadorian military set up shop there in the mid-1940s. The Royal Dutch Shell Company named the oil well it drilled in the area Villano-1; today, the tapped well can be found a mere quarter-hour walk from ARCO's well site. The Ecuadorian military named the installation it built to protect and facilitate Shell's operations the Villano Outpost; today, the small army base sits across the river near ARCO's well site. Consequently, while it makes sense that ARCO named its project Villano given these geographic markers, the name is ironic. Villano is an antiquated term for "villain" in Spanish. Villainous precisely describes how many local Indians view ARCO.

"INVISIBLE PIPELINE"

The second image that I have singled out (see figure 2) is from the center-piece chapter in ARCO's picture book titled "Invisible Pipeline." Rather than a god's-eye view, the photograph offers an intimate, penetrating perspective of a pipeline slinking through the rain forest, as round and smooth and silent as a snake. Far from being an industrial eyesore, the pipeline appears downright natural. Its curves and bends mirror the contours of the terrain, allowing the pipeline to meld into the landscape as it becomes enveloped by nature. The dense foliage appears almost animated, curiously peering out in the brilliance of the noonday sun at its new metallic neighbor. The scene in

2. Photo of pipeline. ARCO 1999, courtesy of BP America Inc.

the photograph is peaceful, vaguely idyllic, and even contemplative, beckoning the viewer to follow the trail as it slips into the distance out of view. There are no signs of bulldozing, uprooting, gashing, or trampling. And while much labor (over five hundred workers for eight months) went into building the pipeline (as other photos in the chapter show), all the pictures of the completed pipeline mirror that above. Once the work is complete, there is not a human in sight. It is an "invisible pipeline," scarcely visible even from that god's-eye view.

Snaking its way through a "green tunnel," ARCO's invisible pipeline carries crude oil from the well site *in* the rain forest to ARCO's processing facilities *outside* the forest. The book's second map (see Map 2) helps locate the pipeline in relation to both. To the right we see the northwestern corner of ARCO's oil concession, Block 10. The clearly marked "Villano Well Pad" is the company's well site. The airstrip indicates where the military installation sits. Two communities (Pandanuque and Santa Cecilia) along the Villano River represent indigenous settlements. The pipeline stretches westward from the well site for thirty-seven kilometers, gaining 300 meters in elevation until it reaches ARCO's Central Processing Facility up an escarpment to

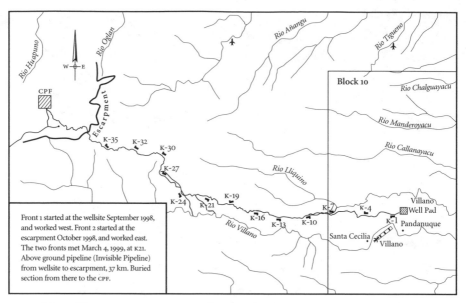

Map 2. Rivers. ARCO 1999, courtesy of BP America Inc.

the left. Small rectangles paralleling the pipeline mark the site of safety shutoff valves and helicopter landing pads.

This is the first pipeline to be built without a road anywhere in the Upper Amazon (an area including parts of Ecuador, Colombia, Peru, and Bolivia). In the northern Ecuadorian Amazon (as is the case elsewhere), oil roads have encouraged poor mestizo farmers to homestead the rain forest, and timber and agrobusinesses to deforest it. Without a doubt, ARCO's decision not to build roads for heavy industrial activity will help prevent Pastaza forests from being further degraded. But the distinction between "in" and "out" is not so sharp or easily defined. "In" or "out," it is all the same biogeography. The area above the escarpment is full of lush vegetation, though substantial portions (but not all) of it have been transformed into a patchwork of pasture and agriculture along with forest. Similarly, however, the Villano area is also a patchwork landscape. The area around the military base is largely pasture, and a handful of mestizo colonists have also home-steaded in the region.[32] Granted, deforestation is not as great in Villano as it is above the escarpment. But the point to keep in mind is that both areas are shifting landscapes where intervention and reforestation continually appear.

One is not "primary" or "pristine" while the other is tainted by human intervention. I noted earlier, even the dense forest through which the invisible pipeline snakes is very much an artifact of human use and manipulation.

Situating the Central Processing Facility "out" of the rain forest was a good design that saved ARCO money. According to a corporate report, ARCO's "small footprint [and] environmentally friendly design cut development costs [for the Villano Project] by 10%."[33] Yet taking special care "in" the forest because it is the realm of biodiversity can legitimize different practices "out" of the rain forest because it is not the realm of pure nature. ARCO spent $5 million on two hundred miles of transit infrastructure in the area the company deemed outside the forest. The company built new roads, reinforced old ones, and retrofitted bridges to carry heavy equipment to the Central Processing Facility. Furthermore, ARCO did build a road along the length of its second and longer pipeline, which carries processed crude oil from the Central Processing Facility to Ecuador's TransAndean pipeline. All of these entailed massive earthworks and ecological disruption to the Upper Amazon. Moreover, locating the Central Processing Facility above the escarpment in an "area designated by the environmentalists [read: environmental impact assessment experts] as 'highly intervened' " does not insulate ARCO from the risks inherent to the oil industry.[34] Oil wells can always explode, pipelines can always rupture or leak, formation waters (the toxic waters that emerge with crude oil from the depths of the earth) and industrial solvents can always seep into and contaminate surface water and subsurface aquifers. No matter how sophisticated, no technology is foolproof, especially in a region subject to earthquakes.[35]

More troubling, however, is the ease with which ARCO's invisible pipeline renders people invisible. Like the aerial photo and site map, the unpeopled pipeline and detailed map imply that the area in which ARCO works is a neutral space of nature. The two communities located near ARCO's well site offer some inkling that people live in the area, but they are never mentioned in the text and consequently represent no competing claims. In fact, the boundaries of Block 10 and the pipeline itself are the only markings on the map and in the text that have a social character. But the area of ARCO's oil operations is very much a lived and contested landscape. Thousands of indigenous peoples—mostly lowland Quichua—live in the province, and hundreds live, manage, and use resources in the area directly surrounding the well site and pipeline. ARCO has figured prominently in creating conflict among them.

When a multinational petroleum corporation acquires an oil concession in Ecuador, it obtains rights to search for and exploit oil. It does not own the land where it works, nor does it own the crude oil it extracts. It only possesses rights to exploit petroleum and obtain a handsome percentage of the extracted crude as payment for its services.[36] At the time that ARCO began its exploratory activities in 1988, none of the forest groups in Pastaza held legal title to their lands. Indeed, the state had repeatedly denied organized attempts by indigenous groups since the mid-1970s to gain title to their claimed ancestral lands in Pastaza. ARCO knew that despite indigenous opposition to its operations, Indians could not legally prohibit oil development. All subterranean resources—of which petroleum is the most coveted—belong solely to the Ecuadorian state, and the state guarantees military backing for petroleum production to proceed if necessary.[37] But the company also knew, from past experience, that indigenous people could interrupt ARCO's oil activities. Within months of ARCO beginning its seismic exploration in 1989, communities belonging to the most representative indigenous organization in the province, Organización de Pueblos Indígenas de Pastaza (OPIP), detained (or "kidnapped," to use ARCO's term) corporate and state officials in the rain forest in an attempt to halt exploratory activity and gain legal title. The OPIP-led action halted ARCO's operation for one year.[38] This interruption, like others to follow, cost the company money.

ARCO reasoned that the best way to secure the company's operations was to build alliances with communities that might be receptive to oil development. Once such a community was identified, ARCO would help it gain legal title to its surrounding lands and then site an oil well there. The company believed that if a community gained legal title as a result of ARCO's intercession, and if the company distributed sufficient gifts among its community members, then ARCO's oil operations would be safe and could proceed unhindered. Pandanuque was such a community. In early 1991, ARCO personnel strategically intervened and secured the legal titling of 10,569 hectares in Pandanuque's name.[39] ARCO drilled its exploratory wells on this land. At the time, Pandanuque was made up of a cluster of houses scattered among planted forest whose residents (estimated at fifty people) lacked the material and cultural capital necessary to pressure the state into giving them land title. Yet through crucial corporate persuasion, the community that ARCO most needed to acquiesce to its practices received communal legal title to the surrounding lands without delay.

ARCO's social and political activity in the area was far from invisible. The company orchestrated the isolated titling of land that it needed to control, and it played a key role in creating and exacerbating divisions among the lowland Quichua in the area. Following the practices of every transnational petroleum corporation working in the Ecuadorian Amazon, ARCO bestowed gifts and small projects on the communities surrounding its work site. When the company began developing the Villano oil field, ARCO gave Pandanuque and Santa Celia an array of material benefits—ranging from candies to blankets, school supplies to a one-room schoolhouse, and airplane rides to high school scholarships—as compensation for letting the company complete its work. For ARCO—a corporation that invested $370 million to explore and exploit oil in Block 10—these were incidental gestures. For a marginalized indigenous community, however, these gifts were near monumental. They became talismans of progress and fetishes of modernity for indigenous recipients.

A leader named Raul from the Villano area put it this way as we chatted one afternoon on the streets of the provincial capital: "We in Pandanuque, we want to work. We have contracts with la compañía. We want to work for a living. We want to develop. If OPIP had its way, we would be naked and barefoot." Raul was taller than most Quichua, and he looked handsome in his slick new haircut and shiny new shoes. "You know," he continued as he leaned against ARCO's Chevy Trooper, "la compañía says that Villano will soon be Ecuador's new petroleum capital. And that land in Villano is ours—we own it and control it."

ARCO's field employees were instrumental in fabricating illusions of imminent betterment for Pandanuque residents: oil operations would catapult Villano hamlets out of their poverty. ARCO had discovered black gold, and Raul, along with others from Pandanuque, were sitting on the jackpot at the end of the rainbow. In order for Pandanuque to take charge of its good fortune, ARCO personnel encouraged the community's residents to form their own indigenous organization, which the company supported financially through the pretense of contracts. Thus was born an anti-OPIP and pro-ARCO indigenous organization. Since ARCO had experienced fierce opposition to its oil operations from OPIP, the existence of an indigenous, anti-OPIP, representative body served the company well. ARCO needed docile and compliant people near its exploratory wells to ensure that its oil operations proceeded unimpeded; a pro-ARCO organization filled this role. In turn, Pandanuque residents jealously guarded their newly titled land and threatened violent retaliation against anyone who sought to intervene in what they deemed to be their affairs.

But the notion that one tiny community could control oil activity on an arbitrarily delineated piece of land was for many Pastaza Indians absurd.[40] As Miquia Magdalena made clear, people's sense of the land and land use did not subscribe to the assumptions of private property. One morning, while I helped her weed her chacra, Magdalena explained, "Those *runaguna* [referring to Quichua Indians in Pandanuque], they are few. We are many. And a few people cannot permit that la cumpañía enter there as it pleases. They think that the *sacha* [forest] is theirs, but it is ours too. Because we might leave from here in one or two days, live for a month there, and return later. Our life is not only here. Our life is not only in one place."

In general, indigenous people throughout Pastaza Province live in and manage the forest in multiple zones that as I learned from my trek with Carmen, could be of great distances apart—crossing river systems and ecological niches.[41] Furthermore, different groups generally have different rights over different resources in the same area. Thus, historically layered overlapping privileges prevent a small cluster of people from having exclusive rights over a large region. Pandanuque was founded in the late 1970s. In the eyes of most Quichua, that was hardly long enough to have sole rights over much of anything other than what one planted. According to most OPIP-affiliated *indígenas*, the community members that formed Pandanuque could not determine the fate of petroleum development for the close to fifteen thousand indigenous peoples who lived in Pastaza's rain forest.

More than simple compensation, ARCO's gifts—be they candies or contracts—served to pry open and transform local senses of self and property. Differences over ways of thinking about land rights and land use led to bouts of conflict between Pandanuque residents and OPIP members. And conflict led to the immediate militarization of ARCO's oil well in the early 1990s, and then the Central Processing Facility and invisible pipeline as soon as their construction began in 1997. In January 1999, special forces from the military armed with weapons and attack dogs threatened and detained groups of Indians protesting an alleged oil leak along ARCO's invisible pipeline.[42] A number of protests in 1998 and 1999 led to the arrest, unlawful detention, and harsh handling of indigenous protesters in July 1999.[43] While to date, confrontations over oil between local people and the military have not led to fatalities, such has not been the case in the northern Ecuadorian Amazon.[44]

By focusing on how to mitigate the effects of oil operations on biophysical processes in the rain forest, ARCO diverted public attention away from how its operations affected local people's lives. A technical, biophysical rationality compelled the company to deny its own practices that actually heightened tensions, conflicts, and inequalities among indigenous groups, and in turn, gave the impression that conflict between indigenous peoples was natural. This was a case of chaotic, irrational Indians fighting among themselves—about which the company could do nothing. ARCO's bioscript granted the company a pretext for not taking indigenous concerns seriously. If ethnic conflicts had created such an unruly mass, then the corporation could easily dismiss indigenous protest as well as delegitimize indigenous claims that were founded in the knotty moral and political questions surrounding resource extraction. Unruliness among indigenous groups justified the use of military force.

"GOOD NEIGHBOR"

The third image (see figure 3) is from the title page of the picture book's final chapter, titled "Community Relations." The photo is a close-up shot (covering two pages) of a small indigenous child peering at the camera through a brightly painted, wooden fence. Her eyes are wide and bright. Her small hand grasps the fence. Like any two year old, this child is full of innocence and charm, and she offers a welcome sense of wonder and hope. We imagine that she is smiling, though we cannot quite tell; her mouth is cut off from view. This photograph differs strikingly from the sweeping panorama on the cover of ARCO's book. Rather than offering expansive depth, this image pulls the child and the fence sharply into focus, while blurring the lived space behind her: the home where she plays and the community that watches over her. Similarly, the indigenous child's narrow view differs strikingly from

3. Girl looking through fence. ARCO 1999, courtesy of BP America Inc.

ARCO's god's-eye view. Far from omniscient, the child's vision is partial. Though she can look out onto the outside world, a fence—that artifact defining borders between neighbors—restricts her sight and confines her movement. She is limited in what she can see (since the wood barrier blocks her view); she is limited in what she can say (since she has no mouth); and she is limited in what she can do (since both the fence and her small size inhibit her).

This child is the first indigenous person to appear in ARCO's book who is not a worker for the company or one of its subcontractors. Likewise, she is the first image in the only chapter where ARCO discusses its relations with local people. Both of these facts are curious in a book as carefully orchestrated as ARCO's. Is the child meant to convey tenderness and humanity among the cold calculus of machines, to suggest that the company too has a heart? Does the child unconsciously reflect ARCO's understanding of Indians as childlike: innocent though mischievous, disobedient though tractable, photogenic though simple? This is the unspoken opposite of everything depicted in the preceding nine chapters in which ARCO appears over and over again as ingenious and rational, sophisticated and developed, and coolly in control. Is ARCO the environmentally minded, economically rational multinational corporation charged with helping the innocent, unsophisticated children of the forest advance?

"La cumpañía," Miquia Magdalena retorted in her chacra, "gives them *kusitaguna* so they will be good and obey, like when you want a *huahua* [small child] to come to you." She was referring to ARCO's practice of bestowing gifts on communities near its operations. Kusitaguna (from the Spanish *cosita*) were "little things" or "trinkets." Like Magdalena, many OPIP-affiliated Indians believe that ARCO treated indigenous peoples as if they were children. ARCO personnel used these trinkets—tin roofing and pencils, blankets and plane rides—to cajole, persuade, and instruct Indians to be appropriate modern subjects: independent, property-owning, rights-bearing, economically maximizing persons.[45] Yet for Miquia Magdalena (as well as others), "those people, they don't know anything. They only want the pay, the money. La cumpañía gives them food, clothes, and they have let them [ARCO] in. Now they believe we [OPIP-affiliated Indians] are being bad to them, telling them not to welcome la cumpañía. They say to us, you don't give us clothes and money; la cumpañía does. But they don't see that it will soon end and they will not know their forest."

Of particular concern was what OPIP leaders called ARCO's strategy of "buying consciences." OPIP leaders condemned ARCO's policies as manipulative and labeled the company's promises of progress disingenuous ploys. A quarter century of petroleum extraction in the northern Ecuadorian Amazon had left the region about one hundred miles north of Pastaza plagued by poverty, political marginalization, and environmental devastation.[46] By keeping a pro-ARCO organization alive through kusitaguna, the company furthered the warped desires of a few small hamlets and denied its own responsibility to the larger, more powerful indigenous communities that the company's operations would also affect. Despite the fact that ARCO's oil operations were relatively sound environmentally, its social policies mimicked those of every other oil company.

Many of OPIP's leaders were articulate, savvy spokespersons who denounced the marginalization, inequality, and exploitation that oil operations produced in the Ecuadorian Amazon. OPIP leaders worked their local and growing transnational networks to make the future of ARCO's operations contingent on meeting broader demands for alternative forms of development throughout Indian territory in Pastaza.[47] Leaders traveled from local protests in Villano to demonstrations in Los Angeles. They mobilized public condemnations at ARCO's Quito headquarters and an anti-oil campaign across the United States. They staged nonviolent occupations of the Ministry of Energy and Mines in Quito, and denounced ARCO at the World Bank in Washington, D.C.[48]

In an attempt to address the growing international pressure, ARCO agreed in 1994 to establish the Technical Environmental Committee, a negotiating forum that included, in the company's words, "all the stakeholders of Block 10"—representatives of ARCO, the state, and three indigenous organizations (OPIP, ASODIRA, and AIEPRA).[49] But while OPIP leaders envisioned the committee as a forum in which to address all aspects of ARCO operations, it quickly became a space only for discussing strictly defined technical, biophysical concerns. The committee focused on issues such as defining the size and location of wells, pipelines, and processing facilities, defining the technology for building these and stabilizing soils, and defining environmental monitoring and waste disposal regimes.[50]

The committee was an important step in reducing the negative environmental effects of ARCO's oil operations, enabling indigenous OPIP representatives to insist that ARCO build the pipeline through their territory without a road. Yet many are disgruntled about the committee's ultimate effect. As Marco noted, "The committee is really a toy that ARCO uses to pretend that it takes indígenas and environmentalists seriously. But little has changed other than the fact that the indigenous members of the committee earn money [salaries]." Ironically, many indigenous people in Pastaza have become skeptical over the years about the committee and see indigenous representatives as sellouts. What such sentiments underscore is how the committee effectively circumscribed what could and could not be talked about. The committee was the space in which indigenous leaders found that they were most legitimately recognized by the company and government. However, in order to be heard, they could only debate what ARCO, and not they, termed their "common ground"—a concern for the environment strictly defined. The result was to grant indigenous peoples a space to "mouth" rather than a forum in which to speak, since their major concerns were only about the environment in a broad social (as opposed to a narrower technical) sense.

On our way down the river to his grandmother's chacra, Marco explained,

"OPIP often carries a banner during its protests that says, 'The Defense of Nature and Social Justice are Inseparable.' Well, the committee undermines that claim." Marco's comment echoed those of others concerned about how the committee was debilitating indigenous political capacity and marginalizing many indigenous demands. Framing oil operations in solely technical, engineering, and scientific terms foreclosed discussion on the social tensions that ARCO's oil operations produced. As in the photograph of the indigenous child, ARCO maintained the Indian representatives in the committee in sharp focus, but the social context in which they lived and their political-economic concerns were blurred from view. More important, it precluded any debate on environmental justice by continually displacing questions of a moral and political nature.

When the Technical Environmental Committee was first created, ARCO said that it would deal with social concerns once the company had garnered revenues from the Villano Project—that is, once oil was flowing. But when that time came in 1999, ARCO changed its strategy. The company once more ignored OPIP and set up a program called Buena Vecindad (Good Neighbor) with the communities near Villano and the rain forest pipeline— approximately five hundred people. The Good Neighbor program was a new name for the company's strategy of handing out trinkets to appease local people. Not surprisingly, the communities along the pipeline route that now received gifts from ARCO renounced their prior allegiance to OPIP and joined the pro-ARCO indigenous organization. But these small, marginalized communities have negligible negotiating power with ARCO. Before ARCO was bought out by BP-Amoco in March 2000, there was growing discontent among people over ARCO's so-called good neighbor policy. Communities accuse the company of not fulfilling its promises and buying off community leaders with *el dólar*.

As Marco steered us around the last river bend, his grandmother's chacra came clearly into view—in fact, its brilliance blinded me as the sun reflected off a shiny new, corrugated-tin roof. Marco's grandmother's home included three raised structures. Two were thatched and semiclosed; the one with a metal roof was newly built without walls. "I don't spend much time there," Marco's grandmother said as I commented on the structure; "It's way too noisy. But it's good for collecting water"—referring to how one of the metal sheets acted like a funnel diverting water from the roof into an empty oil barrel below. That evening, when I slept under the ARCO-donated roof, I quickly saw why Marco and his grandmother did not do the same. The nightly rain ricocheted off it like millions of bullets keeping me awake a good share of the night. The next day, I also came to understand why the collected water was so nice. There were problems in a nearby stream where Marco's grandmother had previously collected water when the main river was too murky. A number of Pandanuque residents had baptized the stream Isma-yacu (Shit River) after it became contaminated by wastes from the well site's encampment and drilling platform. Not all, even those from Pandanuque (ARCO's original supporter), were always in accordance with the company.

SUBTERRANEAN TECHNOLOGIES

How is one to think about the Villano Project? Over a good part of the twentieth century, my father and grandfather worked for the petroleum industry in marginalized parts of Latin America and North Africa. And over a good part of the twentieth century, big oil has wreaked ecological devastation on Third World ecologies. Given the ties between my family history and that of industrial contamination, I feel a personal stake in the emergence of environmentally sensitive oil operations. Many readers will agree with me that it is time that multinational petroleum corporations recognize the environmental repercussions of their operations, especially in an ecosystem as full of biodiversity as the rain forest. Yet I also want oil companies to be ecologically conscientious because though my monthly paycheck comes from elsewhere, I—like any member of our consumption-oriented society— remain as dependent on the industry now as I was as a child. Oil is a game in which we all participate, regardless of how often (or little) we drive our cars, fly in airplanes, or recycle our plastics. This complicity makes it easy to become dazzled by environmentally upright oil operations and makes it difficult to question them.

Most everything I know about oil operations I have learned from my petroleum geologist father. On the basis of what he has taught me, I applaud ARCO for its environmental sensitivity.[51] However, given the contemporary global concern for the environment and the rain forest in particular, I would prefer to see ARCO's ecologically sensitive operations as an expected norm, not as an exceptional concession. Installing ecologically sensitive operations in the rain forest is a positive step. But strictly ecological concerns should not be the only criteria for evaluating an oil company. The critical questions that OPIP leaders and community members are raising about the inequalities inherent in resource extraction are not easily dismissed.

What has concerned me in the preceding pages is not whether ARCO's green technology is sound but "how a technical rationality becomes a surrogate for moral and political rationalities."[52] ARCO's picture book helps in untangling this query. *The Villano Project* embodies the icon (oil and nature harmoniously hand in hand) through which ARCO sought to define itself to the world. Yet when examined carefully, the photos in the picture book contain within them ways of seeing and behaving that reveal both the bioscript and spectacle behind the icon. In particular, they illustrate how ARCO sought to define the rain forest as exclusively a biological and geophysical realm, and how this definition allowed the company to conceive its challenge, solution, and success in purely technical terms. Images, of course, are never innocent. In construing the rain forest as purely a biophysical realm, ARCO celebrated a forest uninhabited by ongoing human historical relations, obscured its own policies so divisive to human life in the rain forest, and limited the terms on which it would engage with local people. Foregrounding environmental concerns marginalized other issues—issues of profound social, political, and economic import.

My aim here is not to bash ARCO. The less degrading and less contaminating technologies that ARCO used in its Villano Project are to be commended. Yet foregrounding ecologically sensitive operations should not blind us from simultaneously seeing or questioning other phenomena. Because ARCO's operations (no matter how ecologically impressive) are still perpetuating the inequalities inherent in resource extraction, I find it hard to be comforted by the company's environmental record. The conditions under which oil operations take place are profoundly inequitable, some might say unjust and unscrupulous. While OPIP leaders are not saints and the organization is far from perfect, it is hard to argue with their insistence that the high social, political, and economic (as well as environmental) stakes of oil operations cannot be exchanged for trinkets. As my now-retired petroleum-geologist-cum-social-justice-activist father would readily agree, the landscapes in which ARCO and other oil companies work are *social* as much as biological and geophysical. Mitigating the social impact of oil operations would mean making compromises that extend a corporation far beyond incidental interchanges with select communities and a few indigenous representatives. Therein lies the *real* challenge.

NOTES

1. Atlantic Richfield Company (ARCO), *The Villano Project: Preserving the Effort in Words and Pictures* (Quito: Imprenta Mariscal, 1999). I am grateful to Teresa Velasquez, my research assistant and dear friend, for sending me ARCO's book soon after it was published. I thank Raymond Bryant, Scott Fedick, Alex Greene, Paul Greenough, Susanna Hecht, Nancy Peluso, Hugh Raffles, Charles Zerner, and especially Jack Putz and Candace Slater for their generous and insightful comments throughout the evolution of this chapter. Research and writing for this project was generously supported by HRI, the Social Science Research Council-MacArthur Foundation Fellowship for Peace and Security in a Changing World, and the National Science Foundation.

2. Ibid., 8.

3. Miquia, while literally meaning "aunt" in lowland Quichua, is an affectionate, honorific title for an older woman.

4. To pump oil successfully from a social forest—one with human residents who have long transformed and been transformed by it—ARCO would have to take into account the highly disruptive and chaotic effects of its operations on social processes. Just as it has for biophysical processes, ARCO would have to devise its oil operations such that they did not tear at the fabric of social relations within the forest. Reconceptualizing oil operations along social lines would raise uncomfortable questions about the social, political, and economic power inequities inherent in how and why oil gets produced. These are questions that neither oil corporations nor those of us who benefit from them care to ask.

5. Other scholars have explored this dilemma. For a broad reflection on the relationship between science and society, see Bruno Latour, *We Have Never Been Modern*, trans. Catherine Porter (Cambridge: Harvard University Press, 1993). For complementary analyses focusing on the environment and rain forests, see W. M. Adams, "Rationalization and Conservation: Ecology and the Management of Nature in the United Kingdom," *Transactions in the Institute of British Geographers* 22, no. 3 (1997): 277–291; Bruce Braun, "Buried Epistemologies: The Politics of Nature in (Post)Colonial British Columbia," *Annals of the Association of American Geographers* 87, no. 1 (1997): 3–31; Peter Brosius, "Green Dots, Pink Hearts: Displacing Politics from the Malaysian Rain Forest," *American Anthropologist* 101, no. 1 (1999): 36–57; and David Demeritt, "Scientific Forest Conservation and the Statistical Picturing of Nature's Limits in the Progressive-Era United States," *Environment and Planning D-Society and Space* 19, no. 4 (2001): 431–459.

6. I thank Susanna Hecht and Andrew Tolan for reminding me of this fact. For other research on petroleum development in Latin America and the Upper Amazon, see Gerard Colby with

Charlotte Dennett, *Thy Will Be Done: The Conquest of the Amazon; Nelson Rockefeller and Evangelism in the Age of Oil* (New York: HarperCollins, 1995); Fernando Coronil, *The Magical State: Nature, Money, and Modernity in Venezuela* (Chicago: University of Chicago Press, 1997); Michael J. Economides, *The Color of Oil: The History, the Money, and the Politics of the World's Biggest Business* (Katy, Tex.: Round Oak Publishing Company, 2000); Judith Kimerling, *Amazon Crude* (New York: Natural Resources Defense Council, 1991); Terry Lynn Karl, *The Paradox of Plenty: Oil Booms and Petro-States* (Berkeley: University of California Press, 1997); Suzana Sawyer, *Crude Chronicles: Indigenous Politics, Multinational Oil Capital, and Neoliberalism in Ecuador* (forthcoming, Duke University Press); and Joel Simon, *Endangered Mexico: An Environment on the Edge* (San Francisco: Sierra Club Books, 1997).

7. The text notes that an "estimated 700-year-old kapok tree" next to the well site became for ARCO engineers the "sacred tree" to stand over and protect the project (ARCO, *Villano Project*, 38).

8. During most of the twentieth century, "green gold" in Ecuador has referred to the wealth of the banana boom on the coast. Since the 1900s, however, the phrase has increasingly referred to the untold wealth of biodiversity in the rain forest.

9. Donna Haraway, "Situated Knowledges: The Science Question in Feminism and the Privilege of Partial Perspective," *Feminist Studies* 14, no. 3 (1988): 576.

10. See Center for Economic and Social Rights (CESR), *Rights Violations in the Ecuadorian Amazon: The Human Consequences of Oil Development* (New York: Center for Economic and Social Rights, 1994); Chris Jochnick, "The Human Rights Challenge to Non-State Actors" (manuscript, 1997); Judith Kimerling, *Crudo Amazónico* (Quito: Abya Yala, 1993), and "Oil, Lawlessness, and Indigenous Struggles in Ecuador's Oriente," in *Green Guerrillas: Environmental Conflicts and Initiatives in Latin America and the Caribbean*, ed. Helen Collinson (New York: Monthly Review Press, 1996), 61–74; Suzana Sawyer, "Fictions of Sovereignty: Of Prosthetic Petro-Capitalism, Neoliberal States, and Phantom-Like Citizens in Ecuador," *Journal of Latin American Anthropology* 6, no. 1 (2001): 156–197, and "Bobbittizing Texaco: Dismembering Corporate Capital and Re-membering the Nation in Ecuador," *Cultural Anthropology* 17, no. 2 (2002): 150–180; and Glen Switkes, "Texaco: The People vs. Texaco," NACLA 38, no. 2 (1995): 6–10.

11. What type of environmental aesthetics and scientific reason make the rain forest appear to be a neutral, ahistoric space ripe for certain kinds of economic intervention? How has an exclusive, narrow understanding of the environment in technocratic and scientific terms authenticated some voices (those of industry and environmentalists) and silenced or off-centered others (those of indigenous peoples)?

12. See Adams, "Rationalization and Conservation"; Braun, "Buried Epistemologies"; Brosius, "Green Dots"; Eric Darier, "Environmental Governmentality: The Case of Canada's Green Plan," *Environmental Politics* 5 (1996): 585–606; Demeritt, "Scientific Forest Conservation"; and Timothy Luke, "On Environmentality: Geo-Power and Eco-Knowledge in the Discourse of Contemporary Environmentalism," *Cultural Critique* 31 (1995): 57–81.

13. ARCO, *The Villano Project*, 8, 10. ARCO's first exploratory well in Block 10 was at a site called

Moretechocha approximately thirty miles south of the Villano area. The next two exploratory wells were at Villano, where ARCO reexplored the geologic structure originally "drilled by [Royal Dutch] Shell in 1950" (ibid., 10). The Villano oil reservoir contained seven hundred thousand barrels of twenty-one-degree API oil—that is, good, light oil.

14. Ibid., 10.

15. Protracted negotiations with the Ecuadorian state delayed the project. Delays resulted from the 1995 Amazonian war between Ecuador and Peru, and disagreements about how to build a pipeline from Villano to Ecuador's main TransAndean pipeline as well as how to expand the carrying capacity of the TransAndean pipeline.

16. See ARCO annual reports from the mid-1970s onward.

17. For scholarship on the history of campaigns decrying oil operations in Ecuador, see Acción Ecológica, *Marea Negra en la Amazonía: Conflictos socioambientalies vinculados a la actividad petrolera en el Ecuador* (Quito: Abya Yala, 1995); Acción Ecológica, *Amazonía por la Vida*; Debate ecológico sobre el problema petrolero en el Ecuador (Quito: Imprenta Mariscal, 1993); Kimerling, *Crudo Amazónico*; Jochnick, "Human Rights Challenge"; and Sawyer, "Fictions of Sovereignty," and "Bobbittizing Texaco." For Nigeria, see Anthony Apter, "The Pan-African Nation—Oil-Money and the Spectacle of Culture in Nigeria," *Public Culture* 8, no. 3 (1996): 441–466; R. Boele, H. Fabig, and D. Wheeler, "Shell, Nigeria, and the Ogoni: A Study in Unsustainable Development," *Sustainable Development* 9, no. 3 (2001): 121–135; S. Cayford, "The Ogoni Uprising—Oil, Human Rights, and a Democratic Alternative in Nigeria," *Africa Today* 43, no. 2 (1996): 183–197; Diana Barikor-Wiwa, "The Role of Women in the Struggle for Environmental Justice in Ogoni," *Cultural Survival Quarterly* 21, no. 3 (1997): 46–49; Michael Watts, "Petro-Violence: Community, Extraction, and Political Ecology of a Mythic Commodity," Environmental Politics Working Paper No. WP 99–1 (Berkeley: Institute for International Studies, University of California, Berkeley, 1999), and "Black Gold, White Heat," in *Geographies of Resistance*, ed. S. Pile and M. Keith (London: Routledge, 1997), 33–67; and C. E. Welch, "The Ogoni and Self-determination—Increasing Violence in Nigeria," *Journal of Modern African Studies*, 33, no. 4 (1995): 635–650.

18. For another case study along these lines, see Judith Kimerling, "Corporate Ethics in the Era of Globalization: The Promise and Peril of International Environmental Standards," *Journal of Agricultural and Environmental Ethics* 14, no. 4 (2001): 425–455.

19. ARCO, *Annual Report*, 1998, 9.

20. A quick look at the website of any multinational petroleum corporation will indicate how environmentally conscientious they claim to be. A click on an "environment" icon often shows the company's ecologically friendly practices and the multiple environmental initiatives that it supports.

   Beginning in the early 1990s, a number of multinational oil companies working in Ecuador have become singularly conscientious, in a way never before witnessed, about how their operations might affect rain forest ecologies. Their concern for the environment represents a dramatic shift for players in an industry that has caused massive contamination in the northern Ecuadorian Amazon since the late 1960s, as well as elsewhere (Kimerling, *Crudo Amazónico*; and CESR, *Rights Violations*). The shift in concern represents the industry response to what one oil executive vice president called the "attack" that "radical environmental activists" have launched against "oil development . . . all across the rain forest" (*Oil and Gas Journal* [*OGJ*] 95, no. 16 [21 April 1997]: 38). Indeed, the doggedly persistent and increasingly transnational campaigns that many environmental organizations (both mainstream and radical) have sustained over the past decade against ecologically degrading petroleum activities have caused numerous corporations to upscale their operations in Third World spaces to near-U.S. environmental standards. In Ecuador, the campaign against Texaco that culminated in a $1.5 billion class action lawsuit filed in the United States by Ecuadorians for alleged environmental contamination in the Ecuadorian Amazon has compelled most corporations to be particularly sensitive about their oil operations. For more detail, see Jochnick, "Human Rights Challenge"; and Sawyer, "Fictions of Sovereignty," and "Bobbittizing Texaco."

21. In addition to *The Villano Project*, ARCO personnel published two articles on its Ecuadorian operations (*Environment* 40, no. 5 (1998): 12–20, 36–45; and *OGJ* 96, no. 15 [14 April 1998]: 22–25). Two other articles in *OGJ* have focused on ARCO's work in Villano (*OGJ* 95, no. 16 [21 April 1997]: 37–42; *OGJ* 97, no. 31 [2 August 1999]: 19–26). ARCO also published multiple reports and updates on its website, http://www.arco.com. Due to the BP-Amoco merger with ARCO in March 2000, this website no longer exists.

22. ARCO *Annual Report*, 1998, 10.

23. Ibid., 6.

24. Ibid., 7.

25. On the role of money in obscuring the social relations of production, see Karl Marx, *Capital: Critique of Political Economy*, trans. Ben Fowkes (New York: Penguin, 1992).

26. ARCO's maps in other articles give a bit more detail but they are still extracted from their social, political, economic, and cultural surroundings.

27. See Daniel Cosgrove and Stephen Daniels, *The Iconography of Landscape: Essays on the Symbolic Representation, Design, and Use of Past Environments* (Cambridge: Cambridge University Press, 1988); J. B. Harley, "Cartography, Ethics, and Social Theory," *Cartographia* 27 (1990): 1–23; Anne McClintock, *Imperial Leather: Race, Gender, and Sexuality in the Colonial Contest* (New York: Routledge, 1995); Mary Louise Pratt, *Imperial Eyes: Travel Writing and Transculturation* (London: Routledge, 1992); Suzana Sawyer and Arun Agrawal, "Environmental Orientalisms," *Cultural Critique* 45 (2000): 71–108; and Peter J. Taylor, "Politics in Maps, Maps in Politics: A Tribute to Brian Harley," *Political Geography* 11, no. 2 (1992): 127–129.

28. Archaeological research and early colonial accounts suggest that the population was substantially denser at and before the Spanish invasion for the entire Amazon Basin. Lowland Quichua are not Quichua peoples from the Andean highlands who migrated to the Upper Amazon. Though their ancient history is somewhat vague, the ancestors of lowland Quichua are believed by most scholars to have lived in the rain forest region for thousands of years. See María Antonieta Guzmán Gallegos, *Para que la yuca beba nuestra sangre* (Quito: Abya Yala, 1997); Elizabeth Reeves, *Los Quichuas del Curaray: El proceso de formacion de la identidad* (Quito: Abya Yala, 1988); Norman Whitten, *Sacha Runa: Ethnicity and Adaptation of Ecuadorian Jungle Quichua* (Urbana: University of Illinois Press, 1976); and Norman Whitten, *Sicuanga Runa: The Other Side of Development in Amazonian Ecuador* (Urbana: University of Illinois

Press, 1985). Achuar also live in the southern part of the province, and Shiwiar and Zaparo Indians inhabit the area near the Peruvian border. For the archaeology, see Anna Roosevelt, "Resource Management in Amazonia before the Conquest: Beyond Ethnographic Projection," in *Resource Management in Amazonia: Indigenous and Folk Strategies*, ed. Darrell Posey and William Balée (Bronx: New York Botanical Garden, 1989), 1–21, and "Paleoindian Cave Dwellers in the Amazon: The Peopling of the Americas," *Science* 272 (1997): 373–384. For a colonial account at contact, see José Medina, *The Discovery of the Amazon* (New York: Dover Publications, 1934).

29. For literature on resource management in Amazonia, see William Balée, *Footprints of the Forest: Ka'apor Ethnobotany—The Historical Ecology of Plant Utilization by an Amazonian People* (New York: Columbia University Press, 1994); William M. Denevan and Christine Padoch, eds., *Swidden-Fallow Agroforestry in the Peruvian Amazon* (Bronx: New York Botanical Garden, 1988); Philippe Descola, *In the Society of Nature: A Native Ecology in Amazonia*, trans. Nora Scott (Cambridge: Cambridge University Press, 1994); Darrel Posey and William Balée, eds., *Resource Management in Amazonia: Indigenous and Folk Strategies* (Bronx: New York Botanical Garden, 1989); and Kent H. Redford and Christine Padoch, eds., *Conservation of Neotropical Forests: Working from Traditional Resource Use* (New York: Columbia University Press, 1992).

30. See Soren Hvalkof, "Outrage in Rubber and Oil: Extractivism, Indigenous Peoples, and Justice in the Upper Amazon," in *People, Plants, and Justice*, ed. Charles Zerner (New York: Columbia University Press, 2000); Marcelo Naranjo, *Etnohistoria de la zona central del alto Amazonas, siglos XVI–XVIII* (Urbana: University of Illinois Press, 1987); Udo Oberem, *Los Quijos, historia de la transculturación de un grupo indígena en el Oriente Ecuatoriano* (Otavalo, Ecuador: Instituto Otavaleño de Antropología, 1980); Udo Oberem and Segundo Moreno Y., *Contribución a la etnohistoria ecuatoriana* (Otavalo, Ecuador: Instituto Otavaleño de Antropología, 1995); John L. Phelan, *The Kingdom of Quito in the Seventeenth Century: Bureaucratic Politics in the Spanish Empire* (Madison: University of Wisconsin Press, 1967); Reeves, *Los Quichuas;* Anne Christine Taylor and Cristóbal Landázuri, *Conquista de la región Jívaro, 1550–1650: Relación documental* (Quito: Abya Yala, 1994); and Whitten, *Sacha Runa.*

31. The Villano River flows into the Curraray River, which flows into the Napo, a tributary of the upper reaches of the Amazon River.

32. After Ecuador's 1981 war with Peru over their shared Amazonian border, the government encouraged individuals to colonize the Villano area. Colono households were given private title to fifty hectares of land just like in the "intervened" area. There are six to eight colono households in the Villano area.

33. *Security Analyst Meeting Report on Upstream Operations*, 1997, http://www/arco.com/ Corporate/reports/SAM97/upstrm6x.htm.

34. ARCO, *Villano Project*, 92.

35. ARCO's operations in the Upper Amazon are set squarely in the seismically active Andean foothills. The area of ARCO's Central Processing Facility is more densely populated, and consequently, toxic pollution could affect more people immediately. Furthermore, any contamination will flow downhill and affect the area ARCO considers to be the rain forest.

36. Legal contracts between multinational oil companies and the Ecuadorian state have varied over the years. The ARCO arrangement was what was called a "risk-service contract."

37. See the Ecuadorian Constitution, art. 46. As defined by the National Security Law (art. 50) and the Hydrocarbon Law (arts. 6 and 8), petroleum is a national security of strategic importance and its production must be guaranteed by military action.

38. For an in-depth description of this process, see Sawyer, *Crude Chronicles.*

39. See the state archives at the Instituto Ecuadoriano de Reforma Agrária y Colonización (IERAC) and interviews with IERAC officials.

40. What ARCO did not anticipate was that a year later, in 1992, the state would adjudicate legal title to a substantial portion (over one million hectares) of the remaining indigenous-claimed lands in Pastaza. An array of historical conjunctures led to this unprecedented land titling; 1992 was a big year in the Americas. On a global scale, the transnational continental campaign against the quincentenary celebration of the "discovery" of the New World as well as the United Nations Conference on Environment and Development in Rio de Janeiro piqued global awareness of indigenous rights and the preservation of the rain forest. But direct pressure to title the land resulted when two thousand OPIP-affiliated Indians marched for two weeks from the Pastaza rain forest to the president's palace in Quito (Ecuador's capital) demanding legal title to ancestral territory. For more detail, see Suzana Sawyer, "The 1992 Indian Mobilization in Lowland Ecuador," *Latin American Perspectives* 24, no. 3 (1997): 67–84; and Norman Whitten, "Return of the Yumbo: The Indigenous Caminata from Amazonia to Andean Quito," *American Ethnologist* 24, no. 2 (1997): 355–391.

It would have been better for ARCO if only Pandanuque Indians had legal title. While the company could not prevent titling, they could intervene in how it was done. The state broke the new million hectares of titled land up into nineteen segments. But since these nineteen newly created juridical divisions were arbitrary and did not reflect actual land-use practices or authority structures on the ground, in most every case, their boundaries were ignored and they were never marked on the landscape.

41. See Guzmán, *Para que la yuca;* Reeves, *Los Quichuas;* and Whitten, *Sacha Runa,* and *Sicuanga Runa.*

42. Confederación de Nacionalidades Indígenas de la Amazonia Ecuatoriana (CONFENIAE), 18 July 1999.

43. OPIP documents, 1 February 1999; and Acción Ecológica, *Informe especial: La ARCO en el Ecuador* (Quito: Abya Yala, 2000).

44. The connection with oil operations and the military is long-standing. The helicopter technology advanced for military warfare in Vietnam greatly enabled oil operations in the rain forest to become more precise and sophisticated. We should not lose sight of the surveillance aspect of this alliance. Another section in ARCO's 1998 *Annual Report* is titled "Monitoring the Environment: In Alaska, ARCO Is Using Some Hot Technology from the Cold War." The article explains how ARCO has adapted an infrared technology that the North American Air Defense Command used in monitoring political hotspots from a satellite hundreds of miles above the earth to keep an eye on its Alaskan pipeline. Once secured to the belly of an airplane, the

infrared technology can detect leaks in the pipeline. Yet the technology is also precise enough to help ecologists conduct wildlife surveys by counting herds of caribou and locating polar bears in their winter dens. The potential for this surveillance technology to also track humans, even in the rain forest, should not be dismissed.

45. On the power of education and persuasion to inform people's sense of self and meaning, see Michel Foucault, *Discipline and Punish: The Birth of the Prison* (New York: Vintage Books, 1995), and *The History of Sexuality, Volume 1: An Introduction* (New York: Vintage Books, 1990).

46. See Acción Ecológica, *Marea Negra*, and *Amazonía por la Vida*; CESR, *Rights Violations*; Kimerling, *Crudo Amazónico*; Hvalkof, "Outrage in Rubber and Oil"; Jochnick, "Human Rights Challenge"; and Sawyer, "Fictions of Sovereignty," and "Bobbittizing Texaco."

47. Though the exact forms that OPIP's indigenous development will take are still in the process of formation, most ideas under consideration represent alternatives to oil dependency. They include improving the solar-powered radio communication systems among communities, strengthening the bilingual education system with more teachers and solar-powered computer technology, increasing the number of solar-powered health dispensaries, and developing a solar-powered monorail transportation system.

48. For more details, see Sawyer, *Crude Chronicles*.

49. The Asociación de Desarrollo Indígena Región Amazónica (ASODIRA) is the name that DICIP, the pro-ARCO Organization, later took. The Asociación Independiente Evangélica de Pastaza, Región Amazónica (AIEPRA) is a small evangelical organization founded in the late 1970s when U.S. Protestant missionaries had a small presence in Pastaza. The majority of AIEPRA communities are located near the border with Peru.

50. For problems with another case study, see Kimerling, "Corporate Ethics in the Era of Globalization."

51. ARCO, *Villano Project*, 16.

52. Braun, "Buried Epistemologies," 9.

ALEX GREENE

THE VOICE OF IX CHEL:

FASHIONING MAYA TRADITION

IN THE BELIZEAN RAIN FOREST

As I awaited an audience with Rosita Arvigo in her gift shop, looking out the window over rows of her bottled Rainforest Remedies, I was struck again by the orderliness of the place. Crews of gardeners kept the leafy stands well pruned, and as with the cabanas of the resort next door, even the thatched roofs seemed opulent. Much of the acreage was a lovingly manicured stage for sharing the teachings of Don Elijio Panti with the world. Panti, a regionally revered Maya healer from the nearby village of San Antonio, was also becoming a global icon of ancient healing wisdom, thanks to Arvigo's books, medicines, and popularity with tourists. I could see why most people born and raised in Belize viewed the place as a bit too sumptuous for someone who claimed to carry on the humble tradition of a Maya healer. Then again, as she later told me, her time for healing was severely limited these days by her speaking tours.

It wasn't my first visit to her farm—she was one of the reasons I kept coming back to this corner of Central America. In 1995, with my graduate studies in cultural anthropology just beginning, I had, like so many others, been searching for a field site in Guatemala when whim led me to stop off in the neighboring backwater of Belize, formerly known as British Honduras.

As I crossed the border, the contrast between the two countries was palpable. Suddenly I could hear almost everyone, from Rastas to Mayas to Mennonites, speaking a rich "kriol" that seemed to blend Jamaican patois with Scottish brogue. The map was littered with the tiny dots of towns named Teakettle, Double Head Cabbage, and More Tomorrow.

At Eva's Restaurant, a backpackers' Grand Central in the western cow town of San Ignacio, I booked the standard canoe trip upriver to a place called Ix Chel Farms so I could see the "Panti Medicine Trail," which Arvigo had blazed in honor of her teacher.[1] She lived in the compound with her family and employees, her only immediate neighbor being the expensive resort of Chaa Creek, and it was here that her life and work was presented on a daily basis to hundreds of tourists like me. I paddled the canoe, read about the plants on the medicine trail walk, and browsed through the gift shop. By the next day I was winging my way home, some half-remembered leaves and trailside signs dancing in my head. In my luggage was a bottle of Traveler's Tonic, one of Arvigo's herbal tinctures, bearing a nubile likeness of Ix Chel, the Maya goddess of healing. It was worth a try when the Central American water didn't agree with you.

One year later, there I was again. Still using that bottle of Traveler's Tonic, I had, after many hours logged in the library, decided that Belize was to be my field site. Though I had not met Arvigo during my ecotour of the year before, I had since learned that she was a magnet for those seeking "traditional healing" in Belize. An obituary for Panti in the *New York Times* earlier that year—he was 103—was my first inkling of how well connected both she and Panti were.[2] Now, a twenty-something North American intern running the gift shop, Ken, made it clear that I was very lucky to meet "Aunt Rose." There was a two-year waiting list of people hoping for a six-month internship like his, he said. Her book, *Sastun: My Apprenticeship with a Maya Healer*, was doing well, complemented by countless articles and a second book written with an ethnobotanist cataloguing the most used healing plants of Belize.[3] And then there were the speaking tours.

Even then, by the time I visited Ix Chel a second time, awaiting an interview, I had learned of a flip side to Arvigo's popularity. Many Belizeans greeted the subject of Arvigo with an ironic smile or comment that seemed to say, "Oh yes, she's doing very well for herself." Some even denounced her in the local papers as a typical U.S. swindler. To be honest, it was the controversy surrounding Arvigo—a mirror image of her new popularity in the North—that drew me to her, more than her work itself.

Finally she appeared, carrying herself like any confident North American professional: a bit more relaxed, but just as busy. She had the loose-limbed posture of a baby boomer, her hair just beginning to gray, her speech ringing with the no-nonsense snap of her native Chicago. We sat on the veranda and drank water laced with fresh peppermint. "Who are you again?" she asked with a frank smile.

If she wanted to know exactly what I was after, it was a tough question. All I had at that point were a series of impressions about medicine, international development, and environmentalism in the late 1980s and 1990s: a new heroic social category, the traditional healer, was capturing the public imagination, applied to persons who once might have been considered backward or even dangerous. Across the globe, pivotal figures like Arvigo were building links between folk healers and modern researchers with modern research budgets, then presenting this marriage to us as an example of our own future. Traditional healers, it seemed, had become living expressions of the global rain forest itself. They were pop images that allowed readers in the North to imagine the storied canopies of the rain forest as a *knowable* landscape. On the other hand, they were more than images. The people, whole communities, were clearly out there, somewhere, beyond our tangle of pavement and gadgetry, using leaves and bark to cure the sick and tormented, walking through, touching, tasting the trees themselves.

Who was I? I was, for one thing, a sympathetic reader. Here was a modern myth (and a true story at that, of mythic proportions) that I could identify with. First, there was the mega-icon of the rain forest, a vital icon, the epitome of all the nonhuman world's beauty and complexity. Because the forest could poison or cure our bodies, or clean the very air in our lungs, it was intimately a part of us.

And under the shadow of the rain forest icon strolled characters (like Panti or the goddess Ix Chel on my medicine bottle), who, through Arvigo, vied with our decadence, exhorting us to make some sacrifice, some change. Here was Arvigo herself, showing us that it could be done, that one could renounce the treadmill of a career and credit cards, and grow collard greens under the palms. One could meet an old man, who with a few leaves and a prayer could care for all those left falling through the cracks of professional medicine. Arvigo was the perfect modest heroine for moderns who, like myself, had become anxious with their own culture's excesses.

The story was even more compelling because its most sinister protagonist was North American culture itself. There was the arrival of television and soda pop in the old man's village, the children's waning interest in the gifts of the forest, and the indifference of Panti's family after they converted to evangelical Protestantism.

And then a baby boomer from Chicago appears, the only person willing to study his life's work.

As Arvigo sat before me, asking who I was, I was deeply ambivalent. Certainly I had little respect for Belize's minister of natural resources, who denounced the Maya as an environmentally unsound society, then quickly opened their common lands to an aggressive Malaysian logging company. But neither could I accept at face value a tale in which all, or most, of Belizean healing and environmental knowledge lived on in one man, give or take a few elderly colleagues. My instinct was to find some *social* element in this tale, which had been reduced by its author to a drama of two heroic individuals.

My curiosity only increased when a biologist who had worked with Arvigo confided to me (back in the United States) that Belize's organization of elderly healers, heralded in *Sastun*'s hopeful final chapter, had splintered apart in controversy and rancor. To me, expecting to find relations of power and struggle in all things social, this had a ring of truth. Though many portrayed Arvigo as doing no more than duly ushering Panti into the daylight for all to see, I began to imagine her instead pinning him in a spotlight, one that selectively scanned the hodgepodge cultural landscape of Belize for an icon appropriate to the pop mythology of North American nostalgia.[4]

Though I appreciated the sheer moxie of Arvigo and the way her apprenticeship confounded traditional Maya gender roles, somehow I would have preferred seeing a Maya granddaughter emerge as the real hero. One could understand why it might rankle a Belizean to read that the mantle of Maya culture had been passed to a white American.

And what about the other cultures of Belize, of which I was only dimly aware? Were their traditions as impoverished as *Sastun* made that of the Maya seem? I sensed that there was more to this story.

I looked in vain for any published works that shared my doubts about

Arvigo's book, but only found glowing reviews, not to mention entire infrastructures of governance, finance, and scientific research in Belize turning on the premises of Arvigo's narrative.[5] Nevertheless, since I had already heard that her credibility was being challenged locally, I wondered what other perspectives on tradition would emerge if I looked beyond her scope.

Yet as I chatted with her on her veranda, not revealing all my doubts, it was a little sad to hear her pick up where her book had left off. In the book, her apprenticeship sent her off on an optimistic arc of success. As the reader learns in the last chapter, Arvigo's vision attracted a good deal of international research and funding, and even inspired a government minister to set aside six thousand acres as a forest reserve for medicinal plants (dubbed Terra Nova by Arvigo). But the woman speaking to me on this day was weary of the tale, not exhilarated.

As for "Don Elijio's knowledge," she still believed she had single-handedly saved it from extinction. But in the moment of relative celebrity that ensued, she was caught up in the tidal force of her own romanticism, her tale meshing perfectly with the stories and values of international research and environmental networks. "Shaman [Pharmaceuticals] sent us money even before we asked for it!" she noted, referring to the company founded on the principle of translating traditional remedies into chemical blueprints for drugs. The generic traditional healer was at the height of vogue then. Idealized beyond any local history, such healers were becoming common heroes in international channels of power. Arvigo too had been caught in this undertow, riding the wave of funding and recognition. "I'm sure I made mistakes. And no one is completely innocent. But I see our main mistake to be that we were too successful too fast," she said, sinking back into her chair.

BELIZE: BEYOND THE MYSTERIOUS MAYA

As Arvigo spoke, I saw that *Sastun*'s international/modern romance of tradition had crashed headlong into local community politics. And in Belize, such conflict takes on a distinctly personal dimension. Even Arvigo, who had lived there over twenty-five years, was not ready for that.

Technically independent from Britain only since 1981 (yet still bearing the Queen Mother on its currency), the country hosts a jumble of intermingling ethnic groups, but is so small that everyone seems to know everyone else.

Even today, there are fewer than 250,000 people living in an area the size of New Hampshire—one reason the land remains heavily forested. Whether in isolated logging camps, trading posts, or the nerve centers of port towns, the many ethnicities of Belize have so influenced each other that to a certain extent, the idea of any stand-alone tradition is absurd.

Since British timber lords settled there in the late seventeenth century, the largest ethnic group until quite recently had always been the black Creole community, descendants of loggers and their slaves who came to dominate governmental service in the twentieth century. In the last decade, however, mestizos who identified more with Central America than the British West Indies have become Belize's largest ethnicity, making for complicated racial relations between the two groups.

In southern Belize, near the coasts, are towns dominated by the Garifuna, or Garinagu, sometimes called Black Caribs, descendants of an eighteenth-century maroon culture (of escaped slaves) that merged with Carib indigenous communities. And as in all British colonies, Belize is sprinkled with other ethnic groups, encouraged to immigrate as merchants or transported en masse as labor: Chinese, East Indians, Lebanese, and descendants of the Miskitu Indians of Nicaragua. Then there are several communities of Mennonites, who began settling Belize in the 1950s.

Predating all of these peoples are the Maya, never fully "conquered," and still the largest ethnicity in neighboring Guatemala. Though pushed from the coasts by British loggers in the eighteenth century, significant numbers of Belize's three Maya groups (the Mopan, Ke'kchi, and Yucatec) began to reoccupy the area in the mid-nineteenth century, during and after the Caste War of Mexico's Yucatán Peninsula. Other Maya and mestizos settled in small communities after fleeing Guatemala, El Salvador, and Honduras during the U.S.-sponsored wars of the 1980s.[6]

Given all this diversity, I wondered, how could a writer represent a single "pure" tradition in terms of its simply living or dying when so many traditions had been cross-pollinating for generations? Such questions became more pronounced once I ventured beyond the pruned gardens and well-marked trails of Ix Chel Farm, and began encountering Belizean herbalists of every stripe.

"Did you know a bat will smoke a cigarette? Yes!" one well-respected herbalist (who worked with Arvigo) remarked, just before he explained that

1. Not ancient enough? The late Mr. Percy Reynolds: plumber, teacher, saw mill worker, and respected healer. While Arvigo praised him privately as one of her biggest influences, he is largely absent from her writings. Photo by Alex Greene.

the best way to treat a bad cavity was to spray perfume on the tooth until it rotted out. Then there was the dreadlocked white man with an Irish accent who usually offered me cashew wine and made a living supplying bars with "bitters" (rum-soaked roots and bark). One Creole man—a great influence on Arvigo—who often plugged a microphone into his stereo at gatherings to sing along to soul cassettes, had learned the basics of chemistry while studying plumbing (figure 1). He eagerly pored over chemical analyses of Belizean plants published by the New York Botanical Garden. An East Indian fellow by the name of Harry Guy had learned his herbalism as a child, growing up in a purely Maya village, but wasn't inspired to earn his living at it until he befriended a Peruvian ethnobotanist and learned of a certain powerful Andean root. Still more cosmopolitan was the Mennonite woman who earned her living selling herbs—all imported, dried, from Canada.

The diverse experiences, ethnicities, and ages of professional herbal healers were staggering, a far cry from the withering vine of knowledge I had come to expect from reading *Sastun*.

Everyone, of course, had a story about Panti, who had been renowned throughout Central America and Mexico as one blessed with supernatural insight. Certainly he had deserved to be singled out in Arvigo's spotlight. But why, I wondered, had the complex realities I was encountering been polished down to the single powerful image of a lonely old indigenous man who was the last of his kind?

THE BALLAD OF ROSITA ARVIGO

As much for what it did not say as for what it did, I was becoming fascinated with Arvigo's story. And if I had realized it when she asked who I was, I might have told her, "I am here to write an annotation, a footnote, to your life's work."

"In 1969," *Sastun* begins, "I had left my hometown of Chicago and a career in advertising to pursue my dream of living closer to the land."[7] Eventually, this led her to live in a Nahuatl village in central Mexico. There, her experiences with local healing techniques led her back to Chicago to study naprapathy, a massage-based form of chiropractic.[8] A three-year course ultimately led to her Doctor of Naprapathy credential. Dissatisfied with the official intolerance of alternative medicine in the United States, by the early 1980s she and her family had settled on an acreage in Belize, where "it was a never-ending struggle to keep the jungle from encroaching."[9]

The landscape may have been a jungle when creeping over her property, but it became a rain forest when Arvigo journeyed a few miles upriver to see Panti: "I remembered why we loved Belize so much. We were far from smog, the roar of traffic, and city grays. I stood above the bank, captivated by this magical glimpse of Mother Nature in her bedchamber, and sighed. . . . I had always considered plants my friends and was anxious to get acquainted with new friends in the Belizean Rain Forest."[10]

She begins visiting Panti at his home: "The room was no more than ten by ten feet. . . . I could see no modern conveniences. It could have been A.D. 800 except for the nearby cement house with its zinc roof."[11] According to *Sastun*, Panti eventually accepts her as his apprentice only because no one else in San Antonio will bother to learn. The main thing we discover about the town itself, except for one reference to an aged couple who collect herbs for Panti, is that evangelical Protestants have convinced Panti's family and neighbors that his healing prayers are evil. And so it is up to Arvigo to take on the role of a more dutiful next generation. This situates Arvigo as the Maya tradition's sole guardian, and as we shall see, the role is hotly disputed today in San Antonio.

The struggle between modern and traditional life is most vividly described in Panti's voice:

> Factory food was ruining people's diets, he scolded. . . . "For 'modern food disease,' " he said, "I give them Balsam bark tea to cleanse the kidney and the liver . . . " He also found grave harm in frozen popsicles.
>
> "People rarely honor the Spirits anymore," continued Panti sadly as he stuffed the leaves in his sack. "They have no respect for the Lord of the cornfield."[12]

Eventually, Panti tells Arvigo of Ix Chel, the goddess "who watches over healers and helps them."[13] After hauling bags of bark and studying Panti's methods on patients for a full ten years, Arvigo finally receives a gift of great honor: Panti presents her with his charm stone, or *sastun*. She begins dreaming of the Maya spirits, and Panti dictates to her the prayers used to address each of them personally. Although she never learns much of the Mopan Maya language except these prayers, the book ends with Arvigo poised to carry on the Maya role of the Maya shaman/healer.

We also read by the book's end that she has begun collaborating with Michael Balick, director of the New York Botanical Garden's Institute for Economic Botany, in a five-year survey of the country's medicinal plants. With an easy grace, she has moved from hacking back the encroaching jungle to tapping the international rain forest network. This leads her to organize the Belize Association of Traditional Healers (BATH), which will manage the newly created 6,000-acre Terra Nova forest reserve for the preservation of medicinal plants and receive royalties from any future drug development.

And finally she describes the birth of her business, which would become such a fixture in the gift shops of Belize. She and her husband had set themselves to rescuing every medicinal plant they could, wherever farmers or loggers were clearing land, and the sheer volume of saved plants became so great that to deal efficiently with the botanical bulk, they were forced to mass-produce tinctures and salves. The Rainforest Remedies were born, not out of a desire for profit but out of a crusade.

Aside from her tinctures, Arvigo's second book of note from this period was also called *Rainforest Remedies*, subtitled *One Hundred Healing Herbs of Belize*. She and Balick first published it in 1993 after questioning eleven healers of various cultural backgrounds. It featured a foreword by Mickey Hart, former drummer for the Grateful Dead, and present-day advocate of forest conservation and global groove awareness. Hart commented that "Drs. Arvigo and Balick light the fires and carry the torch toward preservation of the rainforests and the indigenous cultures that exist within them. There is a chance for us to do something before these cultures vanish, and people like Drs. Arvigo and Balick are leading the way."[14] Arvigo's reputation as the rescuer of a dying tradition was thus cemented, though the work that led to the book was not based on either a single tradition or a dying one.

Rainforest Remedies' eleven healers (briefly noted in the acknowledgments), in contrast to the one hero of *Sastun*, were of Maya, Creole, and East Indian descent. But oddly, the book amplifies the indigenous romance of *Sastun* by bluntly equating "Belizean" with "Maya" in its introduction—a confusion of notions that colors Arvigo's work to this day. One twelve-page section, "Concepts Underlying the Traditional Treatment of Disease in Belize," deals exclusively with Maya medical and spiritual ideas; Creole, Garinagu, or East Indian concepts are nowhere to be found. And yet, only five of the one hundred plants discussed in the body of the book are listed under their Maya names.

Despite this marked favoring of the Maya as the "most authentic," the book was more important than *Sastun* in establishing Arvigo's relations with a variety of Belizean healers—first and foremost because a percentage of net sales were guaranteed to the eleven healers consulted. Indeed, the core membership of BATH has always been these eleven healers.

As I learned from my talk on the veranda with Arvigo, the optimism in the wake of *Rainforest Remedies* and *Sastun* was not sustained for long. For a while, of course, Arvigo's efforts at popularization paid off and there was money rolling in—including a $35,000 grant from the U.S. Agency for International Development for the Terra Nova forest reserve. And the New York

Botanical Garden continued its work, funded by the U.S. National Cancer Institute. But locally, Arvigo had come under increasing scrutiny, to the point where new members joined BATH with the express purpose of wresting control out of Arvigo's hands.

Talk on the street had it that these newcomers were cozy with the newly elected government, especially with the infamous minister of natural resources, Eduardo "Dito" Juan, the disheveled owner of a logging operation known to have designs on the hardwoods of Terra Nova. But when the newcomers published an article in which they accused Arvigo of personally pocketing all the international grant money, they found widespread public support for their allegations. After the newcomers proclaimed themselves BATH's authentic leaders, one newspaper editorial praised them:

> Today, [foreigners] are exploiting Belize of its herbal medicine under the guise as a good Smaritan [sic], giving with one hand and taking away with the other. This time it would seem that they have touched the wrong group of women. These women are Belizean warriors and have worldwide traveling experiences and are bent on ascertaining for Belizeans what is for Belizeans. Having formed themselves into an organization in which—as they put it—they found in their midst an exploiter passing as a do-gooder who was suspected of herding monies from abroad in millions of dollars in the name of the organization, but such monies have never been deposited in the coffers of the account of the name of the association. They told the nation by way of Radio One that not even the true herbal doctors in whose name the government gave 6,000 acres of land for this project, have ever received any recognition let alone compensation for the God-given talent. . . . These ladies have visited this newspaper and have vowed to keep the heat in the pants of those people who believe that they could continue to come to Belize with an assumed professional name and fool and exploit Belizeans.[15]

Clearly, denouncing Arvigo fed into a long history of local frustration at the power wielded by foreigners in Belize. What made this different was that it was Arvigo's representational power fueling the local outcry. No monies were actually missing, but Arvigo's tale of dying Maya tradition had become *the* defining narrative for global institutions investing in traditional healers. However, while those who wrenched control of BATH for themselves succeeded in raising doubts about Arvigo's newfound authority, they would never be able to match the evocative power of her romantic balladry.

After denying the charges of embezzlement in her own letters to the editor, Arvigo resigned from BATH, and soon all of the actual healers, centered on the *Rainforest Remedies* coterie, resigned also. BATH continued on for some years as a "virtual" organization, existing on paper only, with a membership of about three women "warriors." Meanwhile, Arvigo and her dozen or so healers regrouped as the Traditional Healers Foundation (THF), with a new exclusivity and defensiveness. The money is no longer rolling in, but in international circles today—such as in the policies of Shaman Pharmaceuticals—THF is seen as representative of all Belizean herbalists, and Arvigo continues to network during her frequent speaking tours in the North. Balick has joined her and her husband on the board of the Tropical Research Foundation of Ix Chel Farms. Professionals in the medical and biological sciences continue to use Ix Chel Farms as a base of operations, especially visiting ethnobotanists from the New York Botanical Garden. Of course the tourists keep rolling through. And perhaps of greatest significance to most Belizeans, Arvigo is putting the finishing touches on a new house, complete with decorative burglar bars, just up the hill from her medicine trail. Ix Chel is thriving neatly. At the same time, in Arvigo's interwoven narratives, the jungle's transformation into a rain forest echoes her own transformation from antimodern pilgrim into an advocate and crusader for the traditional healer.

BEHIND THE BALLADS: THE MARRIAGE OF NEW AGE AND ETHNOBOTANY, AND THE BIRTH OF THE LIBRARIAN

Arvigo emerges as a familiar figure to any who dabble in U.S. history in her dissatisfaction with the limits of modern life, warm embrace of another culture as an alternative, and romance of the rain forest. Although using the term "New Age" threatens to reify an amorphous, ever shifting cultural movement, it's tempting to describe Arvigo's background with just such a label. The term does capture an ongoing, multifaceted trend in the North of spiritual bricolage (the grafting together of principles from many religious traditions), wherein indigenous religion is given an individualist rather than a communitarian emphasis. At the same time, New Age uses traditions of science just as freely, bringing to light areas of scientific research that occupy a gray area of credibility within the networks of scientific consensus building. Indeed, Andrew Ross, in the midst of a cultural critique of New Age, has remarked on the unrecognized potential of such thinking to forge a more accessible, populist approach to science.[16] Certainly New Age writers have produced some powerful critiques of, say, the U.S. medical establishment.[17]

As a North American herbalist and masseuse who uses Ix Chel goddess imagery, prays to Maya spirits, and yet eschews life in a Maya community, Arvigo is situated solidly within the individualist, entrepreneurial transcendentalism of the New Age tradition. Her qualified embrace of scientific research is also typical of this: not only does she promote the integration of herbalism into conventional biomedicine; she has also served as the key link between Belizean healers and ethnobotanical research.

And of all the sciences, ethnobotany is perhaps the most amenable to New Age audiences. Focusing first and foremost on plants and their uses, an ethnobotanist's typical approach to culture is an all-encompassing relativism. Fashioned more as an observational science than a predictive or interpretive one, ethnobotany has avoided much of the theoretical infighting of anthropology. At times this leads to impressionistic characterizations: as one ethnobotany textbook informs us, "Indigenous peoples are extraordinarily adept at sensing insincerity."[18] But in practice, this may also give ethnobotany an accessibility that theory-bound anthropology may sometimes lack. Thus, ethnobotany is a natural ally of any popularized critique of modern conventions springing from New Age: any exploration of another culture's definitions of disease can become a de facto critique of our own.[19]

Meanwhile, the discipline's basis in the hard facts of botany and chemistry has also made it attractive to pharmaceutical companies (as many of us have read in *Time* magazine's "Heroes of Medicine" issue).[20] New technologies of chemical screening in the 1980s led to the exponential growth of corporate interest in rain forest medicines and ultimately even forest-based healers.[21]

Because the rain forest is now recognized as such a rich source of unique chemical compounds, ethnobotanists often warn that we are "burning the library of Amazonia" and other rain forests.[22] Certainly the razing of forest land is rampant, but the metaphor itself is so popular that the implications of the term "library" bear consideration.

First of all, in ethnobotany, as in all botany, the primary goal is to render the "book of nature" as literal text and make that text a transparent reflection of nature's objective order.[23] It works the other way as well: an ethnobotanist must read the leaf before him and ascribe to it an identity derived from botanical texts—not an easy task. The order of the forest is both discovered and mediated by the researcher.[24] Calling the rain forest itself a library effaces the scientist's role in teasing out the order from experience, and furthermore, portrays this order as a public good, open for the taking (or lending). It also implies a need for a librarian.

Local/indigenous people are, in a sense, coauthors of the information, but the information only becomes "real" in a funding boardroom once the librarians, ethnobotanists, and networks of laboratories to which they are linked textualize it and give it a recognizable identity. The fiction can be sorted out from the nonfiction, the chemical reactivities separated from the rituals. But what is left unsaid in the library metaphor is the power of the librarian to select what is worthy of cataloging, separating the corrupt literature (folk medicine denatured and cheapened by modern influences, like spraying perfume on a rotten tooth) from the religious (living rituals, using plants, defining a cultural identity) and pragmatic (analyses of plant-derived chemicals).

Although the forest-as-library metaphor has been a valuable tool for conservationists, nearly always implying a connection between indigenous peoples and rain forests, we shouldn't avoid asking what the social implications of this deep-seated metaphor might be in terms of the power to allocate recognition and resources. Ethnobotany as a discourse of power deeply informs the controversy surrounding Arvigo in Belize. Thus, her role as an international advocate and publicist of traditional healing brought with it the duties and powers of the librarian: determining where the pure traditions lay and who the credible healers really were. It was her assumption of this power that stirred things up.

MAYA TRADITION AND THE SEARCH FOR PURITY

As a simple thought experiment, one might imagine slightly different circumstances for Arvigo's tale. Would her story have been as successful with allusions to an "ancient" mestizo culture? Or bearing the title *My Apprenticeship with a Creole Plumber*? Obviously, no: these communities are by definition tainted by historical intermingling, just as the purity of ostensibly ancient Chinese or East Indian traditions in Belize are tainted by their geographic dislocation. Of all the communities Arvigo encountered, only the Maya are both ancient and native enough to become iconic in the North.

But even Maya authenticity/purity must be finessed. In the very text of *Sastun*, we see clues that Panti himself emerged from a history of cultural mixing, but it is a matter that Arvigo as narrator quickly passes over without comment. In chapter 4, while telling of Panti's days in the chicle camps of the 1930s (tapping tree sap for the chewing gum industry), Arvigo evokes the iconic image of the mysterious Maya even as she undercuts it with a little-emphasized detail: "That season, the camp was in the Guatemala Rain Forest near the still-untouched ruins of the Maya city of Tikal. There Panti found his teacher, a mysterious Carib named Jerónimo Requeña."[25]

From then on, it is Requeña who schools Panti in the myriad ways of healing and presents to Panti his first sastun, a stone charged with super-natural power. What the reader doesn't learn is that "Carib" or "Black Carib" are Belizean terms for the Garifuna people—a group well-known for its creative blending of African traditions with those of the practically vanished Carib Indians. The remarkable fact is that Panti, considered throughout the region as the paragon of Maya tradition, was taught by a Garifuna man. Of course, the Garifuna man is shrouded in mystery, and therefore seems to exist outside of history or a culture of his own. Arvigo simply notes that Requeña had learned from a Maya, and that Panti's first encounter with him was amid the "untouched" ruins of Tikal. The more complete picture, that these roaming chicle camps regularly thrust workers of several ethnicities together for months at a time, is quietly left unpainted. While Panti even-tually did settle in a "purely" Maya community, his early life in these camps shows just how elusive tradition in its idealized form can be.

Arvigo's inability to confront the sheer messiness of local history does not stem, I'm certain, from a cynical manipulation of the most marketable icons available to her but rather from the deep-seated romantic traditions of our own culture—the fascination with the medicines of "the Indian" discussed above. For many North Americans, and for writers hoping to popularize and promote herbal healing, an imagined Maya authenticity offers a more credible alternative to scientific biomedicine than any jumble of hybrid traditions.

This explains the mixed messages offered by *Rainforest Remedies*, which purports to be a more thorough overview of Belizean herbalism than *Sastun*. When summarizing "Belizean healing" as a system, the authors take recourse in the notion of pure Maya tradition, even when their own experience with healers clearly owes a great debt to hybrid cultural communities (like that of the Creole plumber). Thus, standing in for all the mingled traditions of Belize, we are informed of the Maya preoccupation with "the number nine," "the herbal pharmacy in Maya Healing," or the fact that "long ago, the Maya people recognized the interdependence of one's physical and spiritual being. . . . Our elderly Maya teachers have instructed us in the importance of a healthy and active mind, as well as the need to keep the body in the best condition possible."[26]

This lends an air of coherence and homogeneity to Belizean healing.[27] Indeed, out of seeming chaos, it fashions it into a unified entity. Suddenly, the many cross-cultural patterns in Belizean healing appear as a comprehensive, systematic whole. This is what the bioscript of the traditional healer demands, at the very least: an identifiable tradition, preferably one fit for popular consumption.[28]

Although this tradition is always partly fictional, Arvigo mourns the passing of its idealized form again and again, curiously oblivious to the groundswell of Maya ethnic identity movements that have been a touchstone of all writings on Central America in recent decades.[29] In *Rainforest Remedies*, for example, we read that "tragically, today in Belize, much of the traditional information is in danger of being lost. Many young people are choosing more 'modern' ways of life and showing little interest in the ways of the forest and its treasures."[30] The authors portray themselves in a "race against time" to save the old ways, which ironically, they hope to place alongside Western biomedicine in a hybridized practice.

The greatest irony is that the very media that Arvigo or Balick might indict as the corrupters of tradition—television, radio, fashion, and so on— are also the media at which these writers excel in using. Arvigo left her advertising career behind, but it remains evident in the packaging of her medicines, her prose (especially when working with her ghostwriter, Nadine Epstein), and the spectacle that flowers at her Ix Chel farm. The New York Botanical Garden even produced a video about the need for forest conservation and its link to herbalism, complete with slow-motion footage of Belizean healers associated with Arvigo that is tracked over with synthesized Andean panpipe music—the twenty-first century's universal pop signifier, it seems, for indigenous mystery.[31]

The cumulative effect of Arvigo's and other writers' promotion of authenticity has been a recasting of what traditional healing connotes. While once upon a time, in the realms of health development workers, the term was a euphemism for backwardness and unreasoning superstition, it has now taken on a celebratory glow. In fact, the term's meaning has been reversed. But whether it is used as a pejorative term or its mirror image, tradition remains a reductive concept that denies the gritty history behind healers' lived experience.

INTERNATIONAL AND LOCAL REPERCUSSIONS:
THE SPECTACLE SNOWBALLS

The tendency to confuse Belizean healing with an idealized Maya culture, so clear in Arvigo's writings, is even more evident in North American references to her work. In a sense, the Maya are a highly salable commodity: they have acquired a kind of stardom on international tourist circuits, television docudramas, and in the amorphous cultural movement known as New Age.

Thus we see articles such as "Maya Medicine: Lunch with Rosita Arvigo, Shaman from Chicago," in the *Whole Earth Review*. *Herbalgram* recommends *Rainforest Remedies* as "a window into the sacred world of traditional Mayan healers."[32]

This idealization of the Maya takes on added gravity when the reductions inform the white papers of corporations and nongovernmental organizations (NGOS) conducting research. One recurring example is a series of scholarly articles promoting the work of Shaman Pharmaceuticals. This corporation, which has now declared bankruptcy and is restructuring, based its drug development on ethnobotanically derived chemicals. The company's effort to reward local healers were widely applauded by many. Scholars working for the company or its nonprofit wing, the Healing Forest Conservancy, promoted their work in edited volumes concerned with the question of property rights and botanical materials and knowledge. Several such articles made regular reference to Shaman's support of the medicinal plant reserve in Belize, never failing to mention the "Maya traditional healers" or "herbal remedies using Maya medicine."[33]

Ultimately, the rise of intellectual property rights (and other methods of sharing drug profits with indigenous/local healers) only heightens the influence of northern notions of cultural and traditional purity. In order to attribute intellectual property, an "author" is required—someone who can sign the contract—either in the form of an individual or a community. But in this there lurks the danger of defining authors only in terms of their authenticity. Even if, say, company payments are made to a general local fund (as was the case when Shaman paid out shares to Arvigo's group of healers), questions of representation become crucial. Who represents the association, and to whom are they answerable? And who will be the local librarian, determining who is a real healer and dispensing royalties to them?

One might think this privileging of the Maya tradition, whatever it may be, would be welcomed by the Maya themselves. In some circles, it is. There are Maya who continue to work with Arvigo. One woman of note, from the Creole/mestizo/Maya village of Bullet Tree, whose father, a healer, was too traditional ever to teach his daughter the art of healing, now studies with Arvigo and travels to the United States with her from time to time. And since Ix Chel Farm and its medicine bottles have become a kind of institution in Belize, one often sees the casual appropriation of Arvigo's imagery of the Ix Chel goddess—for example, in the *Maya Atlas*, produced in southern Belize by Maya activists who have been fighting government logging concessions.[34] Arvigo's celebration of Ix Chel and Maya tradition in general is politically useful locally, where the long-marginalized Maya must use every resource available to be heard in the public sphere. And her work in linking herbalists with ethnobotanists has helped to bolster local appreciation of "the bush doctor," whose reputation has by all accounts increased markedly since the 1970s.

Arvigo certainly cannot take all the credit for such transformations, but her work is directly analogous to what Anna Tsing calls the "green development fantasies" of the "tribal elder" in South Kalimantan, Indonesia—fantasies that can take on great political weight. As a romantic icon of "tribal culture" promoted by international environmentalists, the generic "tribal elder," however fantastic, can inspire creative action. As Tsing writes,

To condemn a project, it is not enough to say that it engages in simplifications; all social categories simplify even as they bring us to appreciate new complexities. . . . "Tribal" fantasies . . . lead to collaborations between urban activists and village leaders that offer possibilities for building environmental and social justice. . . . [They foster] moments of creative intervention and the making of new identities.[35]

Nowhere is this more clear than in Arvigo's subtle challenge to the gender relations of Maya culture, both in her very practice of esoteric healing arts and in her explicit critique of Western medicine's treatment of women's health issues. Arvigo's Maya critics in San Antonio will often attack her gall, as a woman, in presuming to be Panti's apprentice, when it is taken for granted that only men can learn the most powerful healing methods. This applies even to techniques such as the external massage of the uterus, at which Panti apparently excelled. Arvigo demonstrates such techniques in many of her talks and promotes them as a holistic approach far superior to the more invasive methods of Western medicine. (At the same time, she calls this therapy "The Arvigo Techniques".)

The repercussions of Arvigo's work, in terms of Maya community gender relations, are still being played out. But the woman from Bullet Tree, studying Arvigo's techniques, has a thriving practice now—one that might not exist if she had been left to her father's tradition.

Arvigo's fascination with rescuing a pure Maya heritage, then, is mostly present on an abstract level. In practice, she will pick and choose elements learned from Panti and numerous other non-Maya healers, blending them further with elements of Western science and naprapathy. She will also boldly ignore the gender restrictions that most Maya themselves see as traditional. Thus, Arvigo usually only resorts to an idealized Maya tradition when she is staging an extralocal spectacle of indigenous wisdom—that is, for tourists or international readers. And it is on this level that Panti himself is reduced to an icon of such purity. Tradition itself, in the heroic form of the solitary Panti, becomes a kind of character in the spectacle, standing alongside the character of modern science, waging rhetorical battle with the forces that would destroy the Eden of the rain forest library. It is the very simplicity of this spectacle that earns Arvigo her international popularity and authority as *the* voice of the Maya. By extension, the voice of Ix Chel is that of the rain forest itself.

Within Maya communities like San Antonio, it is the singular (and nearly monopolistic) power of this voice that seems most offensive. An international spotlight can easily dominate the Belizean landscape. Though the thin charges of embezzlement have not vanished completely, these are only sustained by a handful of individuals. It is more the success of Arvigo's spectacle that disturbs the Maya I spoke with, especially those who studied healing under Panti themselves. For her authority is based to no small degree on the central conceit of *Sastun:* that it was she alone who was willing to learn Panti's prayers and practice amid the apathy or outright hostility of his family and neighbors.

True, the Maya of San Antonio do not uniformly embrace Maya cultural expressions. Over the years, a tightly knit community has formed around evangelical Christian churches, which disdain Maya ritual, healing, and even temples as pagan or even satanic, much in need of being swept away. This Christian community emerges from the pages of *Sastun* as yet another example of U.S. cultural imperialism. But there are at least three practicing healers in the town as well (not to mention several more from nearby villages), and the richness of their tales of learning from Panti since their childhood was just as compelling as the story told in Arvigo's book, if not more so. While their versions are irreconcilable with the voice of Panti in Arvigo's narrative (Panti, so wistfully musing on the lack of interest in his practice), I found these healers highly credible. I began to suspect that Panti had either overstated the case to Arvigo, or that her memory had faltered by the time she assumed her authorial voice.

The three healers from the town of San Antonio make use of Maya prayers alongside the use of healing plants, and regularly treat patients from the area. They are all estranged from those few who focus on the issue of embezzlement.[36] Instead, these three men all draw attention, first, to the incompleteness of Arvigo's knowledge and, second, to the absurdity of a U.S. woman, who lives next to a resort and never visits San Antonio, claiming to carry on the Mayan tradition.

"Well, she's smart," one of them told me. "She can type her own mind, but it's only her mind. She didn't really get the old man accurate. . . . The old man personally told me she didn't learn that prayer. Because when she made her [medicine] trail, she couldn't do it; the old man had to bless her place." Another noted that she had merely written the most important prayers down phonetically, ignorant of the words' meaning. He added that he had learned the prayers by recording Panti's voice on cassettes and listening to them day after day.

Yet pure tradition aside, it was the most "modern" of these three men who was closest to Panti in his lifetime. The principal of the town's public school, Sylverio Canto, who studied education in the United States and is Panti's godson, has been conducting *primicias,* or sacred memorial services, to Panti in the classic Maya style for years. When I visited him, he had summoned the local television and radio stations to his medicine trail to document the proceedings, and then topped off the celebration by hosting a community feast complete with Maya dancing and marimba music. Somehow, as he chatted in the Mopan language with his neighbors, many of whom he had treated, it was easier to believe that whatever tradition could be carried on was living here, in Panti's community.

Later, as I watched the traditional dancing in Panti's honor that day, made surreal by the presence of local television reporters (who later dubbed reggae music over the credits), it occurred to me that these healers from San Antonio were using the same metaphors of pristine tradition as Arvigo, in diverse ways. The script of the ancient Maya had a life of its own, or rather, many lives. Different actors could lay claim to the ancient goddess Ix Chel, pointing out the hybrid nature of others who were thus too compromised to wholly give her voice. Arvigo was being hoisted on her own petard of purity by the healers of San Antonio, as a modern North American woman. Meanwhile, she had rendered them invisible in her own writings. Under the terms of her spectacle, in which Panti-as-icon was either a witting or unwitting participant, these healers simply did not exist.

The differences between the competing spectacles were, to a certain extent, differences of scale and style. Canto's primicia was broadcast on a local public access station, and his say in the editing was nil. A fellow healer from San Antonio may reach as far as the cayes off the coast of Belize, where he sells diarrhea tonic to snorkelers—bags of bark bearing his signature with a hand-scrawled note describing their use. Arvigo can tour the world, publish glossy books, and even write in the professional voice of the grant proposal.

And her command of, or surrender to, the imperatives of these extralocal scripts throws up an insurmountable barrier between her and the communities she wants to celebrate, no matter how lofty her intentions.[37]

Arvigo's fascination with the notion of pristine Maya tradition excludes not only non-Maya (with the exception of a few favored healers that she honors briefly by name) but also Maya who somehow do not embody their ideal tradition enough. While her very hybridity, and the multimedia communication skills that go with it, may ultimately promote all Belizean "bush medicine" in general and the role of Maya women healers in particular, the primary content of her international spectacle promises to remain centered on the lone heroic figure of her mentor. And alongside Panti, Arvigo uses another Maya icon: the image of the goddess Ix Chel.

The authority of voice bestowed by the success of Arvigo's international, Maya-oriented spectacle has translated into her near-monopoly on ethnobotanical research in Belize, with Ix Chel Farm now established as the gateway into Belizean healing for any researchers and students. Beyond the researchers and funds circulating through Ix Chel Farms, Arvigo has established a commercial authority as well, as her tinctures are by far the most widely available in the gift shops and even pharmacies of Belize.

The significance of this power was underscored when a husband and wife who market herbal health blends with a markedly Creole identity (that is, tonics with the creolized labels "Fi Di Man" and "Fi Di Woman," with drawings of Creole "bushmasters" on them) complained to me bitterly of the unfair commercial advantage of Arvigo's Rainforest Remedies tinctures throughout Belize, to the extent that Creole-oriented products are not allowed any shelf space in some shops. This is largely due, one suspects, to the importance in the prepackaged tincture market of international consumers (tourists), who bring with them extralocal expectations of what is authentic and what is "quality" packaging.

The staging of Arvigo's spectacle, then, has a material dimension, and challenging it is not simply a matter of aesthetics or producing a more "accurate" representation of Belize. Many Maya are now busily preparing their own medicine trails to capitalize in their own ways on the script of

Maya-ness, and San Antonio now sits on the edge of the newly created, 5,000-acre Don Elijio Panti National Park.[38] Local voices are now seizing on the global scripts.

And yet Arvigo shows no indication of trying to devolve this symbolic authority over to local actors, even the healers in her circle, as I witnessed clearly during an annual meeting of her THF. On this occasion (July 1998), a consultant from the United Nations Development Program (UNDP) was present, like me, to just observe. Arvigo made explicit comments about how committed to "democracy" she was, yet afterward the UNDP consultant said to me, "Wasn't that something?" and shook his head.

I didn't quite understand. "It was very democratic, wasn't it?" I responded.

"Look," he said. "The first thing a development worker should do is make himself obsolete. It's what we call 'facilitation.' You have to train people to take your place, so the NGO or whatever can live on and grow. And that just is not happening here."

The next time I met with Arvigo, I asked her how she would answer that criticism. She smiled and remarked, "The THF is my baby. Of course I don't want to give it up!" And that was that. A single, charming declaration could do away with the issue. For the fact remained that she had pursued her mission well, meticulously, and had even helped build the legitimacy of her detractors. The powerful icons of untainted Maya tradition, salvaged in the name of the rain forest of our dreams, were for now mostly animated by her. But the modern Maya she had rendered invisible would yet have their hour.

The goddess Ix Chel, it turns out, holds a special place in the hearts of many Maya. She was honored in the Maya Postclassic period by a particularly elaborate shrine on the island of Cozumel, Mexico. Every year, from the thirteenth century on, thousands of pilgrims would gather in the port city of Xcaret to embark for the island. This also happened to be the first place in Mexico on which the Spaniard López de Gómara set foot in the sixteenth century. He later described what he found on Cozumel:

> The body of this great idol [of Ix Chel] was hollow, made of baked clay and fastened to the wall with mortar, in back of which was something like a sacristy, where the priests had a small secret door cut in the side of the

idol, into which one of them would enter, and from it speak to and answer those who came to worship and beg for favors. With this trickery, simple men were made to believe whatever their god told them.[39]

The contempt the Spaniards had for this "trickery" became clearer when Córtes ordered the shrine destroyed and a likeness of the Virgin Mary Most Pure (La Purísima) put in its place. Yet for centuries, Ix Chel survived in the hearts of many, often blended with the image of the Virgin.[40] And the use of "ventriloquism" to give voice to the supernatural never waned: as our rain forest group discovered in the Yucatán, it was a critical feature of Mexico's Caste War of the Maya that stretched from 1847 to 1901, spurred on by the Speaking Cross of Chan Santa Cruz. The Speaking Cross, which articulated prophecies and military strategies alike, was no bit of trickery, but the nuanced performance of reinvented traditions, lending the speaker a mask of timelessness.[41]

Now, five hundred years after Gómara, Westerners like Arvigo are learning to speak through Ix Chel themselves. Perhaps that's what most irks some of the Maya in Panti's hometown. It's as if Arvigo had smoothed over the face of the goddess, had adjusted the image somehow, before sliding behind her eyes. Now Ix Chel is playing to a global crowd—one that includes the Maya still. Arvigo and the icon both grow on the spectacle's power, but others are in the wings, waiting to give voice to other truths, other visions.

NOTES

This essay benefited greatly from a charmed chemistry between the members of the six-month seminar Rainforest(s): Singular and Plural, convened in 2000 by Candace Slater at HRI. Many thanks to Candace Slater, Scott Fedick, Paul Greenough, Nancy Peluso, Suzana Sawyer, and Charles Zerner for their insightful comments and tireless readings of many drafts. Many thanks also to my dissertation committee at University of California, Davis: Carol Smith, Aram Yengoyan, Benjamin Orlove, and Stephen Brush.
1. The medicine trail has become extremely popular on Central American tourist circuits. It is simply a trail through any patch forest containing a number of medicinal or useful plants, with signs indicating each key plant's history and uses. Along a medicine trail, certain plants are highlighted in the landscape by text, making the greenery of the forest less overwhelming. One suspects, though, that few tourists actually remember the information or even have access to most of the plants in their home countries. Such trails are essentially spectacles of the knowable forest.

2. Robert McG. Thomas Jr., "Elijio Panti, 103, Maya Healer with Modern Times" (obituary), *New York Times*, 10 February 1996, 34N, 52L.

3. Rosita Arvigo, with Nadine Epstein, *Sastun: My Apprenticeship with a Maya Healer* (San Francisco: HarperCollins, 1994). Though authored with a ghostwriter, who may have had a significant impact on the imagery, I refer to *Sastun* throughout as though authored by Arvigo only for convenience's sake. See also Rosita Arvigo and Michael J. Balick, *Rainforest Remedies: One Hundred Healing Herbs of Belize* (Twin Lakes, Wis.: Lotus Press, 1993).

4. Readings in colonial and postcolonial history show how often writers from modern, dominant cultures have idealized native traditions, appreciating them in almost aesthetic terms, removed from history. Many a British colonial officer indulged in this "imperialist nostalgia," bemoaning the loss of folkways among his subject—as did many an anthropologist. In literary terms, one might dub it the "romance of the premodern." See "Imperialist Nostalgia," in Renato Rosaldo, *Culture and Truth* (Boston: Beacon Press, 1993).

Today, the influence of the United States on Belize is comparable to that of an empire, having grown steadily since the early influence of the port of New Orleans, and now amped up by mass media. There are a handful of British and U.S. investors, some U.S. expatriates, and former members of the British Army, some of whom own the best resorts. In southern Belize, the expatriate has been a fixture ever since Confederate families settled there after the U.S. Civil War. One can't condemn any of them as imperialists outright, but I've heard many such relocated northerners indulge in nostalgia for a Belize unsullied by the modern world they hoped to escape. And there's no denying that Belize lives in the shadows of Britain and the United States (especially its dollar, to which Belizean currency is tied).

5. The controversy Arvigo has faced in Belize stands in marked contrast with the glowing praise she has received in the United States. The most tempered tribute comes from a review in *American Anthropologist*, which notes that "innocence and idealization creep into this text as Arvigo praises ancient traditions," but concludes finally that it is "an exceptional book" (Hilary Elise Kahn, *American Anthropologist* 97, no. 2 (June 1995): 389). Reviews in *Library Journal* and *Publishers Weekly* refer to its "engaging account" and "enjoyable story" (Teresa Elberson, *Library Journal* 119, no. 6 (1 April 1994): 127; and *Publishers Weekly* 241, no. 10 (7 March 1994): 62.

But it is in the field of ethnobotany that Arvigo has earned the greatest respect. In a 1993 review in *Herbalgram* (a popular herbalist journal with many references to current scientific research), renowned ethnobotanist Wade Davis writes: "There has been in recent years a slew of books built around the theme of a Western traveler seeking knowledge from a traditional healer. The word 'apprenticeship' has been used in so many titles and subtitles that it has lost all meaning. What distinguishes *Sastun* is the fact that Rosita Arvigo is no mail order mystic, and her story is true" (http://www.herbalgram.org/catalog/centamerica/arvigo.html). Davis further characterizes Arvigo's tutelage under Panti as "one of the most astonishing relationships in the recent history of ethnobotany." Michael Balick, director of the New York Botanical Garden's Institute for Economic Botany, who has become a close friend and ally of Arvigo's, writes of watching her work with Panti: "Rosita absorbed every detail that was being presented to her. I was extraordinarily impressed with the commitment that Rosita had made to her teacher and to Maya traditional healing" (foreword to Arvigo, *Sastun*, ii).

6. For a basic overview of Belize, see Tom Barry, *Inside Belize* (Albuquerque, N. Mex.: Inter-Hemispheric Education Resource Center, 1992). For a richer history, see Assad Shoman, *Thirteen Chapters of a History of Belize* (Belize City: Angelus Press, 1994).

7. Arvigo, *Sastun*, 2.

8. Naprapathy is a chiropractic approach to health, involving intense massage of the connective tissues, developed by Oakley Smith in 1908. See http://www.naprapathy.edu/mission.htm.

9. Arvigo, *Sastun*, 5.

10. Ibid., 15–23.

11. Ibid., 17.

12. Ibid., 41, 72.

13. Ibid., 56.

14. Cited in Arvigo and Balick, *Rainforest Remedies*, ix.

15. "Exploitation Continues!" *Labour Beacon* (Belize City), 22 April 1994.

16. Andrew Ross, *Strange Weather: Culture, Science, and Technology in the Age of Limits* (New York: Verso, 1991), 15–74. Other critical analyses of new age include Marianna Torgovnick, *Primitive Passions: Men, Women, and the Quest for Ecstasy* (New York: Knopf, 1997); and Michael F. Brown, *The Channeling Zone: American Spirituality in an Anxious Age* (Cambridge: Harvard University Press, 1997).

17. See Richard Grossinger, *Planet Medicine: From Stone Age Shamanism to Post-Industrial Healing* (Boulder, Colo.: Shambala, 1982).

18. Michael J. Balick and Paul A. Cox, *Plants, People, and Culture: The Science of Ethnobotany* (New York: Scientific American Library, 1997), 42.

19. Given the degree to which the idea of "culture" as a unified entity has been criticized in anthropology since the 1980s, ethnobotany's lack of attention to the problem of defining "culture" may be its strength. Of course, this is a rough generalization, and there are some examples of ethnobotany with richly nuanced cultural analyses. For one such work, see Janis B. Alcorn, *Huastec Maya Ethnobotany* (Austin: University of Texas Press, 1984). And space does not permit a discussion of "cognitive ethnobotany," such as the classic Brent Berlin, Dennis E. Breedlove, and Peter H. Raven, *Principles of Tzeltal Plant Classification* (New York: Academic Press, 1974). Such works grapple with the taxonomic aspects of different languages in depth, often drawing on linguistic anthropology, and often focused on elucidating local taxonomy as a coherent system.

Nonetheless, an analysis of the conjunction of healing and racial politics, as in the anthropology of Michael Taussig (see *Shamanism, Colonialism, and the Wild Man: A Study in Terror and Healing* [Chicago: University of Chicago Press, 1987]) is indeed rare in ethnobotany, as is any cultural-interpretive tour de force such as Jack Goody, *The Culture of Flowers* (Cambridge: Cambridge University Press, 1993). Despite the lack of political-cultural *theory*, however, there have been varying degrees of political engagement by ethnobotanists such as Darrell Posey, Christine Padoch, Gary Nabhan, Wade Davis, and Paul Cox, who have all struggled to combine environmental conservation with advocacy for the rights of forest-based communities. And the institutional legitimacy that ethnobotany lends to once-disdained folk medicine is important in the politics of culture as well.

20. *Time* 150, no. 19 (fall 1997).

21. Now, it seems that "pharmaceuticals" and "rain forests" are associated almost automatically. Recently, as our group entered Mexico's Sian Ka'an coastal reserve, our guide touted the fact that both Bayer and Schering-Plough were doing research there. "They are working to make medicine from some of the rain forest pl . . . or rather, uh, *tropical* plants," he said. The coastal mangrove swamp of the reserve became a rain forest by virtue of the pharmaceutical companies' interest. This interest has led in turn to attempts to formulate local knowledge in terms of Western intellectual property, so that forest-based communities can enter into contracts with research institutions. This legal/economic approach is in marked contrast to colonial botany, when local communities were often kept in the dark about the importance of their medical knowledge, even as plant-derived drugs like quinine became the basis for entire empires. See Lucile H. Brockway, *Science and Colonial Expansion: The Role of the British Royal Botanic Gardens* (New York: Academic Press, 1979). But ascribing property rights to the flow of cultural information has also been roundly critiqued, most famously by Vandana Shiva, *Biopiracy: The Plunder of Nature and Knowledge* (Boston: South End Press, 1997).

22. Richard Evans Schultes, "Burning the Library of Amazonia," *Sciences* 34, no. 2 (March–April 1994): 24–32.

23. See Michel Foucault's historical analysis of this in *The Order of Things* (1970; reprint, New York: Vintage Books, 1994). The "libraries" in this discipline have included botanical gardens, places that Bruno Latour terms "centers of calculation." The former centers of colonial empires, the great cities of the North, remain powerful partly because they continue to collect the most detailed information about every nook of the globe, from botany to anthropology, and aspire to organize it. Patterns, combinations, equivalencies, and contrasts can be arranged on a table or graph, and can be managed from a distance. See Bruno Latour, *Science in Action: How to Follow Scientists and Engineers through Society* (Cambridge: Harvard University Press, 1987), and *Pandora's Hope: Essays on the Reality of Science Studies* (Cambridge: Harvard University Press, 1999).

24. The library metaphor is arising at the very moment when the centers of calculation are becoming more powerful than ever in terms of their command of botanical information. Because of advances in molecular biology, the economic value of collected plant material is reducible to smaller and smaller bits of information. Terming the rain forest a library therefore obscures the growing power of the real libraries that can collect molecular codes easier than ever. For a brilliant analysis of the implications of this, see Bronwyn Parry, "The Fate of the Collections: Social Justice and the Annexation of Plant Genetic Resources," in *People, Plants, and Justice: The Politics of Nature Conservation*, ed. Charles Zerner (New York: Columbia University Press, 2000), 374–402.

25. Arvigo, *Sastun*, 35.

26. Arvigo and Balick, *Rainforest Remedies*, 14.

27. In this light, it is worth mentioning that Arvigo and the ethnobotanists associated with her do not reject modern medicine outright but rather envision a future where both herbal healing

and biomedicine complement each other. By emphasizing the ancient, traditional aspects of local herbal healing, these writers also rhetorically underscore the *systematic* quality of such practices: they are all part of a timeless whole, which may stand side by side with modern science, even as they implicitly reveal science's weakness by their very tenacity. Thus, there is in the connotation of purity a suggestion of a unified system. However, the portrayal of a culture as a homogenized whole, which follows from this rhetorical strategy, is highly suspect in anthropology.

28. The past two decades have witnessed a great deal of writing on the uses of tradition—mostly from the disciplines of social history, sociology, and anthropology. But they have not made much of an impact on popular thought. Even most ethnobotanists, who often allude to the erosion of tradition, resist thinking of tradition as a negotiated, constructed aspect of cultures. While continuities certainly exist in all cultures and can be threatened by various upheavals, one must remain critical of the ways in which tradition is reified in most popular writing.

By both "outsiders" and "insiders" of a culture, tradition is often selectively used, then contested by another party. One anthropologist has written that a regular part of one Indonesian healer's practice is to recite his lineage back to the original "Lord of Omnipotence" who communicated with "Archangel Gabriel" (Peter Worsley, *Knowledges: Culture, Counterculture, Subculture* [New York: New Press, 1997], 223). This sort of reference to the past is a typical legitimizing discourse. For an astute analysis of the use of "ancientness" as a marker of cultural purity and local power, see Arjun Appadurai, "The Past as a Scarce Resource," *Man* 16 (1981): 201–219.

And while historians have popularized the "invention of tradition" as a way to understand how nation-states draw on a mythical past for legitimacy (for example, Eric Hobsbawm and Terence Ranger, eds., *The Invention of Tradition* [Cambridge: Cambridge University Press, 1983]), we should also be aware of how this goes on at a more mundane level within small communities and households. Tradition is creatively mobilized to reinvigorate certain ideas and ways of living within one's culture, or to establish one's authority, and perhaps we should look at Panti's own antimodern nostalgia in this light.

In Belize, Richard Wilk has written a compelling account of how his "search for tradition" in the remotest Maya villages led him to a stunning realization: "The 'traditional' institutions were in many cases increasing in strength and importance in the more developed areas." Ultimately, he concludes that "what looked like directional change [that is, the irreversible loss of tradition] was actually cyclic" (Richard Wilk, "The Search for Tradition in Southern Belize: A Personal Narrative," *América Indígena* 47, no. 1 [1987]). These villages had gone through cycles of development, involving greater and lesser interactions with regional economies. These cycles of revitalization were not confined to "folkloric" matters such as dance or decorative arts but deeply affected the most basic social realms, such as family and household organization. These are the very realms that many countermoderns idealize as the heart of continuity, as in the classic scenario of the elder instructing her grandchildren around the fire before the advent of corrupting radios and televisions.

With his volume *In the Museum of Maya Culture: Touring Chichén Itzá* (Minneapolis:

University of Minnesota Press, 1996), Quetzil E. Castañeda examines the process by which an idealized Maya tradition is presented as a series of consumable signs for modern tourists. In Belize, where the reconstructed ruins of Maya monuments dominate the landscape at every turn, the same process is very much at work, along with the same attendant impact on the self-representation of local Maya.

For a direct analysis of the self-representation of tradition among Maya involved in traditional healing, see Steffan Igor Ayora-Diaz, "Imagining Authenticity in the Local Medicines of Chiapas, Mexico," *Critique of Anthropology* 20, no. 2 (2000): 173–190. As Ayora-Diaz writes, local Maya healers "produce representations and herbal medicines to be consumed by romantic travelers, and attempt to preserve their own medical tradition. However, this tradition involves change and continuous hybridization" (185).

29. For a thorough discussion of a movement that has galvanized Guatemala, see Diane M. Nelson, *A Finger in the Wound: Body Politics in Quincentennial Guatemala* (Berkeley: University of California Press, 1999). Nelson even quotes mestizo Guatemalans who claim that "Maya theories of identity are not their own; they are from the United States" (130). This of course is belied by history (see Carol Smith, ed., *Guatemalan Indians and the State, 1540–1988* [Austin: University of Texas Press, 1990]), but such a backlash is certain evidence of the strength of the Mayan identity movement.

30. Arvigo and Balick, *Rainforest Remedies*, 5.

31. New York Botanical Garden and Ix Chel Tropical Research Foundation, *Diary of a Belizean Girl: Learning Herbal Wisdom from Our Elders*, video with teacher's guide (Bronx, N.Y.: Institute of Economic Botany, New York Botanical Garden, n.d. [ca. 1996]).

32. American Botanical Council, "Herbal Education Catalog," *Herbalgram*, no. 48 (2000): 24.

33. See Steven R. King, Thomas J. Carlson, and Katy Moran, "Biological Diversity, Indigenous Knowledge, Drug Discovery, and Intellectual Property Rights," in *Valuing Local Knowledge*, ed. Stephen B. Brush and Doreen Stabinsky (Washington: Island Press, 1996), 180–181; and Katy Moran, "Biocultural Diversity Conservation through the Healing Forest Conservancy," in *Intellectual Property Rights for Indigenous Peoples: A Sourcebook*, ed. Tom Greaves (Oklahoma City, Okla.: Society for Applied Anthropology, 1994).

34. Toledo Maya Cultural Council and Toledo Alcaldes Association, eds., *Maya Atlas: The Struggle to Preserve Maya Land in Southern Belize* (Berkeley: North Atlantic Books, 1997). See the *Atlas*'s Ix Chel figure (32), which is clearly lifted directly from Arvigo's promotional literature.

35. Anna L. Tsing, "Becoming a Tribal Elder, and Other Green Development Fantasies," in *Transforming the Indonesian Uplands: Marginality, Power, and Production*, ed. Tania Murray Li (London: Harwood Academic Publishers, 1999), 162–163. For similar discussions of international notions of authenticity and their local repercussions, see Peter J. Brosius, "Endangered Forest, Endangered People: Environmentalist Representations of Indigenous Knowledge," *Human Ecology* 25, no. 1 (1997): 47–69; Beth Conklin and Laura Graham, "The Shifting Middle-Ground: Amazonian Indians and Eco-politics," *American Anthropologist* 97, no. 4 (1995): 695–710; and Beth Conklin, "Body Paint, Feathers, and VCRs: Aesthetics and Authenticity in Amazonian Activism," *American Ethnologist* 24, no. 4 (1997): 711–737.

36. One San Antonio healer even exclaimed of the embezzlement crowd, "They don't like me either!" and speaking of one woman in particular, "Her! I trust her even less than Rosita!"

37. See Néstor García Canclini, *Hybrid Cultures: Strategies for Entering and Leaving Modernity* (Minneapolis: University of Minnesota Press, 1995). This is a pivotal work on moderns' romance of the traditional folk, and the diverse ways in which tradition lives and mutates in popular forms of expression. Simply using communications media, or having recourse to international networks, does not invalidate a form of popular expression, even if it is staged as an expression of tradition. Indeed, as he writes, "The supposed process of folklore's extinction did not become more marked, despite advances in mass communications and other technologies. . . . Not only did this modernizing expansion not succeed in erasing folklore, but many studies reveal that in the last few decades traditional cultures have developed by being transformed" (153). García Canclini's work, which helped to spark widespread research into hybrid, nonpristine cultures, still stands as the strongest corrective to forms of modernist nostalgia like Arvigo's.

38. LOVE-FM radio broadcast news, 27 February 2001.

39. Cited in David A. Freidel and Jeremy A. Sabloff, *Cozumel: Late Maya Settlement Patterns* (New York: Academic Press, 1984), 44. This book has several discussions of the Ix Chel shrine.

40. For a description of Ix Chel's usurpation by Mary, see Victor Turner and Edith Turner, *Image and Pilgrimage in Christian Culture* (New York: Columbia University Press, 1978), 50. Ix Chel's melding with her usurper is described in Ronald Wright, *Time among the Maya* (London: Bodley Head, 1989), 314, 345.

41. The tradition of speaking stones is still very much alive among the Maya, as described in Wright, *Time among the Maya;* Paul Sullivan, *Unfinished Conversations: Mayas and Foreigners between Two Wars* (New York: Knopf, 1989). For more on the Caste War (also very much alive in local memory), see Nelson Reed, *The Caste War of the Yucatán* (Stanford: Stanford University Press, 1964); and a special dedicated issue of *Americas* 53, no. 4 (April 1997).

RAIN FOREST AND JUNGLE

SCOTT FEDICK

IN SEARCH OF THE MAYA FOREST

The ruins of soaring, vine-draped temples protrude above a tall, dense, tropical forest—remnants of a mysterious lost civilization of southern Mexico and Central America known as the Maya. This image of ancient temples in the tropical forest has become a familiar icon of the "Mysterious Maya," as they are so often called in popular media (map 1).[1] But behind this single icon are three strikingly different stories about the relationship between the ancient Maya and the tropical forest they inhabited.

The first story behind the icon portrays the ancient Maya in popular culture, film, and the tourist industry as a mysterious civilization that lived in harmony with nature in a primeval rain forest. While this image has been around for quite some time, its current resurgence is largely a New Age vision of the Maya as wise mystics and shamans who respected and preserved nature's sacred paradise. It is an image and a story that teaches the public that temple-cities nestled in an undisturbed rain forest is how it was in ancient times, and how it should be in the future.

The popular image of the Maya forest as nature's sacred paradise stands in stark contrast to the traditional academic image of the Maya as a culture constantly in battle with a hostile, ever encroaching jungle. In this second

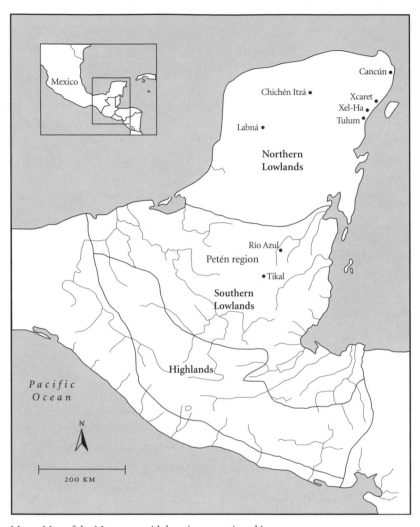

Map 1. Map of the Maya area with locations mentioned in text.

story, the temples in the forest stand as stark reminders that Maya civilization lost a battle that could not be won against an ill-suited environment. Through the five centuries since Europeans first encountered the Maya and the tropical lowland environment in which they lived, the region has been considered a virtual agricultural wasteland. When the forest was cleared, the thin, rocky soil was difficult or impossible to plow. The land was easily exhausted, with erosion and grass invasion a constant threat. How the ancient Maya could have developed an advanced culture in such a hostile and fragile environment was baffling. It was generally thought that environmental degradation was the unavoidable result of the Maya occupation, ultimately leading to the well-known collapse of Classic Maya civilization at about A.D. 900. This "lesson" from the past is often cited as an example of the destructive potential of human occupation in fragile rain forest environments, serving to warn of the need for setting aside such forests in order to preserve them for future generations.

The popularity and persistence of these first two stories is rather puzzling in light of the current evidence indicating that the Maya forest of today is actually a product of many thousands of years of human management, manipulation, and tinkering. It is also a forest that became part of an agricultural system, sustaining astonishingly high populations for a long time without any known loss of species diversity. Yet the contrasting images persist. Were the Maya a rain forest civilization living in harmony with the environment or were they struggling farmers in a constant battle against an advancing jungle? Images, texts, and personal experiences present a dilemma that must ultimately be confronted by public policy on forest conservation. With a growing public concern for "saving the rain forest," should people, including the modern Maya, be denied access to the forested lands of an ecosystem that has, for most of its existence, included intensive human manipulation? Is only the lush, high-canopy, rain-dripping, primordial rain forest worth preserving, and if so, how much of it actually exists in the Maya Lowlands? If efforts are to be made to restore parts of the forest that have been destroyed in recent times, what kind of forest should be restored?

Before attempting to answer these questions, I suggest that there exists behind the icon a third story or interpretation that needs to be told: The forest that now surrounds the ancient temples is a garden gone to seed. In this forest, it is the absence of people that is the aberrant situation. This story offers an image of the ancient landscape as a "managed mosaic," as a complex patchwork of forest, field, and garden in which the Maya were neither perfect guardians of nature nor the foes of a hostile jungle.[2] It is a story in which the Maya were, are, and should be part of the Maya forest.

All three of these stories are important, not just because of the images they create about the Maya past but because of the implications they hold for the current efforts of conservation groups, forest restoration projects, indigenous rights organizations, tourist development boards, and the immigrant farmers that are being drawn to this region in search of new opportunities.

It seems only appropriate that I introduce the currently popular image of the ancient Maya as rain forest dwellers living in harmony with the environment by discussing the encounters I had with this rain forest while viewing the popular animated film *The Road to El Dorado*, visiting the Rainforest Café in Cancún, and touring the "eco-archaeological" park of Xcaret in the so-called Maya Riviera.[3] As an archaeologist who has worked in the Maya Lowlands for many years, I found it enlightening—and at times highly disturbing—to try to view the Maya world as a tourist, with perceptions about the Maya and their forest home being shaped by movies, tourist attractions, and popular literature.

In the animated film *The Road to El Dorado*, as mentioned in Candace Slater's chapter, two young Spaniards find themselves unwitting stowaways on one of Cortez's ships heading for the New World in 1519.[4] The goal of the voyage and the two adventurers is to find El Dorado, the city of gold. Our heroes travel through a forest with trees so large and lush as to make actual rain forests look like groves of broomsticks. They end up in a city of soaring temples and palaces inhabited by people that are clearly portrayed as (but not called) the ancient Maya.[5] The painted temples are nestled within a towering, verdant rain forest, with palaces draped in philodendrons and flowers. The giant trees of the luxuriant forest grow right up to the edges of the plazas in which the Maya gather to dance, witness sacrifices to the gods, and play the sacred ball game. Even in sweeping panoramic views, there is no sign of cleared agricultural fields or farmers inhabiting the countryside surrounding this temple-city.

In the movie theater, I found myself fidgeting in my seat and grumbling to my friends about what an absurd portrayal it was. Where does their food supposedly come from? Did they just gather wild fruit from the first floor like vegan Jain monks in India?[6] Where did the thousands of happy, smiling Maya live—inside the temples? Why do movie producers have to perpetuate misconceptions about the Maya like this? I couldn't help being bothered. I had seen this image of the ancient Maya too many times in recent years, and it made me wonder what implications such popular images might hold for the modern Maya and the future of their forest.

Many viewers of *The Road to El Dorado* would probably recognize the architecture and people portrayed in the film as Maya, and would imagine what a wonderful place the Maya rain forest must be to visit.[7] The most popular tourist destination in Mexico is the beach resort of Cancún, the heart of a thriving coastal development region promoted as the Maya Riviera. For most tourists, mention of the Maya brings to mind a mysterious lost civilization. It is a vision of abandoned temples, palaces, and beautifully carved monuments hidden for centuries by a tangle of vines and the leafy shadows of a lush tropical forest. They remember seeing something about the Maya on a National Geographic television special or reading about them in one of their aunt's picture-filled Time-Life books on ancient civilizations.[8] They know that the Maya built beautiful stone buildings; were masters of mathematics, astronomy, and the calendar; and that their civilization had vanished for unknown reasons at the close of their Classic period (ca. A.D. 250–900), over a thousand years ago.[9]

The image of a rain forest paradise is strongly promoted in the resorts and commercial enterprises of the Maya Riviera. After arriving in Cancún, many tourists are drawn into their first imaginary rain forest encounter by visiting the Rainforest Café, located on the Cancún hotel strip within a short walk of the beach.[10] I took the Rain Forest group there in the spring of 2000 so we could all share the tourist experience. The outside of the mall that houses the Rainforest Café is decorated with a multistory fiberglass and plaster collage of Maya architecture, rain forest foliage, and exotic jungle animals. Inside the café, we found that one could enjoy a "Rainforest Burger" (100 percent beef) or "Mayan Meatloaf" beneath a canopy of plastic vegetation and the watchful stare of motorized fake gorillas and elephants while waiting for the next sound-and-light rainstorm. This must be what the Maya rain forest is really like!

The centerpiece of the café is a fountain surrounded by lush vegetation and a sculpted Atlas figure supporting a globe draped with a banner proclaiming, "RESCUE THE RAINFOREST." In the Rainforest Café Retail Village, we browsed through products that help raise consciousness about endangered species, fed coins to the Rain Forest Meter (every nickel saves eighteen square feet of rain forest), and learned about a free educational program that was offered for groups of children. I purchased a booklet titled *Rain Forest Explorers' Guide* that includes a map placing the entire Maya Lowlands, including the Maya Riviera, within tropical rain forest.[11]

How can the Cancún tourist learn more about the ancient Maya and the currently endangered rain forest in which they lived? Kiosks in the mall and tour hawkers on the street corners of Cancún offered us the answer. Spend a day at Xcaret, a self-described eco-archaeological park that bills itself as "Nature's Sacred Paradise."[12]

Our drive from Cancún to Xcaret took about an hour on a wide, modern highway. Along the roadside were numerous billboards placed by Xcaret to educate us about the fragility of our planet, evils of pollution and littering, and need to protect endangered plants and animals. The entrance road to the park is posted with a 15-kilometer-per-hour speed sign that asks us to slow down so as not to harm any butterflies. A walk through the tree-shaded parking lot led us to the park entrance, a reproduction of a famous Maya archway from the site of Labná—the same archway that protects the hidden entrance to the Maya city of El Dorado in the animated film. We then stepped into the modern El Dorado, the rain forest paradise of the ancient Maya.[13]

Literature produced for Xcaret told us that the park ". . . provides an educational and entertaining forum to interact directly with the region's natural beauty and fascinating Maya culture."[14] The natural setting of the park is described as being "nestled amidst the luxuriant rainforests of the Yucatan Peninsula. . . . In this idyllic paradise, humans, the natural world, the past and the present converge and coexist."[15] It is also explained that "the veneration and preservation of the environment, as much as the legacies and culture of the Mayan peoples, is of paramount importance to Xcaret."[16] The reconstructed Maya temples that are included within the park boundaries provide stages for segments of *Xcaret at Night*, a show that we viewed while strolling through the park after dark.[17] The temples, surrounded by lush vegetation, are illuminated by candlelight as costumed dancers perform to the beat of drums, shrouded by a mist of copal incense. Visitors to Xcaret are constantly reminded that the ancient Maya lived in harmony with nature and that we must do our share to save what is left of the Maya rain forest.

Nature's sacred paradise is presented to the Cancún tourist as an eternal, unchanging forest threatened only by modern development, yet capable of being preserved by responsible ecotourism. The same international

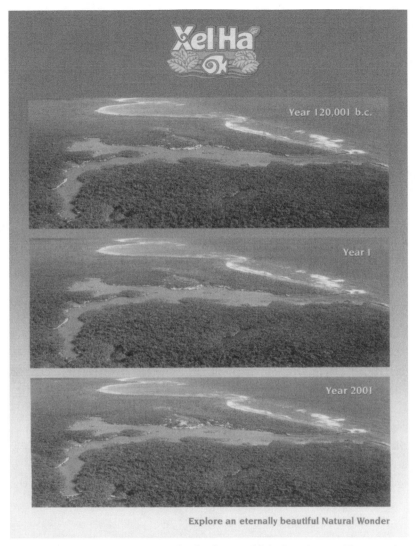

1. An advertisement for the eco-park of Xel-Há, published in *Cancún Tips* magazine, spring 2001. Reprinted by permission of Xel-Há.

corporation that developed Xcaret also manages Xel-Há, another eco-archaeological park in the Maya Riviera.[18] An often seen advertisement for Xel-Há features three identical aerial photographs of the coastal forest (figure 1), one labeled "Year 120,001 b.c.," the second "Year 1," and the third "Year 2001."[19] The only visible difference between the photographs is the unobtrusive thatched roofs of the restaurants, shops, and other facilities of the park in the "Year 2001." While obviously an advertising gimmick, the corporation's iconic ad cultivates a public desire for a Maya landscape of unchanging, eternal beauty—nature's sacred paradise, sprinkled only lightly with ancient temples where the Maya lived in harmony with the enveloping rain forest, the way it's always been and how it should remain in the future.

As with my experience of watching *The Road to El Dorado*, my visit to the Rainforest Café and to Xcaret certainly provided entertainment, but also left me with an uneasy feeling that I was sure the average tourist would not experience. Did people realize that the vast majority of the Maya Lowlands is really not rain forest at all, and that most of the plastic plants and animatronic animals in the Rainforest Café were African species? And what about the Rainforest Burger and Mayan Meatloaf? Even the McDonald's corporation recognizes that cattle ranching is the main cause for tropical forest destruction and vows not to use beef from cleared forest lands. The nature's sacred paradise of Xcaret features an artificial "underground river" that was gnawed out of the natural landscape with dynamite and backhoes, and an artificial beach of trucked-in sand is protected by a bulldozed spit of boulders and gravel. The reef off of Xcaret is now nearly dead, and the park's botanical garden is mysteriously free of insects, just like the resort area of Cancún, which is sprayed in the early morning hours by tanker trucks of insecticide.

And what about the Maya? At Xcaret, the Maya are portrayed as an ancient civilization that apparently spent most of its time conducting strange rituals and dancing. The reconstructed Maya village looks like something out of Polynesia, and gives no clue if it's supposed to represent lifeways of today or a thousand years ago. In fact, as noted in the introduction to this book, a visitor could easily leave Xcaret with the mistaken impression that the Maya are an extinct culture, now exotic and distant except in the theme park world of nature's sacred paradise.

Maybe it's just too easy for me to be sarcastic about what I've experienced as a "tourist" in the Maya Riviera. Yet even for the casual visitor, there is something confusing and contradictory in the descriptions one reads in the Cancún tourist literature about the supposedly tall, lush rain forest that lies inland of the Maya Riviera, and the low, dry, scrubby forest and brush one views out the windows of the bus while riding across the northern Yucatán Peninsula to visit the famous ruins of Chichén Itzá.[20]

The current popular image of the ancient Maya as having lived in harmony with nature in a primeval rain forest is strongly linked to the commercial rise of ecotourism and the public appeal of "saving the rain forest." But where is the Maya rain forest that needs to be saved? Most of the vegetation around the Maya Riviera is low forest and scrub. Many of the trees that tourists see are also nearly bare of leaves since the height of the tourist season coincides with that of the dry season in the area. But aren't rain forests tall, lush, and green? Have people already destroyed the rain forest in this northern part of the peninsula? Is the Maya forest really a rain forest at all?

In the strictest sense, only about one-third of the Maya Lowlands, an area in the southernmost part of the region, would be considered "real" rain forest according to the widely accepted scientific classification.[21] Until recent times, only a few general descriptions of the Maya Lowlands have characterized the region as predominantly rain forest.[22] The northern lowlands, the region most visited by tourists, supports dry tropical forests of low to medium height, with trees that lose most of their leaves during the pronounced dry season. The species composition of the northern forests, however, is similar to forests in the wetter southern lowlands. Likewise, the species composition of the younger "regrowth" forests that predominate in the more heavily populated north also closely mimic the "mature" forest, where it can still be found.[23] From a biodiversity perspective, there is really little justification in valuing the dry, predominantly secondary forests of the north any less than the more majestic ones of the south (map 1). The difference between these forests is really more aesthetic than biological.

Placing too much emphasis on the aesthetic value of the high-stand "undisturbed" forest sets a dangerous precedent in which the public, and government policymakers might tend to write-off the lower, drier, and more "disturbed" forests of the northern Maya Lowlands. This is exactly the pattern that is emerging in the popular media and conservation literature that alerts the public to ongoing destruction of natural habitats in the region, in which the endangered Maya forest is increasingly equated with rain forest.

We can also trace the changing public image of the Maya forest from a hostile jungle to an endangered rain forest through the pages of *National Geographic Magazine*. Thirty-one articles were published in the magazine from 1913 through 2000 that included descriptions of the forest environment of the Maya Lowlands. Before the term "rain forest" was first used to depict the area in 1975, the most common term employed to describe the vegetation (besides the neutral terms of "forest" or "tropical forest") was "jungle."[24] When the term "rain forest" was first used in the magazine in reference to the Maya area in 1975, it was to describe the setting for a murder rather than to offer a sympathetic characterization of the environment.[25] In 1985, the first *National Geographic Magazine* article to highlight the ongoing destruction of the Maya forest made heavy use of the term "rain forest," although "jungle" was still more commonly used in that article.[26] In the magazine, the Maya Lowlands became a rain forest with economic value for ecotourism in 1989 with the publication of an editorial about protecting tropical rain forests and a lead article on La Ruta Maya, a multinational plan to promote tourism in the Maya region.[27] The editorial in that issue alerts the public that:

> Unfortunately, rain forests are falling at an alarming rate, because their timber provides quick cash, and cleared forests become coveted farms for the growing masses of poor people who live near most tropical forests. Even if they wanted to, governments cannot fence the forests and order citizens to stay out.
>
> Only imaginative plans and quick action that provide alternative income can save them. As we explain in the article on La Ruta Maya leading this issue, "eco-tourism" offers one hope. . . . Making some forests into living theme parks with cable-car or monorail systems, as shown above, offers one exciting solution.[28]

An accompanying painting shows a "Maya Route Cableway" in which a Robinson Crusoe–style gondola rides suspended from a cable beneath a high-canopied rain forest while passing by the towering ruins of a Maya temple (figure 2). This "imaginative plan" represents what some think of as the ideal solution: rain forest reserves that exclude people (except tourists and managers), yet provide revenues for local populations living outside the boundaries of the parks. Is this so different than the image of the ancient Maya living in nature's sacred paradise? Apparently not, as illustrated in an earlier issue of *National Geographic Magazine* (figure 3). Accompanying a 1975 article is a painting by a National Geographic artist that reconstructs a bird's-eye view of Tikal, one of the largest ancient cities of the Maya world.[29]

2. Maya route cableway, published in *National Geographic Magazine*, October 1989. Note the ancient Maya temple on the right side. Painting by John Berkey. Reprinted by permission of National Geographic.

3. The ancient Maya city of Tikal, as depicted in *National Geographic Magazine*, December 1975. Portion of a painting by Peter Spier. Reprinted by permission of National Geographic.

In this painting, the temples and palaces of the living city are surrounded by a dense, unbroken expanse of forest that continues on to the distant horizon. As in *The Road to El Dorado*, there is no sign of agriculture nor an indication of human disturbance of the forest besides the ceremonial center. This image is accompanied by a companion article in the same issue of *National Geographic Magazine* that seemingly contradicts the memorable painting. In that text, archaeologist William Coe describes the results of surveying and mapping that extended beyond the major architectural monuments:

> When excavation ended in central Tikal in 1965, we were still curious about how people lived in the countryside. . . . The final map was astonishing. The so-called countryside was heavily populated over an area of 50 square miles. Rarely were family compounds more than 500 yards apart; nowhere was there enough space for the slash-and-burn practices of Maya corn growers. The people must have been backyard horticulturists. . . . Most likely, subsistence was based on highly productive kitchen gardens . . . Utilizing their environment to the fullest, Tikal people perhaps built up low-lying areas, or bajos, into raised fields.[30]

Despite the archaeological evidence to the contrary, the *National Geographic* painting (figure 3) makes it possible for a populated city of temples and palaces to exist in an undisturbed forest—an impossible paradox. Despite Coe's description, it is this image, the painting, that is most likely to stick in readers' minds. To much of this public, the entire Maya forest has now become a rain forest, and it must be preserved and protected, just as was supposedly done by the ancient Maya.

THE HOSTILE JUNGLE

The current popular image of the Maya Lowlands as nature's sacred paradise stands in stark contrast to the common academic vision of tropical environments as hostile and uninviting settings for people to survive, let alone develop an advanced civilization. A negative image of the Maya Lowland environment is a deeply held scholarly perception that had been maintained by outsiders, including explorers, anthropologists, and archaeologists, for nearly five hundred years.

The thinness of the soils in the Maya Lowlands, particularly in the north, has always shocked outside observers. Writing about the northern lowlands, Diego de Landa, the first bishop of Yucatán, wrote in 1566 that "Yucatan is the country with least earth that I have seen, since all of it is one living rock and has wonderfully little earth."[31]

Systematic, large-scale archaeological investigations in the Maya area began in the early twentieth century. The archaeologists who conducted these studies continued to express negative impressions about the environment and agricultural potential of the lowlands that were mirrored in their interpretations of the land and its ability to support civilization.

As Sir J. Eric S. Thompson, one of the most famous twentieth-century scholars of the ancient Maya, observed:

> To me, one of the greatest mysteries is why Maya culture should have reached its greatest peak in this region so singularly lacking in natural wealth, where man, armed only with stone tools and fire, had everlastingly to struggle with the unrelenting forest for land to sow his crops. Moreover, when he had wrestled the necessary area momentarily from the forest's grasp, he usually found a soil so thin and quickly weed infested that after one or two crops, it had to be surrendered to his enemy, who lost no time in covering it once more with dense vegetation.[32]

As Thompson and other scholars of the mid twentieth century were beginning to synthesize the first decades of serious archaeological work on the Maya, a clear consensus emerged regarding the ancient civilization and its tropical forest home. The forest was the enemy, and Maya civilization was in a constant state of struggle against the encroaching, unproductive jungle. Due to the perceived severe limitations imposed on agricultural development in the Maya tropical forest, researchers from Thompson's time held that the only form of agriculture that could have been practiced was slash-and-burn (swidden) cultivation, with maize as the main food crop. They also believed that this cultivation system could be intensified only by shortening the fallow period—the time a field was left unused in order to regenerate forest cover and replenish the soil. The ideal fallow cycle, often said to require ten to twenty years of nonuse, could only be shortened slightly without resulting in erosion, grass invasion, and nutrient depletion of the soil. This has come to be known as the "swidden thesis" of Maya agriculture.[33]

The swidden thesis had profound implications for the interpretation of

Maya civilization, placing severe restrictions on regional population levels and the ability of agriculture to sustain true cities. Archaeologist Sylvanus Morley was perhaps the strongest proponent of the swidden thesis. Writing for *National Geographic Magazine* in the 1930s, and as author of the first textbook on the ancient Maya in 1946, he was fond of citing studies by "corn experts" with the U.S. Department of Agriculture that declared the Yucatán Peninsula was fit only for long-fallow cultivation that could not be intensified without environmental degradation, or cultivated by any other system, even with modern implements.[34] It was believed that swidden agriculture could support only a low population density, and that the soaring temples and grand palaces that were being mapped and excavated represented not true cities, but ceremonial centers that once had only small, resident groups of priests, astronomers, and their attendants. This priestly elite was thought to have been supported by a sparse, scattered population of farmers who congregated at ceremonial centers on occasions ordained by the keepers of the calendar to provide labor and tribute to the priests as well as receive instructions regarding the agricultural cycle of slash-and-burn cultivation.

According to early archaeologists, Maya religion, elaboration of the calendar, and astronomical recordings were all centered around predicting the proper time to clear the forest, burn, and plant corn. Even the alleged astronomical function of buildings was tied to the precarious activity of farming corn in the jungle. For example, a 1931 *National Geographic Magazine* article by Morley is titled "Unearthing America's Ancient History: Investigation Suggests That the Maya May Have Designed the First Astronomical Observatory in the New World in Order to Cultivate Corn."[35] Supposedly, this religious devotion to the agricultural cycle could not prevent either the eventual degradation of the fragile land or the encroachment of grasses that the Maya had no technology to battle.

Perhaps the strongest statement about the limitations of the lowland Maya forest environment is to be found in Betty Meggers's 1954 tour de force of environmental determinism, "Environmental Limitation on the Development of Culture." In an attempt to explain the differential development of cultures in various parts of the world, Meggers classified environments into four types, based on the supposed suitability of environments for food production. Citing only generalizations made by other anthropologists and geographers, she classified the Maya Lowlands as a "Type 2" environment of limited agricultural production:

> Here agriculture can be undertaken, but its productivity is minimized by limited soil fertility, which cannot economically be improved or conserved. When the natural vegetation cycle is broken by clearing, planting and harvesting, the delicate balance between what is taken from and what is returned to the soil is upset. The soil is poor to begin with, and exposed fully to the detrimental effects of the climate, it is quickly exhausted of plant nutrients. The addition of fertilizer is not feasible on a primitive level or economically practical on a modern one.[36]

Meggers saw such an incongruity between the heights of Classic period Maya civilization (ca. A.D. 250–900) and the forest setting in which it is found that she could arrive at only one, rather bizarre conclusion:

> This means that a culture of the level attained by the Classic Maya could not have developed in the Type 2 environment where the archaeological remains are found, but must have been introduced from elsewhere. Furthermore, since Type 2 environments lack the resources to maintain so high a level of culture, the history of the Maya occupation of the tropical forest should represent a decline or deculturation.[37]

Relying on the scant archaeological evidence available in her time, Meggers speculated that Classic Maya civilization had appeared suddenly in the lowlands, without local cultural antecedents or a record of gradual development, only to fall into a long and continuous decline brought on by overuse and degradation of an unsuitable environment.

According to Meggers, the appearance of civilization in the Maya Lowlands was an anomaly; she seemingly viewed the culture as a mistake that was eventually righted by the law of the land: "The level to which a culture can develop is dependent upon the agricultural potential of the environment it occupies."[38] To Meggers and other proponents of the swidden thesis, today's popular icon of ruined temples enveloped in jungle growth would perhaps represent a reminder that humans have no place in the Maya forest: attempts to occupy and develop such a place are thus doomed to failure.

In the 1960s, archaeologists began to explore beyond the grand architecture found at sites such as Tikal, searching the surrounding forest for evidence of the farmers who supported these so-called ceremonial centers. As we learned above from Coe's description of the mapping project at Tikal, what the archaeologists found surprised them. Classic period Maya centers were populated by tens of thousands of people, and the surrounding countryside was filled with innumerable smaller centers and communities. The Maya Lowlands may, in fact, have been one of the most densely populated regions in the world at the time. Estimates place rural population densities of the Late Classic period at 180 persons per square kilometer, and urban densities between 500 and 800 persons per square kilometer.[39] How, then, did they feed themselves in what was still perceived as having been a marginal environment?

In response to revised population estimates for Maya cities and the region in general, researchers began to reassess ancient cropping systems and agricultural technologies.[40] As alternatives to slash-and-burn cultivation of maize, a variety of crops and cropping systems were proposed that would have been less demanding on the soil, such as root and tree crops. Long ignored evidence of terraced hillsides was reexamined, and extensive areas of terracing were newly mapped. Perhaps most significant, the discovery of patterned ground in wetlands of the Maya Lowlands led to speculation that these vast areas had been converted from swampy "wastelands" into intensive raised-field agricultural complexes, similar in form and function to the familiar "floating gardens" or *chinampas* that had been built by the Aztecs in the shallow lakes of the highland Basin of Mexico.[41]

As early as the mid 1970s, a revised academic vision of the ancient landscape of the Classic period was emerging; it was a landscape stripped of forest, heavily populated, and intensively cultivated. Perhaps the only way to tame the jungle was to clear it, while turning to wetlands for the most intensive and productive farming. In a 1975 *National Geographic Magazine* article, geographer B. L. Turner II describes his vision of the Maya Lowlands: "If you could have flown over the Petén at the height of the Classic Period, you would have found something akin to central Ohio today."[42] A 1986 illustration from *National Geographic Magazine* provides an updated image of "A City of Monuments that Keeps the Tropical Forest at Bay" (figure 4).[43]

4. The ancient Maya city of Río Azúl, as depicted in *National Geographic Magazine*, April 1986. Portion of a painting by Roy Andersen. Reprinted by permission of National Geographic.

In this image, the city of Río Azúl and its suburbs have been cleared of forest, with only a scattering of trees remaining.[44] Food was produced in wetland raised-field complexes and permanent dryland plots. There is no sign of slash-and-burn cultivation in the illustration, and no mention of it in the accompanying text by archaeologist Richard Adams. The environment is still described as a "marginal setting," but one that has been cleared, tamed, and intensively cultivated.[45] According to Adams, the forest remained only as "belts of deliberately abandoned wastelands serving as buffer zones between the rival Maya states. There were also stands of forest left, presumably, for hunting and logging, but even these were largely composed of highly selected species of trees."[46] Adams dismisses the notion of the forest-dwelling Maya:

> One of the long-standing misconceptions of Maya archaeology has been that Maya civilization existed within dense tropical forest much the same as the environment of today. This now appears false. . . . Therefore, the popular idea of sophisticated Maya cities set within a primeval wilderness is a romantic fantasy. I think that the Classic Maya themselves would probably have looked upon today's chaotic jungle growth as a reversion to savagery.[47]

Adams's vision of a stripped forest and intensive cultivation of a marginal environment still derives from the long tradition, described earlier, of characterizing the environment as an essentially hostile setting for agriculture and the development of complex civilization. Even when the lowlands are granted the ability to support high population densities, many researchers, including Adams, turn to the wetlands as the primary source of food production, relegating the forest to a marginalized wasteland when not cleared for agriculture.[48]

THE MAYA FOREST UNDER SIEGE

While the swidden thesis of Maya agriculture may have been disproven, the new image of cleared lands and an intensive use of wetlands did not engender an image of the Maya as good farmers or able managers of the environment. Beginning in the 1970s, a number of studies began to compile examples of environmental degradation caused by the ancient Maya.[49] While the swidden thesis had implied an inherent danger of erosion and grass invasion, evidence from some studies indicate that previously unrecognized forms of agricultural intensification did not always necessarily prevent environmental destruction.

What was the driving force behind the documented examples of environmental deterioration? Population levels rose dramatically through the Late Classic period and are often cited as a major factor in environmental degradation.[50]

In the pages of *National Geographic Magazine*, the modern Maya are once again turning against the forest. In a 1992 article, "Maya Heartland under Siege," the modern highland Maya are depicted as destructive frontier farmers laying waste to what is left of the fragile rain forest as they invade the lowlands, fleeing the political upheavals and land shortages of their homeland.[51]

Archaeologist Arthur Demarest worked in the lowland forests of northern Guatemala during the civil war that raged throughout that country during the early 1990s. While witnessing the inherent destructiveness of warfare, Demarest also comments in a 1993 *National Geographic Magazine* article on the invasion of the lands surrounding his archaeological project by displaced slash-and-burn agriculturalists:

I can smell the acrid scent of burning wood. Kekchi Maya colonists from the highlands are burning the forest so they can sow their *milpas*, or fields. An estimated 100,000 acres disappear each year in the Petén region, and some scientists say the forest may last only another generation. Guatemalan and international agencies try to stop the devastation but must also consider the farmers' need for land. . . . Today the rain forest is being destroyed by a population only a fraction the size of the Classic Maya's.[52]

According to Demarest and his colleagues, the inherent environmental pressures of overpopulation along with the destruction caused by rebellion, invasion, and warfare over land rights are destroying the present-day Maya "rain forest," in much the same manner that it did in the remote past. The image of ruined temples in the rain forest—partially obscured by the smoke from exploding mortars and the fires of invading Maya agriculturalists from the overpopulated, poverty-stricken highlands—remind us of the destructive capabilities of people, including governments and the Maya of today.[53]

IMPLICATIONS

What are the implications of all these images for the future of the Maya forest? If the ancient Maya lived in temple-cities as portrayed in *The Road to El Dorado*, nestled in a vast sea of undisturbed rain forest, then that is perhaps how the forest should be preserved—as nature's sacred paradise, lightly treaded by ecotourists or viewed only from suspended gondolas. Assuming we don't want to have people reoccupy the temples and live on food gathered gently from paradise, this would mean keeping human beings—including the modern Maya—out of the forest. But we are also left with the realization that only a small area of real rain forest actually exists in the Maya Lowlands. The popular call to arms to "Rescue the Rainforest" implies taking back what's left of that majestic forest from the Maya farmers and also not worrying about the low, dry forest and secondary growth that actually make up most of today's nonurbanized landscape. As mentioned above, from the perspective of biodiversity and sustainable agroforestry, abandoning this less aesthetically pleasing forest to permanent clearance and development would be a tragic mistake.

Accepting the other visions of the Maya forest as a jungle that had to be cleared and tamed by the ancient Maya, it would seem that human presence in the forest can only lead to land clearance and environmental degradation—a destructive process we see advancing on the forest today. If the ancient Maya couldn't live in the forest without destroying it, then maybe people—with the exception of ecotourists and scientists—should be kept out of it today. The image of modern human occupation of the Maya forest that is paraded by the media before the public eye is one of destruction, burning, and erosion of the land. It is a forest under siege and quickly losing ground. Is the only alternative to fence in what's left of the "pristine" forest, while at the same time fencing out the Maya?

A GARDEN GONE TO SEED

I prefer the alternative way to view the forest that cloaks the ancient temples—as a garden gone to seed.[54] This perspective on the relationship between humans and the forest sees people as an integral component of the forest ecosystem. Viewing the Maya forest as a garden gone to seed is not a tree-hugging variant of nature's sacred paradise. It recognizes that people have been present in the forest since its formation, and that human manipulation of the forest has gone on for millennia. The "Maya forest" is just what the name implies: a forest that has been shaped, modified, and manipulated by the Maya for thousands of years. The absence of people from the forest has occurred only in some areas and for relatively short periods of time. The "natural" rain forest—that which the public is so concerned about preserving—is a result of relatively recent abandonment.

Formation of the Maya Forest

People like to think of rain forests, including the Maya forest, as being primordial in the strictest sense of the term: first, original, unchanging, and extremely old—as in the ad for Xel-Há (see figure 1). Contrary to this notion, we now know that the lowland tropical forest of the Maya region began to appear after the end of the last Ice Age, no more than ten to eleven thousand years ago, and was not fully developed until about nine thousand years ago. There is now evidence that by seven thousand years ago, people were significantly manipulating the tropical forests of Central America through burning.[55]

Subsistence of these early inhabitants was based primarily on hunting and gathering, supplemented (beginning about five thousand years ago) by long-fallow, slash-and-burn cultivation of maize.[56] It has also been suggested that since early times, "tree crops may have been at least as important as maize in early economies," and that the cultivation of root crops, specifically manioc, may have even predated the cultivation of maize.[57] Selective cutting and preserving of certain species of trees while preparing a swidden plot, the planting of desirable trees after milpa cultivation, and the management of productive fallow regrowth are known practices today among Maya groups that have long-established or traditional land tenure—not the invading frontier farmers from the recent pages of *National Geographic Magazine*.[58]

In a sense, slash-and-burn cultivation of maize could be viewed as a step used by the Maya to begin the gradual process of creating increasingly productive forest farms. As botanist Arturo Gómez-Pompa has described the process, "Each abandoned *milpa* is an empirical experiment in directed succession."[59] As a result of these practices, the Maya forest has an amazingly high percentage of food-bearing tree species, and it is apparent that many of the "wild" species found in parts of the forest today may have actually been brought in and planted there by humans in ancient times.[60] While the vast area of the Maya Lowlands contains a great deal of variability in terrain, soils, and rainfall, the majority of tree species are common throughout the region—probably as a result of human selection and management over many millennia.[61] In areas where the Maya of today maintain land rights, we can see the forestry and gardening practices that created the Maya forest still in process. In forested areas that the Maya have abandoned, we see a garden gone to seed.

Garden Cities and Forest Farms

What would this alternative vision of the Maya Forest have looked like during ancient times? In reconsidering the *National Geographic Magazine* illustrations, reality probably fell somewhere between the images of solid forest surrounding the temples of Tikal (figure 3) and the stripped and virtually treeless landscape of Río Azul (figure 4); this alternative landscape would have been composed of garden cities and forest farms.

Although not shown in the *National Geographic Magazine* illustrations of Tikal and Río Azul, descriptions of the landscapes provided by archaeologists Coe and Adams in the same articles emphasize that ancient homesteads were probably set within home gardens, rich with a diversity of trees, root crops, and a variety of other produce. Why are these home gardens not included in the *National Geographic Magazine* illustrations? Why are home gardens not a part of the reconstructed Maya village that tourists visit in Xcaret? The tree-filled homegarden is today as characteristic of Maya agriculture as is the milpa (figure 5).

Walking through a typical Yucatec Maya community of today, one sees family compounds of about a quarter to a half hectare in size, surrounded by low stone walls or living fences of trees and shrubs. Trees make up a significant part of the house lot, shading the structures and open activity areas, and growing in orchards in the back of the lot. Recent botanical studies of modern Maya home gardens show that the cultivated trees include some introduced species, but are mainly comprised of wild trees taken from the Maya forest and domesticated varieties of local species.[62] Home gardens also include a vast array of shrubs and herbs from the forest that are used for food, medicine, dyes, and construction material. In a very real way, these home gardens preserve and perpetuate the botanical diversity that is the Maya forest. The origins of the Maya home garden may go back to the earliest Maya communities of the Preclassic period (ca. 1000 B.C. to A.D. 250).[63]

What would the ancient landscape have looked like outside the garden city? It probably would have appeared as a patchwork of cleared fields, a great variety of plots in various stages of controlled and productive regrowth, and managed stands of forest trees.[64] Wetlands were certainly used, but not in a uniform cookie-cutter pattern of raised fields and canals. The many different types of wetlands found across the lowlands implies that an equally diverse range of cultivation systems would have been used by the ancient Maya.[65]

The garden city hypothesis of ancient Maya settlement and agriculture has been gaining support among academics for some time, but it has not been adopted by the public media, which seem to prefer either a hostile jungle that had to be kept at bay or, more recently, a primeval nature's sacred paradise that the Maya barely touched. The real Maya forest, as noted earlier, has included humans for the majority of the time that forest has existed in the region. Sometimes well managed, sometimes mismanaged, but sustainable enough that after numerous regional abandonments—both ancient and

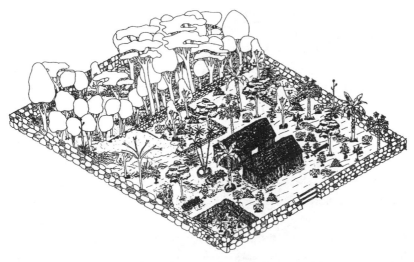

5. Schematic representation of a modern Maya home garden in X-uilub, Yucatán, Mexico. From Natividad Delfina Herrera Castro, "Los Huertos Familiares Mayas en el Oriente de Yucatán," in *Etnoflora Yucatanese*, monograph 9 (Mérida: Universidad Autónoma de Yucatán, 1994). The drawing is by Armando Cerdenares Domínguez. Reprinted by permission of Universidad Autónoma de Yucatán.

following European contact—the Maya Forest was able to rebound with no known loss of species diversity.

RECOGNIZING THE REAL MAYA FOREST

The real Maya forest is a forest with Maya people in it, a managed mosaic of differing land resources and wetlands. It is neither a primordial rain forest nor a hostile jungle.

When it comes to public policy on forest preservation, I think there is a distinct difference between the goals of rescuing the rain forest and preserving the Maya forest. Stands of tall, majestic forest are certainly a thrill to witness and can be a big attraction for ecotourists, as perhaps can best be seen in Costa Rica, where numerous forest reserves generate an income from tourism that drives that nation's economy.[66] Costa Rica has virtually no indigenous population remaining, however, and ecologists such as Daniel Janzen point out that the beautiful rain forests of the country are also representative of long-abandoned forest gardens.[67]

It seems to me that areas of uninhabited, high-stand forest are certainly worth preserving in the Maya Lowlands, while recognizing that the preservation of such habitats is really based more on the aesthetic value of such forests rather than being representative of the "natural" environment. Many forest reserves have been established in the countries that today divide the Maya Lowlands, and are, with varying degrees of success, saving stands of high forest for future generations.[68]

I suggest the Maya forest that is really in need of being rescued is the traditional, sustainable mosaic of forest farming and home gardens that represents the true Maya forest—one that is actively used and maintained by stable communities of indigenous people.

Unfortunately, the loss of lowland forest that so alarms the public is falling to pioneering agriculturalists—many of whom are Maya people moving out of the crowded highlands—who practice slash-and-burn cultivation without applying the techniques of selective cutting, productive fallow management, tree planting, and establishment of home gardens that characterize Maya land use where people hold established land rights.[69] After a few years of cultivation, pioneering farmers often either convert the ravished land to cattle pasture or sell the land to ranchers. The resulting landscape of degraded soils and tenacious grasslands has not been a significant part of the Maya forest in the past, and represents an ecological transformation from which the forest is not likely to recover.

Fortunately, there are many grassroots programs, often organized by established Maya communities, to preserve traditional knowledge of sustainable agriculture while establishing markets for forest products that can generate income and improve living conditions.[70] Other programs, frequently initiated by ecologists, anthropologists, and even archaeologists, are helping to reintroduce forgotten intensive cultivation practices and to transfer cultivation technologies used in some parts of the Maya Lowlands to other areas where they could be appropriately adapted. There are even a few programs in which local Maya farmers teach traditional forms of sustainable gardening and forest management to immigrants from other regions.[71]

I am rather dismayed that with few exceptions the field of conservation biology has focused on preserving or restoring allegedly natural landscapes that are based on the absence of people, when landscapes such as the Maya forest have, and should, include people as part of the natural environment. I don't see anything contradictory in an alternative approach to conservation that puts people back into the forest, managing and manipulating it in a way that conserves biodiversity rather than just saving a few stands of very tall trees.[72]

Can ecotourism be compatible with a notion of the Maya forest that actually contains and is managed by Maya people? I think it can be, but it will require a major change in attitude by the tourist industry. Ecotourism in the Maya forest does not need to be restricted to gondolas suspended from cable routes through the primeval rain forest. Over the last couple of years, I have visited numerous Maya communities that have developed a variety of programs in agricultural intensification, home gardening, controlled logging, reforestation with economically important local species, and harvesting of various forest products.[73] Many of these visits have included walks through beautiful forests where the Maya guide points out all the examples of forest management and manipulation that most tourists would not notice. Interestingly, these walks often wind around and over the low mounts that represent the remains of ancient houses.

The popular image of the Maya forest as nature's sacred paradise does great injustice to Maya culture and its achievements. The Maya were, and still are, a brilliant culture, with one of their greatest achievements having been the ability to sustainably manage a vast lowland forest in a manner that preserved biodiversity over many millennia while feeding so many generations. Promoting the notion that the only forest worth saving is the uninhabited, majestic rain forest is an overly exclusive and impractical approach to conservation. There is much practical knowledge we can learn from Maya agriculture and forest management, both ancient and modern, that can be applied to help save the real Maya forest—a forest with people.

NOTES

This chapter has benefited from the helpful comments of the HRI rain forest group, as well as the visitors to our group while in residence at UC Irvine, Francis Putz, Claudia Romero, David Baron, and Robert Patch. I have also benefited from comments and conversations with my colleagues Eugene Anderson, Mary Baker, Arturo Gómez-Pompa, and Dominique Rissolo. I thank all of these friends, while taking responsibility for my use or abuse of their suggestions in this chapter. The HRI group visit to the Yucatán Peninsula was made possible in part through the generosity of our hosts Michael Baker and Teresa Martelon during our stay at the beautiful resort of Rancho Santa Maria, in Quintana Roo, Mexico.

1. For example, George E. Stuart and Gene S. Stuart, *The Mysterious Maya* (Washington: National Geographic Society, 1977). For an analysis of the National Geographic portrayal of the Maya, see Peter Hervik, "The Mysterious Maya of National Geographic," *Journal of Latin American Anthropology* 4 (1999): 166–97.

2. Scott L. Fedick, ed., *The Managed Mosaic: Ancient Maya Agriculture and Resource Use* (Salt Lake City: University of Utah Press, 1996).

3. The Maya Riviera is a tourist-industry designation for the coastal strip of land and beaches that begins at Cancún and extends about 160 kilometers south to the archaeological site and modern town of Tulúm. Much of this area has been, or is being, developed with hotels, resorts, and so-called eco-parks. The development of Cancún and the Maya Riviera has resulted in striking contrasts and conflicts between modern development and the "traditional" Maya way of life. For an analysis of this phenomenon, see Alicia Re Cruz, "The Thousand and One Faces of Cancun," *Urban Anthropology* 25 (1996): 283–318.

4. *The Road to El Dorado* (DreamWorks Pictures, 2000).

5. On the Web page for *The Road to El Dorado* (www.roadtoeldorado.com) it is acknowledged that the ancient Maya is the culture portrayed in the film.

6. Jains are the most vegetarian of vegetarians, and cannot prepare their own food or do violence to any form of life. For a description of Jain ascetics in regard to food and eating, see James Laidlaw, *Riches and Renunciation: Religion, Economy, and Society among the Jains* (Oxford: Clarendon Press, 1995); and Lawrence A. Babb, *Absent Lord: Ascetics and Kings in a Jain Ritual Culture* (Berkeley: University of California Press, 1996).

7. Numerous tour companies offer "temples and rain forest" tours in the Maya area.

8. For example, see Stuart and Stuart, *The Mysterious Maya; Lost Kingdoms of the Maya* (Washington: National Geographic film/video, 1993); Time Life Books, *The Magnificent Maya* (Alexandria, Va.: Time Life Books, 1993).

9. For a solid introduction to the ancient Maya, see Michael D. Coe, *The Maya*, 6th ed. (London: Thames and Hudson, 1999); John S. Henderson, *The World of the Ancient Maya*, 2d ed. (Ithica: Cornell University Press, 1997); and Robert J. Sharer, *The Ancient Maya*, 5th ed. (Stanford: Stanford Univ. Press, 1994).

10. Rainforest Café, Inc., headquartered in Minnesota, has twenty-five cafés throughout the United States as well as twelve cafés in seven other countries (according to their website as of February 2001). Besides the restaurants and retail shops, the company offers educational tours of the cafés for classes and other children's groups that include lectures on the environment, conservation, and endangered species. To help support conservation efforts, the retail shops sell Rain Forest Preservation Kits for $10 each, with funds being administered by The Nature Conservancy. Our rain forest research group from the University of California's Humanities Research Institute visited nearly identical Rainforest Cafés in Costa Mesa, California, and Cancún, Mexico. More information on Rainforest Café, Inc. is available at www.rainforestcafe.com.

11. Steven Blaski, Mary Heaton, and Ann Kline, *Rain Forest Explorers' Guide* (Rainforest Café, Inc., 1998).

12. Xcaret is a major tourist attraction built on the coast of Quintana Roo, Mexico, about eighty

kilometers south of Cancún, in the Maya Riviera. The park is often characterized as a Maya Disneyland and has received mixed reviews in tour books. Besides the negative aspects of Xcaret, they do offer a number of programs and services to benefit the public and conservation causes: retail shops at Xcaret that buy and sell handicrafts from local communities to help improve the local economy; two treatment plants for water used in the park; programs to compost organic waste and recycle other refuse generated by the park and its visitors; an environmental education program for students; and breeding and propagation programs for endangered species. Information on the park is available at www.xcaretcancun.com. Our HRI group visited Xcaret in March 2000. An interesting perspective on the park is offered by Angela M. H. Schuster, "Faux Maya," *Archaeology* 52, no. 1 (1999): 88.

13. Labná is a major archaeological site situated in west-central Yucatán. Most of the architecture dates to the Terminal Classic period, about A.D. 800–1000. An excellent guidebook to Maya archaeological sites of Mexico is Joyce Kelly, *An Archaeological Guide to Mexico's Yucatán Peninsula* (Norman: University of Oklahoma Press, 1993).

14. "Editorial," *Xcaret Magazine*, March–June 1999, 7.

15. Eliza Llewellyn, "Mejorando el Mundo Xcaret: Improving the World," *Xcaret Magazine*, March–June 2000, 51.

16. Ibid.

17. Prior to the opening of Xcaret, archaeological research was conducted at the site by Mexico's National Institute of Anthropology and History.

18. Xel-Há is located about thirty kilometers south of Xcaret. Information on Xel-Há is available at www.xelha.com.mx.

19. *Cancun Tips Magazine*, spring 2001, 51.

20. The site of Chichén Itzá is a popular day-trip destination for Cancún tourists.

21. According to the widely used Life Zone classification system based on the work of Leslie Holdridge, rain forest (wet, tropical, perennial forest) in the Maya Lowlands is restricted to the southern Petán of Guatemala, the Lacandon forest area of Chiapas and southeast Tabasco in Mexico, and the southwestern portion of Belize that borders on the Petén. For the classification system, see Leslie Holdridge, *Life Zone Ecology*, rev. Joseph A. Tosi Jr. (San Jose, Costa Rica: Tropical Science Center, 1967).

22. See, for example, Norman Hammond, *Ancient Maya Civilization* (New Brunswick, N.J.: Rutgers University Press, 1982), 81, fig. 3.6.

23. In a sense, the notion of mature versus regrowth forest is a false dichotomy in the Maya Lowlands; all of the forest represents regrowth from human disturbance of varying ages.

24. Of ninety-seven references to the vegetation used in *National Geographic Magazine* from 1913 through 1974, 47 (48 percent) were "forest" or "tropical forest," 35 (36 percent) were "jungle," and 15 (16 percent) were a variety of other terms such as "bush," "brush," "woods," and "wilderness."

25. The first use of the term "rain forest" is in George E. Stuart, "The Maya Riddle of the Glyphs," *National Geographic Magazine* 148 (1975): 768–791. A single use of "rain forest" also appears in Norman Hammond, "Unearthing the Oldest Known Maya," *National Geographic Magazine* 162 (1982): 126–40.

26. S. Jeffrey K. Wilkerson, "The Usumacinta River: Troubles on a Wild Frontier," *National Geographic Magazine* 168 (1985): 514–543. In Wilkerson's article, of 46 references to the vegetation, 22 (48 percent) were "forest" or "tropical forest," 11 (24 percent) were "jungle," 10 (22 percent) were "rain forest," and 3 (6 percent) were "wilderness."

27. Wilbur E. Garrett, "Editorial," *National Geographic Magazine* 176 (1989): 422, and "La Ruta Maya," *National Geographic Magazine* 176 (1989): 424–479. From 1989 through 2000, there were 109 references to the vegetation of the Maya Lowlands in *National Geographic Magazine*; 67 (62 percent) were "forest" or "tropical forest," 7 (6 percent) were "jungle," 31 (28 percent) were "rain forest," and 4 (4 percent) were other terms, "ecological cornucopia," "wilderness" (2 references), and "scrubland" (to describe the Yucatán Peninsula; it was the only reference distinguishing the northern forests from those of the south).

28. Garrett, "Editorial," 422.

29. Alice J. Hall, "A Traveler's Tale of Ancient Tikal," *National Geographic Magazine* 148 (1975): 799–811. Tikal is located in the northern Petén of Guatemala. For a recent summary of the site and its history, see Peter D. Harrison, *The Lords of Tikal: Rulers of an Ancient Maya City* (London: Thames and Hudson, 1999).

30. William R. Coe, "The Maya: Resurrecting the Grandeur of Tikal," *National Geographic Magazine* 148 (1975): 792–95.

31. Alfred M. Tozzer, *Landa's Relación de las Cosas de Yucatan*, Papers of the Peabody Museum of American Archeology and Ethnology, 18 (Cambridge: Harvard University, 1941), 186.

32. J. Eric S. Thompson, *The Rise and Fall of Maya Civilization* (Norman: Univ. of Oklahoma Press, 1954), 26. Thompson is referring to the central/southern lowlands, an area with deeper soils and generally better agricultural conditions than the northern lowlands.

33. See B. L. Turner II, "The Development and Demise of the Swidden Thesis of Maya Agriculture," in *Pre-Hispanic Maya Agriculture*, ed. Peter D. Harrison and B. L. Turner II (Albuquerque: University of New Mexico Press, 1978), 13–22.

34. The first textbook on the Maya is by Sylvanus G. Morley, *The Ancient Maya* (Stanford: Stanford Univ. Press, 1946). For more on "corn experts," see Sylvanus G. Morley, "Yucatán, Home of the Gifted Maya," *National Geographic Magazine* 70 (1936): 598, 615, and *The Ancient Maya*, 141–50, 448.

35. Sylvanus G. Morley, "Unearthing America's Ancient History: Investigation Suggests That the Maya May Have Designed the First Observatory in the New World in Order to Cultivate Corn," *National Geographic Magazine* 60 (1931): 99–126.

36. Betty J. Meggers, "Environmental Limitation on the Development of Culture," *American Anthropologist* 56 (1954): 803.

37. Ibid., 817.

38. Ibid., 822.

39. Don S. Rice and T. Patrick Culbert, "Historical Contexts for Population Reconstruction in the Maya Lowlands," in *Precolumbian Population History in the Maya Lowlands*, ed. T. Patrick Culbert and Don S. Rice (Albuquerque: University of New Mexico Press, 1990), 1–36.

40. Peter D. Harrison and B. L. Turner II, eds., *Pre-Hispanic Maya Agriculture* (Albuquerque:

University of New Mexico Press, 1978); Kent V. Flannery, ed., *Maya Subsistence: Studies in Memory of Dennis E. Puleston* (New York: Academic Press, 1982); and Fedick, *The Managed Mosaic*.

41. The first published report on the discovery of patterned ground in wetlands of the Maya Lowlands is by Alfred H. Siemens and Dennis E. Puleston, "Ridged Fields and Associated Features in southern Campeche: New Perspectives on the Lowland Maya," *American Antiquity* 37 (1972): 228–239.

42. B. L Turner II, quoted in Howard La Fay, "The Maya, Children of Time," *National Geographic Magazine* 148 (1975), 733, 735.

43. Richard E. W. Adams, "Archaeologists Explore Guatemala's Lost City of the Maya: Río Azul," *National Geographic Magazine* 169 (1986): 420–451.

44. The ancient Maya site of Río Azul is located in northeastern Petén, Guatemala. For a summary of the site and recent archaeological research conducted there, see Richard E. W. Adams, *Río Azul: An Ancient Maya City* (Norman: University of Oklahoma Press, 1999).

45. Adams, "Archaeologists Explore Guatemala's Lost City," 435.

46. Ibid., 444.

47. Ibid., 443.

48. Richard E. W. Adams, "Swamps, Canals, and the Location of Ancient Maya Cites," *Antiquity* 54 (1980): 206–214.

49. For summaries of major projects, see Don S. Rice, "Paleolimnological Analysis in the Central Petén, Guatemala," in *The Managed Mosaic*, ed. Scott L. Fedick, 193–206; and John Wingard, "Interactions between Demographic Processes and Soil Resources in the Copán Valley, Honduras," in *Managed Mosaic*, ed. Scott L. Fedick, 207–235. For a more recent study of Maya impact on the environment, see Nicholas Dunning, David J. Rue, Timothy Beach, Alan Covich, and Alfred Traverse, "Human-Environment Interactions in a Tropical Watershed: The Paleoecology of Laguna Tamarindito, El Petén, Guatemala," *Journal of Field Archaeology* 25 (1998): 139–151.

50. See references cited in note 49.

51. George E. Stuart, "Maya Heartland Under Siege," *National Geographic Magazine* 182 (1992): 94–107.

52. Arthur Demarest, "The Violent Saga of a Maya Kingdom," *National Geographic Magazine* 183 (1993): 109, 111.

53. For a quantitative analysis of differences in "folkecology" knowledge and forest management between different Maya groups and recent Spanish-speaking immigrants to the Petén of Guatemala, see Scott Atran, Douglas Medin, Norbert Ross, Elizabeth Lynch, John Coley, Edilberto Ucan Ek, and Valentina Vapnarsky, "Folkecology and Commons Management in the Maya Lowlands," *Proceedings of the National Academy of Science* 96 (1999): 7598–7603.

54. The notion of the Maya forest as a garden gone to seed is partly in response to ecologist Daniel Janzen's suggestion that tropical forests should best be considered as wildland gardens that have always been subject to human manipulation, and will stand the best chance of surviving if we recognize and cultivate that relationship. See Daniel Janzen, "Gardenification of Wildland Nature and the Human Footprint," *Science* 279 (1998): 1312–1313, and "Gardenifica-

tion of Tropical Conserved Wildlands: Multitasking, Multicropping, and Multiusers," *Proceedings of the National Academy of Sciences* 96 (1999): 5987–5994.

55. See David J. Rue, "Archaic Middle American Agriculture and Settlement: Recent Pollen Data from Honduras," *Journal of Field Archaeology* 16 (1989): 177–184; Barbara W. Leyden, Mark Brenner, David A. Hodell, and Jason H. Curtis, "Late Pleistocene Climate in the Central American Lowlands," *Climate Change in Continental Isotopic Records: Geophysical Monograph* 78 (1993): 165–78; Dolores R. Piperno, "Non-Affluent Foragers: Resource Availability, Seasonal Shortages, and the Emergence of Agriculture in Panamanian Tropical Forests," in *Foraging and Farming: The Evolution of Plant Exploitation*, ed. D. R. Harris and G. C. Hillman (London: Unwin Hyman, 1989), 538–554; Dolores R. Piperno, "On the Emergence of Agriculture in the New World," *Current Anthropology* 35 (1994): 637–639; Dolores Piperno, Mark B. Bush, and Paul A. Colinvaux, "Paleoenvironments and Human Occupation in Late-Glacial Panama," *Quaternary Research* 33 (1990): 108–16.

56. Rue, "Archaic Middle American Agriculture and Settlement." See also Dolores R. Piperno and Deborah Pearsall, *The Origins of Agriculture in the Lowland Neotropics* (San Diego: Academic Press, 1998), 297–308.

57. Quote is from Rue, "Archaic Middle American Agriculture and Settlement," 177. For crops predating maize, see Piperno and Pearsall, *Origins of Agriculture*, 299–308; and John G. Jones, "Pollen Evidence for Early Settlement and Agriculture in Northern Belize," *Palynology* 18 (1994): 205–211.

58. See Arturo Gómez-Pompa, "On Maya Silviculture," *Mexican Studies/Estudios Mexicanos* 3, no. 1 (1987): 1–17; James D. Nations and Ronald B. Neigh, "The Evolutionary Potential of Lacandon Maya Sustained-Yield Tropical Forest Agriculture," *Journal of Anthropological Research* 36 (1980): 1–30; and Scott Atran, "Itza Maya Tropical Agro-Forestry," *Current Anthropology* 34 (1993): 633–700.

59. Gómez-Pompa, "On Maya Silviculture," 9.

60. Ibid. See also Arturo Gómez-Pompa, José Salvador Flores, and Victoria Sosa, "The 'Pet Kot': A Man-Made Tropical Forest of the Maya," *Interciencia* 12, no. 1 (1987): 10–15; Arturo Gómez-Pompa and Andrea Kaus, "Traditional Management of Tropical Forests in Mexico," in *Alternatives to Deforestation: Steps Toward Sustainable Use of the Amazon Rain Forest*, ed. Anthony B. Anderson (New York: Columbia University Press, 1990), 45–64. For an alternative view on forest composition and structure in the northern Maya Lowlands, see Victor Rico-Gray and José G. García-Franco, "The Maya and the Vegetation of the Yucatan Peninsula," *Journal of Ethnobiology* 11 (1991): 135–142.

61. See references in notes 59 and 60.

62. For comparison of plant species cultivated in Maya home gardens with those found in the forest, see Natividad Delfina Herrera Castro, Arturo Gómez-Pompa, Luís Cruz Kuri, and José Salvador Flores, "Los Huertos Familiares Mayas en X-uilub, Yucatán, México. Aspectos Generales y Estudio Comparative entre la Flora de los Huertos Familiares y la Selva," *Biotica*, n.s. 1 (1993): 19–36. For other recent studies of Maya home gardens, see José Salvador Flores, "Observaciones Preliminares sobre los Huertos Familiares Mayas en la Ciudad de Mérida, Yucatán, México," *Biotica*, n.s. 1 (1993): 13–18; Luz María Ortega, Sergio Avendaño, Arturo Gómez-

Pompa, and Edilberto Ucán Ek, "Los Solares de Chunchucmil, Yucatán, México," *Biotica*, n.s. 1 (1993): 37–51; James W. Stuart, "Contribution of Dooryard Gardens to Contemporary Yucatecan Maya Subsistence," *Biotica*, n.s. 1 (1993): 53–61; E. N. Anderson, "Gardens in Tropical America and Tropical Asia," *Biotica*, n.s. 1 (1993): 81–102; and particularly, Natividad Delfina Herrera Castro, "Los Huertos Familiares Mayas en el Oriente de Yucatán," *Etnoflora Yucatanense*, monograph 9 (Mérida: Universidad Autónoma de Yucatán, 1994).

63. Archaeological studies have identified space likely to have been used as home gardens through analysis of spacing between household residential units, mapping of boundary walls that define space around residential units, and chemical enrichment of soils resulting from household waste being spread in apparent garden areas. For examples, see Robert D. Drennan, "Household Location and Compact Versus Dispersed Settlement in Prehispanic Mesoamerica," in *Household and Community in the Mesoamerican Past*, ed. Richard R. Wilk and Wendy Ashmore (Albuquerque: University of New Mexico Press, 1988), 273–293; Guillermo Antonio Goñi Motilla, "Solares Prehispanicos en la Peninsula de Yucatan" (thesis, Escuela Nacional de Antropología e Historia, Mexico City, 1993); Thomas W. Killian, Jeremy Sabloff, Gair Tourtellot, and Nicholas P. Dunning, "Intensive Surface Collection of Residential Clusters at Terminal Classic Sayil, Yucatán," *Journal of Field Archaeology* 16 (1989): 273–294.

64. For examples of land use surrounding modern Maya communities, see Toledo Maya Cultural Council and Toledo Alcaldes Association, *Maya Atlas: The Struggle to Preserve Maya Land in Southern Belize* (Berkeley: North Atlantic Books, 1997).

65. For a sampling of various ancient wetland cultivation systems used in the Maya Lowlands, see B. L. Turner II and Peter D. Harrison, eds., *Pulltrouser Swamp: Ancient Maya Habitat, Agriculture, and Settlement in Northern Belize* (Austin: University of Texas Press, 1983); Stephen R. Gliessman, B. L. Turner II, Francisco J. Rosado May, and M. F. Amador, "Ancient Raised Field Agriculture in the Maya Lowlands of Southeastern Mexico," in *Drained Field Agriculture in Central and South America*, ed. Janice P. Darch (Oxford: British Archaeological Reports, BAR International Series no. 189, 1983), 91–110; T. Patrick Culbert, Laura J. Levi, and Luís Cruz, "Lowland Maya Wetland Agriculture: The Río Azúl Agronomy Program," in *Vision and Revision in Maya Studies*, ed. Flora S. Clancy and Peter D. Harrison (Albuquerque: University of New Mexico Press, 1990: 115–24; Mary Deland Pohl, ed., *Ancient Maya Wetland Agriculture: Excavations on Albion Island, Northern Belize* (Boulder: Westview Press, 1990); Scott L. Fedick, "Ancient Maya Use of Wetlands in Northern Quintana Roo, Mexico," in *Hidden Dimensions: The Cultural Significance of Wetland Archaeology*, ed. Kathryn Bernick (Vancouver: University of British Columbia Press, 1998), 107–29; and Scott L. Fedick, Bethany A. Morrison, Bente Juhl Andersen, Sylviane Boucher, Jorge Ceja Acosta, and Jennifer P. Mathews, "Wetland Manipulation in the Yalahau Region of the Northern Maya Lowlands," *Journal of Field Archaeology* 27 (2000): 131–52.

66. Costa Rica has chosen the path of preserving large areas of forest in reserves, while clearing the rest of the landscape. For a history of conservation in Costa Rica, see Sterling Evans, *The Green Republic: A Conservation History of Costa Rica* (Austin: University of Texas Press, 1999).

67. Daniel Janzen, "Gardenification of Tropical Nature: The Imperative and the Hurdles," paper presented at the ninety-seventh annual meeting of the American Anthropological Association, Philadelphia, 1998.

68. For a summary of ecological reserves and conservation areas in Mexico, see Arturo Gómez-Pompa and Rodolpho Dirzo, *Reservas de la Biosfera y Otras Areas Naturales Protegidas de México* (Mexico City: Instituto Nacional de Ecología, 1995). See also Lane Simonian, *Defending the Land of the Jaguar: A History of Conservation in Mexico* (Austin: University of Texas Press, 1995).

69. Atran et al., "Folkecology."

70. Examples of community efforts to preserve and promote traditional Maya agricultural practices include a series of pamphlets produced about agriculture in the village of Tekantó, Yucatán, by Sergio Medellín and María Mercedes Cruz, *Reverdezcamos Tekantó* (3 issues in 1994, 1997, and 1998, published by Grupo Xunan Kab, Tekantó, Yucatán). An example of community based, government supported forest management is the Plan Piloto Forestal of Quintana Roo, Mexico; see Michael J. Kiernan and Curtis H. Freese, "Mexico's Plan Piloto Forestal: The Search for Balance between Socioeconomic and Ecological Sustainability," in *Harvesting Wild Species: Implications for Biodiversity Conservation*, ed. Curtis H. Freese (Baltimore: Johns Hopkins University Press, 1997), 93–131. See also various chapters on forest management programs in Richard B. Primack, David Bray, Hugo A. Galletti, and Ismael Ponciano, *Timber, Tourists, and Temples: Conservation and Development in the Maya Forests of Belize, Guatemala, and Mexico* (Washington: Island Press, 1998).

71. Notable examples of training programs in traditional Maya agriculture and demonstration gardening include the School of Ecological Agriculture (Uyits ka'an) directed by Juan J. Jiménez-Osorio of the Universidad Autónoma de Yucatán, and the demonstration forest-garden developed by Anabel Ford at the El Pilar Archaeological Reserve for Maya Flora and Fauna in Belize. See also Juan J. Jiménez-Osorio and Véronique M. Rorive, eds., *Los Camellones Chinampas Tropicales* (Mérida: Ediciones de la Universidad Autónoma de Yucatán, 1999).

72. Encouragingly, there is a recent trend in conservation biology to view agro-ecosystems as a means for biodiversity conservation. See John Vandermeer, "The Agroecosystem: A Need for the Conservation Biologist's Lens," *Conservation Biology* 11 (1997): 591–591; Susan K. Jacobson and Mallory D. McDuff, "Training Idiot Savants: The Lack of Human Dimensions in Conservation Biology," *Conservation Biology* 12 (1998): 263–267; Michael K. Steinberg, "Neotropical Kitchen Gardens as a Potential Research Landscape for Conservation Biologists," *Conservation Biology* 12 (1998): 1150–1152; E. N. Anderson, "Biodiversity Conservation: A New View from Mexico," in *Ethnobiology and Biocultural Diversity: Proceedings of the Seventh International Congress of Ethnobiology*, ed. John R. Stepp, Felice S. Wyndham, and Rebecca K. Zarger (Athens: International Society of Ethnobiology, 2002), 113–22. There are others, however, who argue that the goals of conservation and supposedly sustainable extraction of natural resources are incompatible; see Thomas T. Struhsaker, "A Biologist's Perspective on the Role of Sustainable Harvest in Conservation," *Conservation Biology* 12 (1998): 930–932; John Terborgh, *Requiem for Nature* (Washington: Island Press, 1999).

73. For examplee, the Mexican communities of Tekantó in north-central Yucatán and El Naranjal in west-central Quintana Roo.

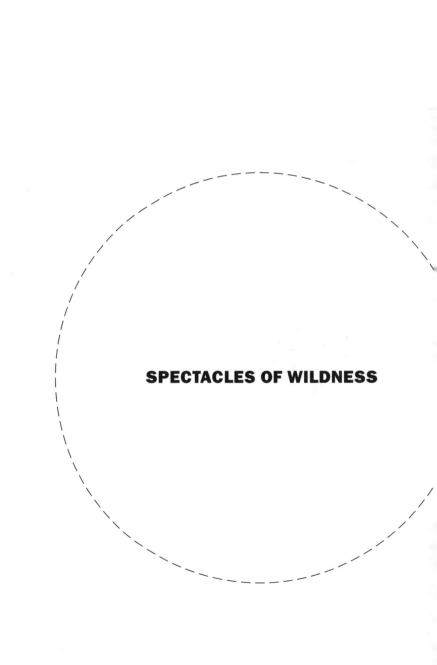

SPECTACLES OF WILDNESS

PAUL GREENOUGH

BIO-IRONIES OF THE FRACTURED

FOREST: INDIA'S TIGER RESERVES

Troubling realizations about India's wild tiger program crystallized for me in
Mexico in April 2000 when I joined a nighttime crocodile hunt in the Sian
Ka'an Biosphere Reserve. Though similar to a national park, a biosphere
reserve differs in that it sets aside a "core" where animals and plants are
sovereign, and from which humans (except scientists) are banned. Like a
bull's-eye, the core is surrounded by a "buffer" where limited human activity
is allowed, and the buffer in turn is surrounded by a "transitional" forest
with somewhat milder constraints (figure 1).[1] The Sian Ka'an Biosphere
Reserve in Yucatán contains half a million hectares (1.24 million acres) of
mangrove forests, freshwater and saltwater lagoons, marshes, and tropical
wet savanna. Two kinds of crocodile, the American and Morletti, are found
in this vast area and are the focus of an ongoing census.

ARCHIPELAGOES OF TEMPTATION

My Mexican hosts "Juan Carlos" and "Ricardo" and I set out at dusk in a
skiff from Muyil, a Mayan ruin near the Caribbean coast.[2] Despite the bright
stars hanging overhead, the lagoon was darkening and the depths were no

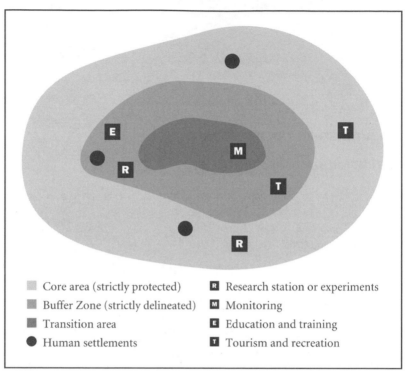

Core area (strictly protected)   **R** Research station or experiments

Buffer Zone (strictly delineated)   **M** Monitoring

Transition area   **E** Education and training

● Human settlements   **T** Tourism and recreation

1. Model arrangement of biosphere reserve. From J. R. Vernhes, "Biosphere Reserves: The Beginnings, the Present, and Future Challenges," in *Proceedings of the Symposium on Biosphere Reserves, September 11–18, 1987*, ed. William P. Gregg et al. (Atlanta: U.S. Department of the Interior, 1987), 7.

longer transparent. Only hours earlier I had been moving over the same water—part salt, part fresh—as a thrilled tourist, photographing waterbirds, monkeys, and tropical plants like the dwarf mangrove, the brilliant pink plumeria, and the lavender-leaved wild pineapple. In the daylight, the up-beat script for conservation had played out smoothly without a hint of contradiction: natural diversity is best preserved when scientists watch over sequestered woods and wetlands.[3] But now, as dusk turned to ink, I sensed other possibilities, and was unexpectedly eager for the hunt.

Standing one at a time in the boat's bow, and waggling a spotlight back and forth to catch their reflective eyes, Juan Carlos, Ricardo, and I spied crocs poised here and there on the lagoon's edge and along the adjacent canals. The brilliant white beam, powered by a twelve-volt battery, stupefied the crocs from fifty yards, allowing us to cut the motor and glide forward on their glassy-orange stares; we slipped wire nooses over their heads and pulled them hooting and honking aboard. With jaws banded and tails immobilized, and then flipped upside down, the crocs quieted down and meekly submitted to indignities. We took their measure with ruler and scales, probed them with fingers to determine their sex, clipped permanent code marks into their scaly tails, and heaved them back with a splash. The work was absorbing, and columns of data slowly accumulated in our notebooks. But sometime after midnight, when we had drawn close to the lagoon's eastern shore, we spied several cars in the distance, headlights lit and engines running. Juan Carlos and Ricardo cried out, "Drug traffickers!" They doused the spotlight, jerked the tiller aside, and we sped away in the darkness. The danger had been fleeting, but what suddenly struck me was that the enormous, empty biosphere reserve was a perfect setting for criminals to do their dark business. All I could get my companions to say was that nighttime powwows of the sort we had just seen were familiar events in the reserve. "Look," they said, "the *narcotraficantes* stay over here [gesture to the east], and the army and police stay over there [gesture to the west]." It was at this point I realized I had heard variants of this story in India.

Mexico is hardly unique in fostering crime in its natural parks and biosphere reserves; the same happens elsewhere, including in the United States.[4] Indeed, for the last century, governments around the world have been setting aside huge swaths of forest to form "protected area networks" (PANs) where nature is encouraged to take its course once human residents have been expelled.[5] PANs now make up more than 6 percent of the surface area of 174 countries, and the establishment of new parks, sanctuaries, and reserves continues apace.[6] In setting aside zones of exclusion and protection, conservationists imagine that the chief result is the continuous survival of valued plants, animals, and whole ecologies. This of course happens, and is one of the distinctive pieties of our age, but not infrequently there is another, hugely contradictory result: PANs become archipelagoes of temptation—giant tracts of emptied terrain that pull in outsiders who have secret projects requiring isolation. Instead of embedding zones of tranquillity into the landscape, PANs not infrequently attract the chaos of criminal gangs, such as the narcotraficantes in Sian Ka'an. PANs also tempt poachers, who strip out the most cherished species, and they give cover to insurgents and terrorists who boldly contest state authority. All these groups are totally indifferent to the fate of rare and endangered species.

This theme—that the modern practice of conservation expels established residents of forests, and so on, only to foster the intrusion of criminals, poachers, and rebels—is the first of several *bio-ironies* that need to be thought through everywhere.[7] In India, where tiger conservation has been a prominent project, there are related ironies. The first is that where there is initial success in protecting tigers, they reproduce so successfully that they later spill out of the reserves to attack cattle and humans, thereby triggering violent peasant retaliation; the second is that in at least one tiger reserve—Sariska in Rajasthan—the most promising success with tiger conservation depends on humans *flooding into the reserve from the outside* to establish a human-tiger détente based on—hold onto your seat—mutual respect and face-to-face dialogue.

Let me review briefly the tiger situation in South Asia. The Indian subcontinent—India, Pakistan, Bangladesh, and the Himalayan states of Nepal, Sikkim, and Bhutan—is a region of enormous biological diversity. India alone is home to one-fourteenth of the total animal species and one-eighth of the total plant species in the world.[8] The western mountains of India (called Ghats), along with the island of Sri Lanka, form one of the "biodiversity hotspots" that are a focus of global concern.[9] Forest cover across South Asia is extensive, although what remains is less than half of what stood at the beginning of the twentieth century. Forest types offer a range from evergreen rain forests in the Andaman and Nicobar Islands, Western Ghats, and hilly states in the northeast, to the mangrove swamps of the Sunderbans and lower Indus, to dry alpine scrub high in the Himalayas. In between these extremes are deciduous monsoon forests, thorn forests, subtropical pine forests, and temperate montane forests. Strikingly, tens of thousands of tigers once trooped over all these wooded settings, but as forest cover disappeared, tigers declined, and only a few thousand are now found in widely separated patches. These still sizable forests—separated by agriculture, roads, and other human uses—now isolate tiger breeding populations and may spell the end of genetic health.

Tigers are valued by conservationists not only as exemplary "top predators" but also as "indicator species," whose presence in the wild indicates an intact ecosystem. Where tigers flourish, so do tiger prey species such as deer and wild pig, along with hundreds or thousands of other animals, plants, insects, birds, and so forth. Tigers are simultaneously "charismatic mega-fauna," a sardonic phrase from the 1980s that references the middle-class public's enthusiasm for large, beautiful wild animals, and "accidental he-roes," who arouse awe and stimulate the public's desire to protect tiger habitats. It's not surprising, then, that biologists, state agencies, and conser-vation organizations have been working together for decades to stabilize the tiger's remaining foothold in South Asia.

In the early 1970s, the World Wide Fund for Nature (WWF), a nongovern-mental organization based in Switzerland but with global involvements, identified the Indian tiger (aka the Royal Bengal tiger) as one of a handful of large felids (cats) facing extinction. Logging, expanding agriculture, and uncontrolled hunting throughout the Old World had eliminated many tiger habitats and depleted their numbers, yet two-thirds of the remaining tigers in the world still roamed in South Asia's forests.[10] With WWF support, the Indian, Nepali, and Bangladeshi governments agreed to initiate conserva-tion programs. The prime minister of India, Indira Gandhi, took a personal interest in setting up "Project Tiger" on a grand scale, beginning in 1973 with nine reserves deliberately sited in vast state forests in different ecological settings.[11] Many of the reserves were connected by "corridors" of state for-ests, which allowed tigers to move back and forth to hunt and breed. By 1995, the tiger network had expanded from nine to twenty-five reserves (map 1). Operational costs for Project Tiger were divided between the central Minis-try of Environment and Forests and state-level forest departments, and it was state forest officers who actually policed and administered the reserves. A Project Tiger directorate was set up in New Delhi to provide overall coordination and work with foreign donors on special initiatives. These arrangements continue to the present day.[12]

The forested areas set aside for Indian tigers are large, and Project Tiger began on premises that tended to push up reserve size. One premise, for example, was that every adult tiger required at least ten square kilometers of undisturbed forest for hunting and breeding. Another was that every reserve should support at least three hundred tigers, a number that it was thought would ensure reproductive success and genetic variation. These ideas con-verged on a minimum reserve size of 3,000 square kilometers (1,800 square miles) or a total of 270,000 square kilometers (168,000 square miles) for the

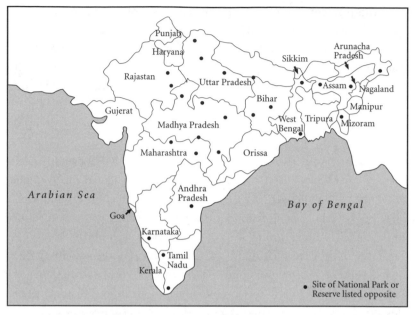

Map 1. Project Tiger reserves in India with sites of disputes and incursions. Adapted from Valmik Thapar, "Tragedy of the Indian Tiger," in *Riding the Tiger: Tiger Conservation in Human-Dominated Landscapes*, ed. John Seidensticker, Sarah Christie, and Peter Jackson (Cambridge: Cambridge University Press, 1999), 298.

initial nine-reserve network. This was an immense area for so densely populated a country, and the Project Tiger task force bowed to political realities by opting for more modest figures—most of the original reserves were established on less than 1,500 square kilometers (900 square miles), some were only half that, and only a few were larger.[13]

The establishment of big tiger reserves in South Asia in the mid-1970s reflected a serious debate among conservation biologists: How large, what size, and how numerous should reserves be to ensure a species' long-term survival? Experimental work by E. O. Wilson of Harvard University and his student, R. H. MacArthur, sparked interest in a theoretical field they called "island biogeography." The new theory quantified the optimal ratios between biodiversity (number of species present) and area (in hectares) in stable ecosystems. From these ratios flowed practical recommendations for conservation set-asides that have been summarized as "bigger reserves are better, the closer they are is better, the more circular [they are] the better, and [they] should be linked by habitat corridors."[14] In addition to establishing tiger reserves, India and the other regional governments set aside large tracts for other big mammals (for instance, rhinos, lions, and elephants), rare and endangered migratory birds (such as the Great Indian Bustard), and sheer biodiversity (for example, Silent Valley). While island biogeography has had its critics, conservationists inside and outside the Indian government pushed to set aside the largest possible protected area network.[15] By 1996, India had piled up a system of 75 national parks and 421 sanctuaries covering more than 3.2 million square kilometers or 4.5 percent of the country's surface. Of this area, the Project Tiger reserves comprised about one-third.[16] Almost as an afterthought, and unsupported by research of any kind, India's 1972 Wildlife Protection Act obliged the police to remove all persons living inside the forest tracts slated for Project Tiger on the assumption that conservation requires the strict separation of tigers and humans. As a result, tens of thousands of tribal and other forest-dwelling persons have been removed.[17]

BIO-IRONY ONE: THE "PROTECTED AREAS" OF INDIA
HAVE ATTRACTED A LEGION OF INTRUDERS

Criminal gangs, illicit loggers, smugglers, poachers, drug runners, and armed subversives roam through many Indian forests (see map 1), and police and troops have mostly failed to drive them out. At times the intruders are more at home than the police; they are either natives of the forest, or buy support and information from villagers, forest guards, and other petty officials, who fear their weapons and benefit from their largesse. The Indian state's own history of using rough measures against forest-dwelling locals biases villagers against the police. The intruders' trump card, however, is the state's own irresolution: an all-out effort with vehicles, helicopters, and large weapons would destroy the cherished flora and fauna. As a result, pursuers and pursued have parried in small-group encounters, sometimes for years.

Poaching wild animals is the most widespread intrusion into tiger reserves.[18] Most poaching in India is the work of poor village hunters, who use bows and arrows, homemade guns, and poison baits. In the 1970s and 1980s, well-heeled "sportsmen" with high-powered rifles and four-wheel-drive vehicles appeared, often using money and connections to get into reserved forests to hunt. Yet poaching never threatened Project Tiger until the 1990s, when single-minded agents offered large sums to all and sundry to poach and cut out tiger skins, bones, and body parts, all of which are valued ingredients in East Asian folk and traditional medicine.[19] The global conservation community responded with alarm, and in the last ten years India, Nepal, and Bangladesh all obtained funding for increased forest guards, vehicles, and two-way radios for patrolling. (In Nepal, the army protects national parks and reserves.) NGOs' interest in these measures is intense and has led to more accurate censuses, more expert research, and even a specialized online newsletter.[20] Meanwhile, multilateral coordination among the affected countries, including China, has choked off much of the international market in tiger parts.

Poaching nonetheless sticks like a bone in the throat of the Indian state. The stakes are higher than ever because poachers nowadays shoot to kill and the police have been authorized to do the same. A vivid example of the present crisis is the failure to stop the notorious criminal/poacher Veerappan, who has plundered the forests of Karnataka and Tamilnadu in southern India since the mid-1960s.[21] Veerappan and his gang of ten to twenty have killed hundreds of elephants and cut down uncounted numbers of sandalwood trees; the estimated street value of the poached ivory and timber is US$50 million.[22] He has sporadically shot his way out of ambushes and is said to have killed more than 120 people including 20 police officers. In one instance, his gang beheaded a state forest officer. The confidential police profile of Veerappan (leaked to the press) is interesting in that it acknowledges his complexity, and attributes his elusiveness to guile and a complete familiarity with the forest milieu:

> Veerappan staunchly believes in gods, deities, and omens, does not fail to worship any stone or tree enroute, [to] which is ascribed any divine status. . . . After every major operation he goes to deities and fulfils his vows. He does not believe in devils, though [he] has faith in ancestral worship. After killing [District Forest Officer] Srinivas he offered a big brass bell to his favourite deity. He is fond of ritual folk dances, which he does for hours, at times playing the cassettes of folk songs. He has a fairly good knowledge of herbal medicine and applies it on his gang mem-

bers. . . . Veerappan is a sharp shooter and possesses abundant presence of mind. He takes meticulous care every day and every minute so as not to be traced by the [Special Police] Task Force [STF]. He is a good imitator of bird and animal calls, which he uses for hunting animals and for communication with his gang members. His continued stay in the forest since childhood has made him acquire certain instinctive behavioural traits of animals, which he effectively uses for jungle survival and guerilla strategy. He can thus be called a wild animal in human form. . . . Veerappan has an excellent communication system using traditional sources. Besides his active sympathisers and supporters whom he contacts in an unscheduled manner, he uses the cattle grazers of the Pattis and Doddis who are available in plenty in almost all the seasons. . . . The gang sends one of its members to talk to these people and gather as much information as possible about the movement of STF and police informants. They always administer threats of severe consequences quoting the previously known instances of massacres lest anyone inform the police.[23]

Over the years Veerappan has kidnapped dozens of naturalists, officials, merchants, quarry owners, and notables. His most famous victim was the much-beloved Karnataka film star Rajkumar, whose abduction and concealment during the summer and fall of 2000 nearly paralyzed government and disrupted local commerce. This exploit shredded the reputation of the Karnataka and Tamilnadu police as well as central paramilitary units, and Indian newspapers openly mocked the combined police search effort as the "Bungle in the Jungle."[24] Veerappan's repeated escapes from custody (probably involving bribery of his jailers), impudent challenge to high-level politicians and senior police inspectors to "catch me if you can," and summoning of a journalist to record his autobiography (dictated onto videotape and broadcast on regional television in its entirety) have made him India's most famous desperado.[25] But Veerappan is only the most media savvy of a host of other forest intruders, as a half a dozen other examples will show.

In the northeastern Indian border state of Assam, the tribal Bodos are a community constituting about a third of the population. Militant Bodo organizations have been struggling for autonomy or independence since 1989.[26] In February of that year, armed members of the All Bodos Students

Union (ABSU) invaded Manas National Park and Tiger Reserve (2,800 square kilometers), killing a dozen forest guards and one staff member and destroying thirty-six administrative buildings. Park managers and the rest of the guards fled. The victorious ABSU then proclaimed an autonomous Bodo homeland north of the Brahmaputra River. Other Bodo liberation groups subsequently entered the field, and a running struggle has ensued among them, the ABSU, the Assam state police, and the Indian army. Early in 1998, the security situation improved when a police battalion was deployed in the national park, and government control was reasserted in seven of forty-three guard camps by the spring of 1999. All efforts to dislodge insurgents from Manas have been unsuccessful, however, and the Manas Tiger Reserve remains in Bodo hands to this day.[27]

The ABSU rebels established their headquarters in the tiger reserve's core (400 square kilometers). Prior to the 1989 attack, the tribal population living in the core had been evicted in order to protect an estimated 80 to 120 tigers. Because the northern edge of the core opens directly across the international border onto Bhutan's Royal Manas National Park, Bodo rebels now move freely between two protected areas—and mock the sovereignty of both countries. The rebels are known to raise operational funds by poaching tigers and one-horned rhinoceroses; they sell the carcasses for huge prices to smugglers of rare animal parts (for instance, tiger skins, tiger bones and teeth, and rhino horns) used in traditional Chinese medicine.[28] Armed encounters between reserve staff and rebel Bodos caught in the act of poaching have been a feature of the park for years. While many tigers in Manas have been killed, their actual number is unknown, and they are rarely seen. Ominously, not a single rhinoceros has been sighted since 1998. Meanwhile, timber merchants have taken advantage of the disruption to illegally cut trees inside the park.

In May 1996, Bodo rebels robbed a bank and escaped into the Buxa Tiger Reserve (759 square kilometers), which lies to the west of Manas in the adjacent Indian state of West Bengal. One gang member was caught by forest staff and turned over to police. Increased police attention to Buxa in the mid-1990s revealed that Bodo militants were involved in smuggling timber from the reserve, and the Indian army was called in to carry out joint patrols with police. In August 1996, the army demolished six sawmills operating illegally in the reserved forests and nearby tea estates. Because the police fear that they will fall into the hands of rebels, the Project Tiger staff's firearms and radios in Buxa have been sequestered.[29] Manas and Buxa are thus premier examples of disruptive intrusion.

Another kind of intrusion is found in central and peninsular India, in the states of Madhya Pradesh and Andhra Pradesh, where a broad belt of dry deciduous forest supports numerous tigers and a large tribal population. The same forest is the hiding place of small bands of armed maoist revolutionaries who are collectively referred to as Naxalites (but often divided by matters of doctrine and tactics). The Naxalites, protected by their tribal supporters, evade capture by moving frequently and by terrorizing landlords, forest guards, and merchants, who they fall on in sudden ambushes. Thirty-seven forest employees were killed by Naxalites in Madhya Pradesh alone between 1993 and 1999. In Bastar district on the border between Madhya Pradesh and Andhra Pradesh, sixteen police officers were killed in an ambush in October 1998 when a rebel group blew up their jeep.

Several of India's most important tiger reserves, such as Kanha in Madhya Pradesh and Srisailam-Nagarjunasagar in Andhra Pradesh, lie in this densely forested region and are regularly encroached on by rebels. The Kanha reserve, located inside Kanha National Park (1,940 square kilometers), was one of the first protected areas to be established in independent India. Kanha's tiger-, monkey- and bear-filled forest was immortalized in Rudyard Kipling's fable, *The Jungle Books*. Kanha was also where U.S. ecologist George Schaller carried out a classic study of predator-prey relationships, *The Deer and the Tiger* in the 1960s.[30] In 1990, a series of devastating fires wasted large areas of Kanha, and Naxalites were accused of arson. Large police units moved in to counter the rebels, but an Indian journalist commented that "from the environmental point of view, probably the presence of huge police forces is causing more damage than the fire that grabbed the headlines." He speculated that Naxalites would make Kanha their stronghold.[31] In fact, rebel groups in Andhra Pradesh and Madhya Pradesh have continued their peripatetic wandering from one forest base to another, while the police, pursuing an old colonial tactic, have helped to establish an armed vigilante group—known ironically as the Green Tigers—whose mission is to combat Naxalite groups.[32]

Similarly violent disruption has occurred in the Palamau Tiger Reserve (928 square kilometers) in Bihar, where Naxalite rebels killed a Project Tiger driver and a forest guard in February 1998 by exploding a mine beneath their jeep. The real target of the attack was a divisional forest officer who had luckily stepped out of the vehicle just before it rolled over the mine. The forest officer radioed for help, but police were unwilling to come to his assistance, fearing a direct encounter with the Naxalite force. The forest officer walked ten kilometers for help and then returned, on foot (fearing other mines), to collect the scattered remains of the victims. Belinda Wright, executive director of the Wildlife Protection Society of India, observed that timber smugglers were being protected by the Naxalites, and that there was a close relationship between the planting of the mine and the illegal timber trade. She added that "without protection [and] support, no member of the forest staff dares to go into the field. For now, Palamau and its tigers are unprotected."[33]

I could go on, of course.[34] But perhaps these examples from Manas, Buxa, Kanha, and Palamau tiger reserves, and the bizarre career of Veerappan, prince of poachers in Karnataka-Tamilnadu, are enough to show that large, depopulated, and (barely) protected reserves—the reigning orthodoxy in biodiversity conservation—function as magnets that draw in a range of illegal actors: criminals, rebels, poachers, or simply poor people looking for subsistence. As is the case in Yucatán, so it is in the tiger reserves of India: law-lessness takes sanctuary in an emptied forest, where the powers of the state seemingly melt away. Indian politicians have read this situation with worldly realism and some have decided that protected areas simply aren't viable; they have begun to hand out chunks of parks and reserves to their clients right and left through the legal maneuver of "denotification."[35] The fact is that after setting up one of the world's more elaborate protected area networks in the 1970s and 1980s, India's reserves are now being gnawed on by criminals, interlopers, and in some instances, elected representatives of the state.

BIO-IRONY TWO: ACTUAL SUCCESS IN
INCREASING TIGER NUMBERS LEADS DIRECTLY
TO A PUBLIC RAGE TO KILL THEM

Not all tiger reserves have been invaded or fiddled with by politicians. Project Tiger has had many successes and has protected many hundreds or thousands of tigers and overseen their numerical increase. By the middle of the 1980s, the tiger population in several reserves was at a peak, and hapless young adults began to spill out of the reserves into nearby, densely inhabited agricultural districts; inevitably, some of them attacked humans and their domestic animals. The public response was deeply hostile—local people began shooting, trapping, and poisoning the tigers—and a gradual decline in numbers set in that has continued ever since. This little-known trajectory—breeding success leads to attacks on cattle and humans, which leads to fatal retaliation—is another bio-irony working its way through rural life around several important reserves. To understand how this happened, we need to know more about how local communities fared at the hands of state-led tiger conservation.

When Project Tiger began in 1973, police pushed thousands of villagers from their homes—a practice that accompanies reserve development to this day.[36] Economic activity was banned in the cores and buffers, including cattle grazing and foraging, and peasant and tribal access to vital forest products like honey, herbs, thatch, and firewood needed for day-to-day subsistance was ended. Only by giving bribes (and taking beatings) could access continue. At the same time, forest officers fostered an increase in prey species, such as deer, antelope, and wild pig, by digging water holes and sowing cereals in forest clearings. While varying from reserve to reserve, these pro-tiger measures often worked. Tigers slowly increased in number after 1975.[37] Recall also that the tigers inside Project Tiger reserves were only half of India's total tiger population. The other half roamed the state forests outside the reserves. In the 1980s, however, these state forests began to melt away; intense pressure from cultivators, insatiable demand from wood and paper pulp contractors, and unstoppable development activities such as mines, dams, and roads were eating them up. Earlier, the state forests beyond the buffer zones had formed "corridors" linking many of the tiger reserves together, but by the mid-1980s sugarcane and paddy were being cultivated in some places right to the edge of the buffers. Because adult tigers were reluctant to cross cultivated fields, they no longer migrated to find appropriate mates. Isolated age- and sex-mismatched tigers are not viable; either they fail to reproduce or their cubs do not thrive.[38] Thus India's free-ranging "outside" and reserve-based "inside" tigers were marooned from each other, and they have had separate destinies since the late 1980s.

The downward drift of South Asia's tiger population *outside* the Project Tiger reserves is well-known to naturalists and fuels their fear that the Indian tiger will crash as did the Bali, Caspian, and Javan tigers in earlier decades. What is less widely known is that successful tiger conservation *inside* the reserves has sparked a devastating hatred of tigers among nearby villagers. The reason for such hate is that young tigers wandering away from the reserves not infrequently kill villagers and their domestic animals. Although "cattle-killers" and "man-eaters" have always been present in rural India, both had been declining over the course of the twentieth century until they suddenly flared up right around the reserves in the 1980s.[39] In the Corbett, Dudhwa, Ranthmbhore, and Sunderbans reserves in India, and in the Royal Chitawan reserve in Nepal, tigers increased in such numbers that young adults could obtain only limited access to breeding and hunting territories; at the same time, villagers pressed their cultivation closer and closer to the buffers, clandestinely entering protected areas to remove traditional forest products like thatch, grass, and wood. The resulting mix was explosive. Villagers typically disturbed tigers at night when the animals were hunting; alternatively, tigresses lying in sugarcane fields, a setting they favored for raising cubs, attacked startled field-workers. Either way, sudden human intrusions and then headlong flight triggered fatal attacks. Dozens of lives were lost in and around reserves each year, disturbing proof that Project Tiger was succeeding beyond all expectations.[40]

Project Tiger staff followed politically unrealistic procedures in coping with "success." When tigers killed villagers, Project Tiger rules obliged forest officers to compensate victims' families, and then to track and tranquilize the offending tigers—no easy task. The drugged tigers were then supposed to be released inside the core. But anguished villagers demanded that offending tigers be shot. In retrospect, killing man-eaters would have been the best course—a drastic step, but one that the International Union for Conservation of Nature and Natural Resources, the premier international conservation agency, recommended just because of the public relations disaster that a human death precipitates.[41] The struggles that erupted between Project Tiger staff and villagers after a failure to destroy killer animals continued for years. As the naturalists Ramesh and Rajesh Bedi observed (in reference to the Dudhwa Tiger Reserve in Uttar Pradesh state):

The district [of Lakshimpur-kheri] was infested with tigers relishing human flesh. No one seemed to be safe. As soon as one man-eater was killed another tiger would soon replace it. . . . Hardly had the *sansis* [peasants] settled down with relief when a fresh wave of panic and terror spread through them. Sometimes as many as five man-eaters [took] their toll of human lives. It [was] like an epidemic. . . . Baits were provided regularly and the man-eaters started taking them. But the experiment [that is, tranquilizing and relocating tigers] suffered a setback. It was possible to control the movements of local people inside the park; but the reserved forest outside the park, where grazing and a limited felling of trees is permitted, continued to attract the villagers. . . . People ignored the warnings of the officials attempting reformation of the killer tigers; and, after every [human] kill, the pressure to eliminate the offenders rose to such heights that it became a problem for conservationists. . . . Local political leaders deprecated the critical situation and fanned the agitation for getting compensation from the government. They pressed for the arrest of officials, whom they accused of being heartless. In one of the public meetings they incited the people to promise that they would tie up the forest guards and put them up as bait for the man-eaters. . . . It was a golden opportunity for the big farmers to demand the lifting of the ban on big-game hunting. They said they were unable to harvest the sugarcane crop because of the active marauders in the fields.[42]

By the mid-1980s, then, tiger conservation had triggered a war on tigers by furious villagers living close to the Dudhwa reserve. Arjan Singh has described how revenge was exacted in Kheri, a district that lies just north of Dudhwa:

The public took retaliatory measures into their own hands, and tigers were slaughtered by poisoning, electrocution, trapping and shooting. Poisoned tigers floated down canals where they had collapsed with a raging thirst. . . . Twenty-five known cases of destruction occurred in 1985–86. The local legislative member went on record saying that tigers were committing suicide.[43]

The legislator's remark was a macabre joke. But it was a joke that agilely bridged the contradiction between a successful state-supported tiger conservation program and the hatred of hothouse tigers around the reserve. Stabilizing a cherished species turned out to be more challenging politically than anyone had ever imagined.

BIO-IRONY THREE:
TIGERS THRIVE ON HUMAN ATTENTION

In July 2000, just a few months after my nighttime encounter with narcotraficantes in Yucatán, I found myself in India and resolved to visit another Indian tiger reserve. It wouldn't be my first such visit—I'd been to the Corbett, Sunderbans, and Royal Chitawan (Nepal) reserves before—but I now had some definite issues in mind. Where to go? A good choice would have been the Ranthambhore Tiger Reserve (1,334 square kilometers) in the state of Rajasthan, one of the most celebrated tiger sites in India. Famous persons and Indian and foreign media frequently go to Ranthambhore, and with reason: good science and social conscience flourish together there. In 1988 Fateh Singh Rathore, the Ranthambhore reserve's field director, and Valmik Thapar, an independent photographer-conservationist, launched an unusual foundation to foster the health, education, and democratic decision making of villagers living near the reserve. Their sincerity won over many locals, and in many ways the Ranthambhore Foundation's work exemplifies successful "eco-development," an approach to conservation that tries to wean villagers from hostility to wild animals and protected areas by offering employment, social services, and a share in the benefits generated by research and tourism.[44] For eco-development to succeed, however, substantial external funding is required. A further assumption behind eco-development— that villagers living near reserves require the tutelage of outsiders in conservation—is puzzling to local tribal and peasant leaders, who say in effect, "Why are *you* evicting *us* from the forest? *We* are the real forest guards." Hence, many Indian conservation activists ask, Is eco-development just? They also ask Is it sustainable? As long as outside funds continue to flow, Ranthambhore tigers can rest easy. But will the necessary conditions be met in perpetuity—*should* they be met in perpetuity? Is there perhaps an alternative model based on insiders and local knowledge rather than outsiders, paternalism, and exogenous funding?[45]

With these questions in mind I opted to visit the Sariska reserve, also in

Rajasthan, instead of Ranthambhore. Sariska (800 square kilometers) is one of the most unlikely tiger reserves in India. For one thing, there are *two* highways running through it with car and truck traffic both day and night. For another, 200,000 pilgrims enter the core every year in September to worship at ancient temples that can neither be removed nor closed.[46] Then, during the dry and cool seasons, foreign tourists eager for a glimpse of tigers roar in and out of the core in jeeps and even motorcycles. Finally, migratory graziers and herders called Van Gujars ("forest Gujars") have moved into Sariska by the thousands in recent decades and have joined fellow Van Gujars already residing there. There are now a dozen old and newer villages in core areas, each with herds of free-roaming buffalo, while several dozen other villages are spotted around the buffer zone (map 2). All these disruptions violate the logic of a biosphere reserve.

While the tourists and pilgrims are left alone, Project Tiger personnel have tried to eject the Van Gujars; these efforts have been publicized, and as a result, the Van Gujars of Sariska have gotten a measure of public sympathy. One leading conservation biologist has expressed exasperation with what he considers to be the Indian public's soft-headed indulgence:

> There was no such thing [before] as Van Gujjars [*sic*]. They have come in the last few decades. They were nomadic earlier, but they now [have] become permanent residents. NGOs have helped them in this. This kind of activity will ultimately result in destruction. . . . I know what I say is not very palatable to many people, but there it is. . . . [You] cannot marry the long-term conservation interest of the protected area with the exploitative interest of the people.[47]

Not only is Sariska thronged by pilgrims, tourists, traffic, and in-migrants but politicians have begun denotifying chunks of it to allow marble quarrying and other economic activities to continue.[48] Sariska would thus seem to offer tigers a scant promise and bleak future. Surprisingly, however, tiger numbers have probably doubled in the last decade, and the population is as large and stable as Ranthambhore's: the quadrennial 1998 tiger census found twenty-five to twenty-nine tigers in Ranthambhore and twenty-five to twenty-eight in Sariska. The reason for this unexpectedly healthy situation is not obscure: local opinion and published accounts agree that the Van Gujars watch over the tigers, and practice a way of life underpinned by religious and

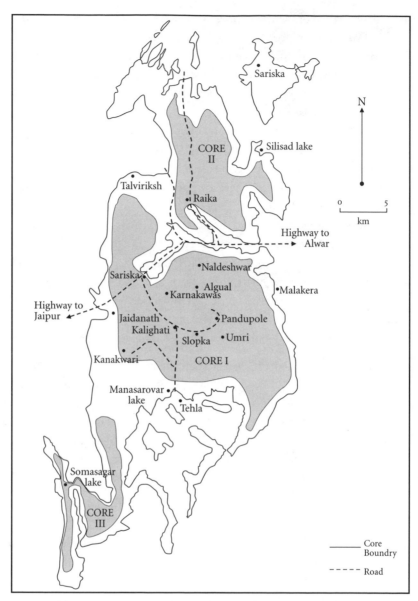

Map 2. Sariska Tiger Reserve, Rajasthan. From Brij Mohan Sharma, assistant conservator of forests, Sariska.

economic interests that are beneficial to the tigers and reserve forest. In going to Sariska I wanted to see for myself a type of conservation that didn't require the Ranthamborean mix of exceptional leadership, public relations flair, and external funding.

A flight to Jaipur and five-hour bus ride on narrowing highways led into the spur valleys of the Aravalli range and dry deciduous forests of Sariska. It was July, the height of the monsoon season, and the trees and grass glistened after daily downpours. Small dams along the Ruparel River had raised the water table on all sides, and field crops glittered. Macaques raced beside the bus in overhanging trees, and hordes of wild peacocks cried out, batting their tails in urgent seasonal display. I was unprepared for the clear air, animal clamor, and natural beauty.

The bus set me down by the forest officer's bungalow near a gate into the reserve. Just as had been the case with the biologists in Yucatán, I found "Brij Mohan" and "Daulat," the tiger wardens at Sariska, eager to explain their work and the difficulties they faced protecting the animals.[49] While they held dogmatically to the view that the tiger reserve should be cleared of humans, they admitted that numerous Gujar villages in the core zone of Sariska had not led to a decline in tiger numbers and that the Gujars readily responded if a fire in the forest threatened to get out of control. Further, the push to relocate villages had lost its momentum, partly because the Sariska tigers are so clearly not threatened by villagers, and partly because the state government is reluctant to pay for rehabilitation and other amenities of relocation that are required by law.[50] On the other hand, the settlements inside Sariska are technically unlawful, so Gujar villages receive no government services such as electricity, telephones, primary schools, public transportation, or health care. It is in everyone's interest—including the tigers—to leave the Sariska Gujars just as they are.[51]

Van Gujars live simply and exert light pressure on the protected forest. They collect small quantities of deadwood—estimated at ten kilograms per day per family—for making hearth fires, and their cows and buffalo graze on the forest floor. Gujars are vegetarians whose diet consists of milk products, cereals, and vegetables. Most Gujar families own buffalo and sell milk, which brings them between 2,000 and 4,000 rupees (US$40 to $80) per month. They harvest the thorny branches of *khejuri* trees and weave them into low corrals that surround their stabled herds at night. Occasionally they forage for wild grains, fruits, and berries when these are in season, but they never hunt the abundant deer, antelope, wild pig, and peacocks that are found everywhere in the reserve.[52]

Approximately twenty to twenty-five wild tigers live in the forest core; this number has apparently been stable in Sariska for a decade.[53] The Gujars report intruders and poachers to the forest offices, but they themselves never hunt, poison, or trap tigers. In turn, and somewhat remarkably, Sariska tigers rarely harm the Gujars who live in the reserve; only one episode of man-killing and a single mauling incident have occurred in the last twenty years. Why? What differentiates Sariska from, say, the Dudhwa Tiger Reserve, which has been plagued by tiger attacks each year?

For one thing, the Van Gujars in the Sariska reserve speak positively of the predatory habits of tigers, and regard the occasional loss of a buffalo as a necessary sacrifice for the health and success of the rest of their animal herds. For another, Gujars consider themselves to be the protectors of tigers, mediating between them and other, hostile humans. More remarkably still, Gujars prepare themselves for face-to-face encounters with tigers, which they regard as a privilege, whereas villagers in most other reserves—who see tigers only exceptionally—consider them innately destructive and expect the worst from every encounter. Thus Van Gujar proximity to, admiration of, and familiarity with tigers turn out to be critical features of the ecosystem in Sariska. We learn about these matters from João Pedro Galhano Alves, a Portuguese biologist-ethnographer.[54] According to Galhano Alves,

> The villagers explain that, "If the tigers almost never attack people in the region it is because we don't kill the tiger and so the tiger does not kill us. We respect the tiger and it respects us. The tiger is a friend of men. We respect the forest; we don't kill the animals that the tiger needs to eat, and so it has plenty of food." The villagers also manifest a deep admiration for the tiger. Being Hindus, they believe that the tiger is the sacred "vehicle" of the Goddess Durga, being an avenger who maintains the world's harmony by fighting against the demons' chaos. But they also respect the tiger because of cultural beliefs and for aesthetic and empirical ecological reasons. For example, they say: "The tiger is beautiful, very beautiful" and "Because of the tiger the buffalo is not sick too often. Tiger breath is good for the buffalo," and "If there were no tigers or panthers, there

would be too many *sambar* (deer) and *nilgai* (antelope) and so one day there would be no grass or trees, and there would be no forest." Finally, a local folk proverb says: "The tiger and the cow must drink from the same pool."[55]

The Gujars admit freely that crossing paths with wild tigers is dangerous: "If one lives here, he must accept to die."[56] However, they also assert that there can be arresting moments of recognition and interest on both sides. As Galhano Alves explains,

> When a villager meets a tiger, which happens to many villagers (and mostly to the buffalo herders) once, twice or more a year, the tiger usually just looks at the person for some seconds and then goes away. In some cases, it keeps looking at the person until *they* go away, without threatening them. . . . Many testimonies concerning man and tiger encounters refer to the most impressive effect that the tiger produces on man, saying that meeting a tiger in the wild gives such an intense feeling that it has a paralysing effect. Usually, this paralysis only lasts a few seconds, but, in some cases, it may be completely dominant, blocking any other kind of reaction in the presence of the tiger. But usually, after this short period, the person is dominated by fear that lasts until they feel that there is no more danger of being attacked. . . . Many villagers are so used to meeting tigers from their childhood that the presence of the animal does not cause fear any more. For example, an old villager says: "I am not afraid of the tiger. I see it all the time. For me, it is not possible to be afraid of the tiger as I live here." But, in general, they feel some degree of fear. As another villager says: "When I see a tiger I am afraid. Of course, the tiger can kill me."[57]

This normative account of exchanged glances suggests that connections other than violence are possible across the species barrier. In such moments, tigers and humans are neither opposites nor enemies but mutually curious and conscious beings. Interestingly, Sariska Gujars interpret these encounters in terms of a familiar political hierarchy, insisting that tigers are kings and humans must conduct themselves as if standing in the presence of royalty. Hence Gujars have worked out a protocol for meeting tigers that includes appropriate verbal greetings, eye contact, gestures, and movements (see appendix at end of this chapter). It is unlikely that this protocol was elaborated only within Sariska in the last twenty-five years; a more likely explanation is that Gujars—the historically preeminent graziers of India— have come to terms with tigers over many centuries.[58]

During my visit to Sariska in July 2000, I spoke with herders in three Gujar villages (Kundelka, Umri, and Haripura) and found that some, and especially those whose regular task is to escort buffalo to and from the forest, have more frequent contact with tigers than others. But the essential points of Galhano Alves's account (in the appendix) were confirmed: Gujars make eye contact with tigers and greet them verbally, and they expect peaceable denouements from encounters. I also found that their tiger views are under-pinned by a matrix of beliefs that exalt tigers as gods or gods' emissaries.[59] No ethnographer other than Galhano Alves has, as far as I know, explored the ecological significance of offering respect to wild tigers, although there are descriptive reports of wild animals' forming an entente cordiale with humans elsewhere in India, Asia, Europe, and Africa.[60]

It may be that there are specific ecological features in the Sariska reserve— for example, an unusual abundance of prey animals or a favorable age and sex ratio among the tigers—that will one day explain more fully the tigers' tolerance of humans.[61] On present evidence, however, the conclusion is irresistible that the Sariska reserve depends on the Gujars' watchful, wor-shipful presence to maintain the conditions under which tigers survive and successfully reproduce. Galhano Alves refers to this situation as a "human-ised ecosystem," and its most striking element is the truce on interspecies violence. If we compare Sariska's humanized ecosystem with tiger reserves elsewhere in South Asia, we come to a third bio-irony: the larger, depopu-lated, and more aggressively policed reserves see most of the poaching, poisoning, and popular rage against tigers, whereas smaller, road-riven, Gujar-invaded, buffalo-cluttered, and pilgrim-crowded Sariska exhibits an enviable truce between humans and tigers. What is the key condition for such success? Can Sariska's results be repeated elsewhere? Several transpos-able elements are evident. For one thing, the simple fact of Gujars living *inside the reserve* has discouraged rebels and criminals from setting up shop in the core; Gujars serve, in their own way and for their own reasons, as the eyes and ears of Project Tiger. For another, Sariska suggests that *a host of intermediaries*—officials, scientists, activists, NGOs, philanthropists, and bureaucrats such as eco-development requires—*may not be necessary.* A few villages of Gujars scattered about in the core may suffice, and their full-time, religion-infused desire to see tigers flourish is so much more intense than that of all but the most dedicated conservation official. Finally, Sariska suggests that animal conservation can succeed on the basis of *interspecies accommodation* rather than of exclusionary separation—a proposition that Project Tiger officials still reject as an unsustainable fantasy.

Is it possible, though, that what is unsustainable is actually the received model of conservation, based as it is on "exclusion and enclosure"? On the one hand, the sheer injustice of using force to establish and maintain national parks and reserves—of depriving rural peoples of lands and lifeways without offering real compensation or meaningful alternatives—is a ticking bomb. If the United States, for example, where state conservation coincided with the consolidation of the western half of the country, local antagonism against officials has roiled agrarian and forest politics since the nineteenth century; continuing state control of forests and prairies is always and everywhere disputed, especially by Native Americans.[62] On the other hand, conservation science is beginning to come around to the view that it misread the ecological significance of humans residing in wild terrains when it pushed for exclusionary laws in the 1970s and 1980s. There is agreement nowadays that "getting people out of parks" should give way to a growing (if still hotly debated) view that certain kinds of humans—those with acceptable low technologies and benevolent motives—may be necessary for the long-term conservation of animals, forests, and biodiversity.[63] Indeed, some visionary tiger conservationists argue that the main goal should be to persuade villagers living near reserves to regard tigers as "neighbors" rather than enemies—something that has been achieved in Sariska with scant external support and despite state hostility. In a yet more controversial vision, the future of tiger conservation depends on an altered set of human values and incentives in which the whole landscape of South Asia would be treated as an unbounded terrain shared between humans and wild animals. In such a future, parks and reserves would no longer be necessary, and the present-day core, buffer, and transitional-forest structure will be viewed as an unworthy exercise in state force and imperial science.[64]

Not every rural community will be easily reconciled to tigers living next door—that's not my argument. And Indian villagers who own only a few plow bullocks, whose culture emphasizes meat-eating rather than vegetarian diets, and who transmit local lore in which tracking and killing tigers are heroic undertakings will not be persuaded. They are unlikely to take up hymn-singing to the goddess Narsati. Moreover, even if Sariska were adopted as a model for future tiger conservation, there are aspects of the reserve that are less than paradisiacal. The punitive policy of denying electricity, schools, health care, and public transportation to Gujars, for example, is clearly intended to drive them out and is a political scandal. Tigers, too, have much to complain about. In 1997, a speeding truck on the Jaipur-Alwar highway ran down a Sariska tiger standing at night in the middle of the road: transfixed by the truck's oncoming headlights, the gaping cat was no more able to avert its gaze than a New World crocodile—both species helpless against a technology that stupefies them below the level of evolutionary caution.[65]

AS THE CURTAIN FALLS

In an earlier age, tigers were living icons. Although rarely seen, they radiated power, majesty, and ruthlessness—all the virtues of aristocratic prowess. Or so many of the archaic poems, rituals, and graphic representations of tigers from South and Southeast Asia tell us.[66] In contrast, tigers in the present are on the edge of extinction and are best known for needing protection.[67] One by one, the eight species known to biology at the beginning of the twentieth century are being hunted to death or deprived of habitat, or both.[68] But a new icon has appeared: this is the Tiger in the Reserve, an unexpected figure of the once condemned getting a reprieve that has been possible only after rearranging the forest to exclude humans. There—in the sanctuary, national park, or biosphere reserve—the tiger plays his role: no longer king of the jungle, he is now a "keystone species" anchoring tropical "biodiversity." This new figure—not the focus of popular ritual but an indexical "top predator" whose well-being shelters a hundred thousand other species in a protected area network—is a peculiarly modern emblem of hope in our age of extinctions. Correspondingly, the reserve is where certain privileged humans stash pure Nature in order to adore it from a distance. Besides hope and adoration, though, there is considerable suspense and hand-wringing: the tiger is now in a witness protection program, and we wonder if the biology police will fend off the mob (hunters, loggers, and poachers) as civilization goes on trial.

In truth, the strategy of bunkering tigers away from humans, after which they can be expected to thrive in perpetuity, is already eroding in India. Numerous causes are at work and have been resolved here into three bio-ironies or knotty ecological contradictions. First, expelling substantial numbers of humans from tiger reserves has simply tempted reckless intruders to take their place; the ensuing struggles between forest guards, police, and the army, on the one hand, and rebels, poachers, and criminals, on the other, has put tigers (and elephants as well as rhinos) at risk and shredded the credibility of government. An obvious example is found in the saga of Veerappan. Second, even where there's been success in protecting tigers and increasing their numbers, the excess animal population has spilled out to attack cattle and humans, triggering in turn peasant retaliation. The result has been a net loss of tigers to an orgy of poisoning and poaching, for example, at Corbett and Dudhwa tiger reserves. Third, the big, emptied-out tiger reserves seem to have fared badly—the loss of control of Manas National Park in Assam to the Bodo rebels comes to mind—whereas smaller, divided, and overrun Sariska reserve has sustained an enviable equilibrium between animals and humans.

Significantly, Project Tiger no longer excites conservation biologists as it did in the 1970s and 1980s, while some of the benefits claimed for reserve-based conservation—for instance, that it enhances biodiversity by sheltering numerous humbler species within tiger habitats—are no longer so scientifically persuasive.[69] Further, Indian conservationists have figured out that in their drive to establish pristine, people-free protected areas, they have engendered the greatest threat to tiger survival: the hostility of rural cultivators and their political representatives.[70] Understandably, then, conservation theorists are exploring new forms of protection that go beyond parks and reserves to envision protective regimes that encompass the whole landscape where tigers are found. The key to this newer approach lies in forming "partnerships" with whole rural populations, not just those villagers living on the lip of reserves.[71] Peasant claims on the reserves are also being considered in a new light, and this implies that human needs and social justice can no longer be dismissed as blips on the radar of bureaucracies and global science. Urgent efforts are now being made in India to broaden the tiger reserves' popular support and secure the collaboration of rural populations.[72] Social activists take grim satisfaction at this turn of events; they've been trying for

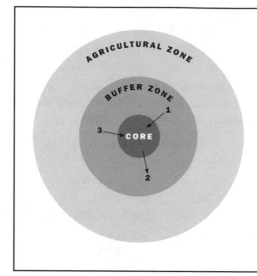

External Pressures

1 Animals moving from the buffer zone to the core for grazing.
2 Animals moving from the core to the buffer zone for hunting and grazing.
3 External pressures created by human movement from the buffer zone to the core

Other External Pressures

- Illegal: Smugglers
- Legal or need-based: Bio-mass collectors, biodiversity collectors, animal cullers, industry and mining.

2. Proposed NGO model Indian wildlife reserve acknowledging popular claims on the forest. From Anil Agarwal, *Protection of Nature Parks: Whose Business?* (New Delhi: Centre for Science and Environment, 1996), 11.

years to redefine protected areas in ways that acknowledge the justice of local claims on forest cores and buffers alike (figure 2; compare with figure 1). As this newer approach gains adherents in India and globally, there has been a backlash; some biologists refuse to give up on the classic biosphere reserve, which they consider the last bastion of biodiversity.[73]

EPILOGUE

September 2001. "Brij Mohan" and "Daulat," forest officers at Sariska, still work every day to keep poachers at bay, while on the other side of the world "Juan Carlos" and "Ricardo" still go out on the lagoon to count and measure crocodiles. There have been some favorable recent developments. For one, in mid-2002, the Mexican police rolled up the leadership of the notorious Gulf drug cartel responsible for cocaine trafficking along the Yucatán coast. And for another, the Jaipur-Alwar road in Rajasthan that used to run through the reserve was diverted around the park boundary in late 2001 to avoid further collisions with tigers. From one perspective, then, the cherished animals are tucked safely into their sanctuaries, looked after lovingly by their attendants. Rumbling beneath and behind this icon, however, is the spectacle of ungovernable forces—crime, "development," grinding poverty, the rough-and-tumble of interest group politicking—all of which erode solemn promises of wildlife protection and the promise that extinctions can be halted. Is this too bleak, too cynical? The annual Indian tiger census, based on pugmark counts, hasn't yet been published, but it is expected to show that less than three thousand tigers remain. Meanwhile, the 2001 decennial Indian census has just been completed and shows more than one billion humans.[74] The key professional newsletter for South Asian conservationists, *Protected Area Update*, reports monthly a doleful record of poachings, incursions, and "denotifications" in an evidently downward spiral. The momentum clearly lies with our own species and our all-too-human needs, and it may now have become more difficult to look a landless cultivator in the eye, without flinching, than a wild tiger.

APPENDIX: RULES FOR MEETING
TIGERS FACE-TO-FACE

According to João Pedro Galhano Alves ("Men and Tigers in Sariska Tiger Reserve, India," *Cat News* 30 [spring 1999]: 12–30), the following rules govern tiger encounters:

1. A person meeting a tiger must stop and not move. They must never run, or turn their back to the tiger. They must not squat. They must remain standing and must never make abrupt gestures or move in the tiger's direction.

2. The person must look fixedly into the tiger's eyes, never taking their eyes off those of the tiger.

3. The person should "talk to the tiger" without aggressiveness, making the following exclamations, with a loud voice: "Ahuah! Ahuahu! . . . ," or other variants, but, in all cases, alternating the U and A sounds. According to the villagers, these vocalisations "act on the tiger's mind and calm it." As a villager says: "We say 'Ahuahu" and the tiger stops. 'Ahuahu' has some effect on its mind." However, some old villagers, such as Baxi Gurjar, from Aripura, make a sound that seems a mix of a snake's hiss and purring, which they say also "keeps the tiger calm."

4. If the person has a walking stick or a crook (as villagers normally do) they should bang it on the ground, making a noise, while making the sounds.

5. If the tiger takes a step forward, the person should take a step back; one step back for each tiger step.

6. Making the same exclamations and banging the ground with the stick, the person should keep moving slowly backwards.

7. The person should never walk in the direction that the tiger appears to have gone. As in many encounters the tiger and the person meet face to face, this means that the person should not walk straight back but at an angle to the tiger's path.

8. When the person has distanced themselves about 20 meters from the tiger, they can turn their back to the tiger and move away more quickly, but always keeping visual contact with the tiger.

9. Even if the tiger usually goes away, the person should move away some hundreds of metres from the meeting point. If possible, the person should enter a safe place, such as a village or a house.

NOTES

I thank my fellows in the spring 2000 residential seminar on contemporary rain forests at HRI for their helpful suggestions and critical comments on numerous earlier drafts. Staff members at HRI were wonderful hosts and made visitors feel like prodigies. Careful readings of this essay by Jack Putz, Claudia Romero, David Baron, and Vasant Saberwal clarified points of confusion, while commentators at the decennial conference of the Agrarian Studies Program held at Yale University in May 2000, the South Asian History and Culture Studies Colloquium at the University of California-Los Angeles in June 2000, and the Clyde Kohn seminar of the University of Iowa's Department of Geography in February 2001 have tried to set me on the right path. I am grateful to Candace Slater, Jim Scott, Vinay Lal, Ramchandra Guha, Raman Mehta, and Gerry Rushton for special efforts on my behalf. Others who helped in specific circumstances are acknowledged below. All the residual confusion and errors are my own.

1. The common layout and legal restraints on human use of cores and buffers in biosphere reserves derive from a global model scripted in the 1970s by the United Nations Educational, Scientific, and Cultural Organization (UNESCO), Food and Agriculture Organization of the United Nations, and the United Nations Environmental Program (UNEP). See *The Biosphere Reserve and Its Relationship to Other Protected Areas: Commission on National Parks and Protected Areas* (Gland, Switzerland: International Union for Conservation of Nature and Natural Resources, 1979), and International Biosphere Reserve Congress, *Conservation, Science, and*

Society: Contributions to the First International Biosphere Reserve Congress (Paris: UN-ESCO-UNEP, 1984). In May 2002, there were 408 biosphere reserves in 94 countries participating in a global network coordinated by UNESCO. See the UNESCO Man in the Biosphere site at http://www.unesco.org/mab/wnbr.htm.

2. I am grateful to the Amigos de Sian Ka'an, a private Mexican research NGO, and in particular, Gonzálo Merediz for his instructive conversation and the opportunity to see the crocodile project close-up. Juan Carlos and Ricardo are fictional names of real biologists.

3. I use the terms "conservation" and "preservation" interchangeably in this essay, although a stricter use would distinguish between conservation for human use and preservation for the sake of wild nature.

4. It's no secret that forests and beaches on the Caribbean coast have been wide open to crime and drug trafficking for years. Parcels of cocaine are dropped on the eastern cayes or on shore by speedboats, or are brought in by vehicles from the south. See Molly Moore, "Vast Mexico Drug Crackdown Targets Top Officials," *Washington Post*, 8 April 1999, 13A; Sam Dillon and Tim Golden, "Drug Inquiry into a Governor Tests Mexico's New Politics," *New York Times*, 26 November 1998, 1A; and Juanita Darling, "Belize Caught in Middle of Drug War," *Los Angeles Times* 25 July 1997, 5A. The United States has its share of park-based crime, and demand from the United States for heroin, cocaine, and marijuana drives the Mexican drug industry. See Traci Watson, "National Parks an Escape—for Drug Smugglers," *USA Today*, 10 December 1999, 21A; Andrew Murr, "Prime Evil Forest," *Newsweek*, 4 November 1996, 69; Richard and Joyce Wolkomir, "Drug Outlaws in Our National Forests," *Reader's Digest* 133 (October 1988): 193–194; and Helen Moss, "Big Bend: Where Mountains Seam the Sky," *National Parks* 58 (November–December 1984): 22–27.

5. PANS comprise national parks and wilderness areas; bird, tree, and animal sanctuaries; and biosphere reserves and other demarcated zones of protection and conservation. While mostly forested areas, PANS also include sizable desert, wetlands, and alpine terrains.

6. By 1996, 174 countries had set aside 9,869 protected areas comprising 931.8 million hectares worldwide. A handful of countries—for example, the United Kingdom, Venezuela, Norway, Greenland, Panama, and New Zealand—have set aside more than 20 percent of their territories for conservation. See "1996 Global Protected Areas Summary Statistics," World Conservation Monitoring Centre (WCMC) website, at http://www.wcmc.org.uk/protected_areas/data/summstat.html. In 2001, for a more recent example, China established the Huangnihe River National Park in Jilin Province for the preservation of Siberian tigers; see Environmental News Service, 2 January 2001, at http://ens.lycos.com/aboutens.html.

7. A bio-irony is the outcome of the application of contradictory conservation/preservation principles or policies. A familiar bio-irony in the United States concerns the management of national forests to attain "pristine wildness," which flatly contradicts the management of the same forests to "preserve biodiversity"—that is, what is most "wild" is not necessarily what is most "diverse," and vice versa. See David M. Graber, "Resolute Biocentrism: The Dilemma of Wilderness in National Parks," in *Reinventing Nature? Responses to Postmodern Deconstruction*, ed. Michael E. Soulé and Gary Lease (Washington: Island Press, 1995).

8. Ministry of Environment and Forests, *National Action Plan on Biodiversity* (New Delhi: Government of India, 1997), 32.

9. Twenty-five "biodiversity hotspots" are defined by Conservation International as those parts of the planet's surface that while comprising only 1.4 percent of its area, contain more than 60 percent of its animal and plant species. See Russell Mittermeyer, *Hotspots: Earth's Biologically Richest and Most Endangered Terrestrial Ecoregions* (Chicago: University of Chicago Press, 2000).

10. Of the eight subspecies of *Panthera tigris*, three—the Bali, Caspian Sea, and Java tigers—became extinct in the 1940s, 1970s, and 1980s, respectively. Today, only the Indian and Indo-Chinese tigers exist in viable numbers, the other three subspecies being "completely isolated, critically endangered and facing a bleak future" (John Seidensticker, Sarah Christie, and Peter Jackson, eds., *Riding the Tiger: Tiger Conservation in Human-Dominated Landscapes* [Cambridge: Cambridge University Press, 1999] xv, table 0.1).

11. Conservation efforts in Bangladesh and Nepal, while important, were considerably smaller. The major tiger reserve in Bangladesh is the Sunderbans Animal Sanctuary (32,000 hectares); in Nepal, the two major tiger reserves are sited within the Royal Chitwan and Royal Bardia national parks (190,000 hectares in total). For the political context of tiger conservation during Indira Gandhi's brief experiment with autocracy, 1975–1977, see Paul Greenough, "Pathogens, Pugmarks, and Political 'Emergency,' " in *Nature in the Global South: Environmental Projects in South and Southeast Asia*, ed. Paul Greenough and Anna L. Tsing (Durham: Duke University Press, 2003).

12. Foreign donors' support has been important for maintaining the reserves in recent years. It had been hoped in the 1970s that tourists would flock to see Indian tigers and pay dearly for Africa-like safari experiences, but Indian tigers (unlike African lions) are solitary and secretive, and thus almost impossible to see. Forests in India with their dense undergrowth usually are traversed on the backs of elephants rather than in open cars, as on the savannas of East Africa.

13. See Indian Board for Wild Life, *Task Force, Project Tiger: A Planning Proposal for Preservation of Tiger (Panthera tigris tigris Linn.) in India* (New Delhi: Ministry of Agriculture, 1972). Some of the original reserves have had their territories augmented since the early 1970s.

14. C. R. Margules and R. L. Pressey, "Systematic Conservation Planning," *Nature* 405 (11 May 2000): 243. The authors go on to say that "in the real world of conservation planning, the opportunity to apply such guidelines is constrained by costs and patterns of land-use history." For criticism of the Wilson-MacArthur position, see A. J. Higgs, "Island Biogeography and Nature Reserve Design," *Journal of Biogeography* 8 (1981): 117–124.

15. See Michael L. Lewis, *Inventing Global Ecology: Tracking the Biodiversity Ideal in India, 1947–1997* (Hyderabad, India: Orient Longman, 2003), 200–211.

16. For a factual overview and numerical indicators for all national programs, see "1996 Global Protected Areas," WCMC Website.

17. Despite a policy of relocation, not all peasant and tribal villagers have been removed from the core and buffer areas of tiger reserves in India. This is due, in some cases, to a scarcity of compensatory lands on which to resettle the displaced communities; in other cases, there are

vested interests in mining and timbering that exploit the cause of forest villages to block closure of core and buffer areas. In recent years, there has been a movement of villagers in the opposite direction—from outside the buffers into the core of the reserves—which infuriates Project Tiger personnel. While conservationists and biologists continue to press for village relocations, NGOS and human rights groups defend the interests of residents and newcomers who rely on forest resources. See Anil Agarwal, *Protection of Nature Parks: Whose Business? Proceedings of a Debate on Who Should Protect Nature Parks: The Bureaucracy or the People?* (New Delhi: Centre for Science and Environment, 1996).

18. For a good account, see Belinda Wright, *India's Tiger Poaching Crisis* (New Delhi: Wildlife Protection Society of India, 1997. For poaching statistics, see Wildlife Protection Society of India, "Tiger Poaching Statistics, January 1994–August 1997," Tiger Information Centre website at http://www.5tigers.org/ConservationOrganizations/WPSI/wpsi_stat.htm. Recent policy moves now commit Provincial Armed Constabulary and Border Security Forces to the hunt for poachers and militants, establish new police stations inside protected areas, and alter the criminal law to immunize forest officers from lengthy trials when they shoot forest intruders. See Ajay Suri, "Paramilitary Forces Will Now Hunt for Poachers," *Hindusthan Standard*, 20 May 2000.

19. Asian folk myth and traditional Chinese medicine ascribe curative, aphrodisiac, and pain-killing powers to tiger bones and organs. See Ginette Hemley and Judy A. Milles, "The Beginning of the End of Tigers in Trade?" in Seidensticker, Christie, and Jackson, *Riding the Tiger*, 217–229.

20. See, for example, a brief account of the Tiger Information Center in Seidensticker, Christie, and Jackson, *Riding the Tiger*, 210–212.

21. The most detailed account of Veerappan's career is the "Status Report on the Task Force Constituted to Nab the Notorious Killer Veerappan and His Associates," by K. Arkesh, superintendent of police, Karnataka, 22 May 2000; cited in *India Today* (New Delhi), at http://www.india-today.com/ntoday/extra/rajkumar/ (alternative website: http://www.dalitstan.org/tamil/stfreprt.html). Other key articles include M. S. S. Pandian, "The Moral World of 'Sandalwood' Veerappan," *Economic and Political Weekly*, 25 December 1999 (online); S. Rohini, "Veerappan Mythology," *Economic and Political Weekly*, 11–17, March 2000 (online); and Ravi Sharma, "On the Veerappan Trail," *Frontline*, 3 August 2001, 37.

22. The gang's principal areas of operation are in the reserved forests of Dankanikotai, Pennagaram, Bargur, Guttiyelathur, Talamalai, and the Nilgiri Eastern Slopes—all in Tamilnadu; and Heggadadevanakote, Bandipura, Biligiri Rangana Betta, Chamarajanagar, Edeyarahalli, Mahadeswarana Malai, Chikkailur, Hanur, Dhangur, Cowdally, and Doddasampige—all in Karnataka.

23. Arkesh, "Status Report on the Task Force." Undoubtedly, Veerappan threatens villagers, but he is also one of them; he comes from the rural Padayachi/Vanniyar caste of Tamilnadu.

24. Veerappan's release of Rajkumar in November 2000 was described in Celia W. Dugger, "To Great Joy in India, a Bandit Frees Movie Idol He Kidnapped," *New York Times*, 16 November 2000, 16; Ravi Sharma and T. S. Subramanian, "The Release and After," *Frontline* (Chennai),

25 November–8 December 2000 (online). Subsequently, police sources admitted that a ransom assembled from public and private sources of "at least" 120 million rupees (approximately U.S. $2.5 million) was paid to Veerappan. See Sharma, "On the Veerappan Trail," 37.

25. Abraham Verghese, "The Bandit King and the Movie Star," *Atlantic Monthly*, February 2001. As of 2003, three full-length studies by journalists have been published by Indian publishers.

26. See Shin-wha Lee, Aamena Malik, and Deepa Khosla, "Assamese and Bodos in Assam, India," in *Minorities at Risk Project* (College Park: University of Maryland, 1999); for background information, see http://www.bsos.umd.edu/cidcm/mar/indbodo.htm.

27. W. Hussain, "Bodo Militants on Shooting Spree Inside Manas Tiger Reserve," *Telegraph* (Calcutta), 5 April 1989 (online); J. Gavron, "Marauders Slaughter Assam Tigers," *Sunday Telegraph*, 10 September 1989, 16; Peter Jackson, "Critical Situation for Wildlife in India," *Cat News* 13 (July 1990) (online); Vivek Menon, "Militants!! Continuing Concern about India's Manas Tiger Reserve," *Cat News* 23 (fall 1995) (online); United Press International, "Tribal Militants in India Kill Tigers," 10 October 1995 (online); World Conservation Monitoring Centre, at http://www.wcmc.org.uk/; "Guerrilla Groups Disrupting Indian Wildlife Reserves," *Cat News* 30 (spring 1999) (online); Valmik Thapar, "Obituary Sanjoy Deb Roy (1934–1999)," *Cat News* 31 (fall 1999) (online).

28. A bowl of tiger penis soup sold for $350 in the markets of Southeast and East Asia in 1998. See Hemley and Milles, "The Beginning of the End," 219.

29. See "Militants!! Problems in India's Buxa Tiger Reserve," *Cat News* 24 (spring 1996) (online); Mian Ridge, "Indian Tigers in Danger of Extinction," Reuters Service, 25 November 1999 (online).

30. Rudyard Kipling, *The Jungle Book* (London: Macmillan, 1894); and George B. Schaller, *The Deer and the Tiger: A Study of Wildlife in India* (Chicago: University of Chicago Press, 1967).

31. "The Naxalites champion tribal peoples, who, they say, are oppressed and cheated. Some reserve officials were beaten up and accused of not handing over to tribal labourers their full wages. Some animals were found poisoned near saltlicks. Several hundred police were moved to Kanha to combat the Naxalites, but none has been caught" (Peter Jackson, "Critical Situation for Wildlife in India," *Cat News* 13 [July 1990], citing *Sunday Magazine*).

32. See Ramachandra Guha and Madhav Gadgil, *Ecology and Equity: The Use and Abuse of Nature in Contemporary India* (New Delhi: Penguin India, 1995), 96. See also Bittu Saghal, "Tigers and Terrorism," *Sanctuary* (June 1999) (online); and Federation of American Scientists, "People's' War Group," Intelligence Resources Program (May 2000), at http://www.fas.org/irp/world/para/pwg.htm.

33. Cited in *Cat News* 28 (spring 1998) (online).

34. Disturbances and occupations have also been reported in the tiger reserves at Namdapha (Arunachal Pradesh), Dampa (Mizoram), Dachigam, Overa (Kashmir), Gadichiroli (Maharashta), Panna, Indravati (Madhya Pradesh), and Srisailam (Andhra Pradesh).

35. Regarding denotification, see Samar Halarnkar with Ruben Banerjee, Rohit Parihar, and Amarnath K. Menon, "National Parks: The Last Stand," *India Today*, 9 March 1998, 66ff.; and "Wildlife Protection and People's Livelihood: Building Bridges," *Economic and Political Weekly*,

14–20 August 1999 (online). A steady drumbeat of denotifications can be found in issues of *Protected Area Update*, an online newsletter published by Kalpavriksh, a New Delhi–based environmental NGO; see http://131.103.239.3/pa0801.htm (passim).

36. The number of persons displaced to establish the tiger reserves is uncertain, and expulsion of residents of the reserves has never been complete. One author suggests that as many as eight million persons, mostly tribals, were displaced, while the Project Tiger directorate suggests a more modest figure of less than twenty-five thousand. See "Relocation of People from Indian Tiger Reserves," *Cat News* 26 (spring 1997) (online). Forced removals of villagers for development projects in India since the 1950s is a major theme in civil society's criticism of the government in India; see, for example, the special issue on displacement and rehabilitation, in *Lokayan Bulletin* 11, no. 5 (March–April 1995) (online). See also "Wildlife Protection and People's Livelihood."

37. How much of an increase was never clear, and a press debate erupted in 1978 over the accuracy of tiger censuses. The debate was the first public discussion of Project Tiger. See Greenough, "Pathogens, Pugmarks, and Political 'Emergency.'" The tiger censuses were based on pugmark (paw print) counts, and depended on the unproven assumption that every tiger had unique and distinguishable pugmarks. See K. Ullas Karanth, "Tigers in India: A Critical Review of Field Censuses," in *Tigers of the World*, ed. R. L. Tilson and U. S. Seal (Park Ridge, N.J.: Noyes Publications, 1987).

38. These dilemmas were documented with great clarity in families of tigers at the Royal Chitawan Park in Nepal and the Ranthambhore tiger reserve in India. See Mel Sunquist and Fiona Sunquist, *Tiger Moon* (Chicago: University of Chicago Press, 1988), 75–78, 123–133. See also Eric Dinerstein et al., *Framework for Identifying High Priority Areas and Actions for Conservation of Tigers in the Wild* (Washington: WWF-U.S. Wildlife Conservation Society, 1997).

39. For a review of the dangers to humans and domestic animals in colonial and postcolonial India, see Paul Greenough, "*Naturae Ferae*: Wild Animals in South Asia and the Standard Environmental Narrative," in *Agrarian Studies: Synthetic Work at the Cutting Edge*, ed. James C. Scott (New Haven: Yale University Press, 2001), 141–185.

40. Between 1979 and 1984 there were at least 485 recorded fatalities from animal attacks in 39 of India's national parks and 167 sanctuaries; this is an average of 32 deaths each year. Unrecorded fatalities, and fatalities from attacks occurring outside the parks and sanctuaries, are not reported, but may have equaled these figures. See Ashish Kothari, Pratiba Pande, Shekhar Singh, and Dilnavaz Variava, *Management of National Parks and Sanctuaries in India: A Status Report* (New Delhi: Indian Institute of Public Administration, 1989), 45–46. A total of 22 persons were killed by tigers in and around a single reserve, Royal Chitwan National Park, in Nepal between 1979 and 1996 (private communication with B. N. Upreti); see also Sunquist and Sunquist, *Tiger Moon*. Human deaths from tiger attacks in the Sunderbans region of West Bengal and Bangladesh varied between 20 and 80 per year between 1993 and 1996 (see *Cat News* 26 [spring 1997; online]), for an average of 50 deaths annually. It does not seem unlikely that the average annual number of deaths from tiger attacks in India, Nepal, and Bangladesh between 1980 and the present has exceeded 100, suggesting at least 2,200 fatalities in 22 years.

41. International Union for Conservation of Nature and Natural Resources, *Conservation of the Tiger in India: A Report to the Chairman of the Tiger Steering Committee on a Mid-Term Study of Project Tiger, March/April 1976* (Morges, Switzerland, 1976), 25.

42. Ramesh Bedi and Rajesh Bedi, *Indian Wildlife* (New Delhi: Brijbashi, 1984), 78. The Bedis probably exaggerated the frequency of "man-eaters" around Dudhwa reserve. Tigers in India normally hunt deer and pig; they may attack and kill unlucky humans who threaten or startle them without necessarily "relishing human flesh." Much of the hatred directed against man-killing tigers is premised on the widespread assumptions that they are all potentially man-eaters.

43. Arjan Singh, "A Brief History of Project Tiger" (manuscript, 1995), 11. I am grateful to Ramchandra Guha for showing me this document. Poisoning tigers by salting their kills with pesticides is the safest (yet most painful) means of killing them. See "Poisoning of Tigers and Leopards in India," *Cat News* 28 (spring 1998) (online). For further discussion of the struggle over man-killers and man-eaters, see Greenough, "Pathogens, Pugmarks, and Political 'Emergency.'"

44. The Ranthambhore Foundation provides mobile health care and family planning services; a tree nursery, regreening, seed collection, and afforestation services; dairy development and animal husbandry services; promotion and propagation of appropriate technology; and an informal nature education program for local youth and children. It also works to strengthen community decision-making organizations. See "Ranthambhore Foundation Initiatives around Ranthambhore National Park and across India," *Tiger Information Centre* (summer 1999), at http://www.5tigers.org/ConservationOrganizations/rantham.htm. For an example of the highly effective publicity about Ranthambhore, see Valmik Thapar and Fateh Singh Rathore, *Wild Tigers of Ranthambhore* (Delhi: Oxford University Press, 2000).

45. The "Why are you evicting us?" statement was made by Santosh Mohanty, Social Research and Development Council, Mayurbhanj, at a 1996 roundtable on forest protection organized by the Centre for Science and the Environment (cited in Agarwal, *Protection of Nature Parks: Whose Business?* 20, emphasis added). The limitations of eco-development in general are reviewed in Vasant Saberwal, Mahesh Rangarajan, and Ashish Kothari et al., *People, Parks, and Wildlife: Towards Coexistence* (Hyderabad, India: Orient Longman, 2001): 83–87. For the drawbacks of eco-development projects around a tiger reserve in particular (Buxa National Park in Assam), see B. G. Karlsson, "Ecodevelopment in Practice: Buxa Tiger Reserve and Forest People," *Economic and Political Weekly*, 24–30 July 1999 (online).

46. See A. J. T. Johnsingh, K. Sarkar, and Shomita Mukherjee, "Saving Prime Tiger Habitat in Sariska Tiger Reserve," *Cat News* 27 (fall 1997) (online). According to this article, "The Alwar-Thanaghazi-Jaipur State Highway passes through the reserve, and 2,000 vehicles ply on it every day. Of these, 1,000 are trucks, which use the road mainly at night. Although there are two check posts, one at the entrance and the other at the exit point of the reserve, it is impossible for the Reserve authorities to check all the vehicles on this highway. . . . Another road through the reserve, heavily used by people, is the Sariska-Kalighati-Pandupole road. On Tuesdays and Saturdays pilgrims visit Pandupole temple of the Hindu deity, Hanuman, driving for 20 km

through Kalighati, the only habitation-free wildlife-rich valley of the Reserve, in 100–200 vehicles."

47. M. K. Ranjitsinh, former director of the wwf Tiger Conservation Programme, "Wildlife Mismanagement in India, Mutually Assured Destruction," interview by editors, *Down to Earth* 8, no. 20 (15 March 2000) (online). See also Johnsingh, Sarkar, and Mukherjee, "Saving Prime Tiger Habitat in Sariska Tiger Reserve."

48. See *New Scientist*, 26 August 1995, 10.

49. I am grateful to "Brij Mohan" and "Daulat" (pseudonyms), forest officers at Sariska, for hours of informative conversation. Like most Forest Department tiger guardians, they work every day around the year. I am also obliged to Mangaldas Singh, a Forest Department tracker, for accompanying me on foot to several Van Gujar villages and the pilgrimage village of Narsati.

50. Police have removed only a single Sariska village, Kanakabal, from the core to the buffer in the last twenty-five years. In contrast, twelve villages were removed from the core of Ranthambhore Tiger Reserve between 1976–1979 and twenty-six from the core of Kanha Tiger Reserve. See "Critical Review of Project Tiger (1993)," Project Tiger Directorate website, http://envfor.nic.in/pt/status.html.

51. There are other ethnicities besides Gujars living in the Sariska Tiger Reserve—for example, Meenas, Jogis, and Bauriyas. Gujars, however, are the predominant settled group.

52. For details on Gujar income, diet, and so on, see Joäs Pedro Galhano Alves, "Men and Tigers in Sariska Tiger Reserve, India," *Cat News* 30 (spring 1999): 10–12.

53. This is approximately the same number that inhabit Ranthambhore. According to Brij Mohan, the forest officer in charge of the Sariska reserve, the 1999 pugmark-based census data showed a population of twenty-six to twenty-eight tigers. Ten years ago the figure was probably around eleven, although a census in 1988 had suggested more than forty. The tigers in Ranthambhore reached a numerical peak of forty-five in 1991; since then, there has been a marked decline. See "Critical Review of Project Tiger (1993)," Project Tiger Directorate website.

54. João Pedro Galhano Alves, "Tigers and People: Strategies for Tiger Conservation in Sariska Tiger Reseve, India," *Cat News* 29 (fall 1998): 9–11; and "Men and Tigers," 10–12. See also João Pedro Galhano Alves, "Of Large Carnivores and Humans," *Down to Earth* 8, no. 4 (15 July 1999): 27–31; and "Vivre en Biodiversite Totale. Des Hommes, des Grands Carnivores et des Grands Herbivores Sauvages. Deux etudes de cas: Loups au Portugal, Tigres en Inde (Ph.D. diss., Universite d'Aix-Marseille III, Laboratoire d'Ecologie Humaine et d'Anthropologie, 2000).

55. Galhano Alves, "Men and Tigers."

56. Cited in Galhano Alves, "Tigers and People," 101. In July 2000, I interviewed Ram Karan of Umri Village, Sariska, who had been mauled by a tiger in September 1998. He showed me his scars, but insisted that the injuries were his own fault because he and his companions had slept on a forest track that was well-known as a tiger route and thus bound to antagonize passing tigers. Our conversation took place among a crowd of children, for whom this lesson was also presumably intended.

57. Galhano Alves, "Men and Tigers," 11.

58. Historical studies of Gujar ecological practices are limited, but see D. D. Dangwal "State, Forests, and Graziers in the Hills of Uttar-Pradesh: Impact of Colonial Forestry on Peasants, Gujars, and Bhotiyas," *Indian Economic and Social History Review* 34, no. 4 (October–1997): 405–435; Darshan Singh Manku, *The Gujar Settlements: A Study in Ethnic Geography* (New Delhi: Inter-India Publications, 1985); and Ram Parshad Khatana, *Tribal Migration in Himalayan Frontiers* (Gurgaon, Haryana, India: Vintage Books, 1992).

59. Gujars openly profess beliefs about tigers and their presiding deity, who tigers serve at the temple of Narsati ("tiger sati") in the village of Narsati located in the Sariska core. There I saw high-caste goddess worship conducted by a *pujari* (officiant), accompanied with a liturgy that relates the brave deeds of the sati (self-immolating widow) who obtains the tiger's protection for her child. Illustrations of some elements of the mythography were painted on the temple walls. It seems likely that a fusion of the sati cult, goddess worship, and protective tiger motifs in Narsati's ritual stabilizes local attitudes of respect for tigers in the wild.

60. See Galhano Alves, "Vivre en Biodiversité totale."

61. Considerable research is underway to better understand the dynamics of tigers' hunting and reproductive success under less-than-ideal conditions. Recent research suggests that the greatest threat to tigers' survival may not be habitat destruction, poaching, and retaliatory poisoning, as has long been assumed, but the destruction of tiger prey by villagers and market hunters living in and around the reserves. See K. Ullas Karanth and Bradley Smith, "Prey Depletion as a Critical Determinant of Tiger Population Viability," in Seidensticker, Christie, and Jackson, *Riding the Tiger*, 100–113.

62. See Mark David Spence, *Dispossessing the Wilderness: Indian Removal and the Making of the National Parks* (New York: Oxford University Press, 1999). Related studies of a later era are Robert H. Keller and Michael F. Turek, *American Indians and National Parks* (Phoenix: University of Arizona Press, 1999); and Philip Burnham, *Indian Country, God's Country: Native Americans and the National Parks* (Washington: Island Press, 2000). See also James McCarthy, "States of Nature and Environmental Enclosures in the American West," in *Violent Environments*, ed. Nancy Lee Peluso and Michael Watts: Cornell University Press, 2001), 117–145. The phrase "exclusion and enclosure" comes from Peluso and Watts, *Violent Environments*, 53.

63. See the debate among Stephan Schwartzman, Adriana Moreira, Daniel Nepstad, John Terborgh, Kent Redford, Marcus Colchester, and Avecita Chicchón in the "Conservation Forum" section of *Conservation Biology: Journal of the Society for Conservation Biology* 14, no. 5 (October 2000): 1351–1374.

64. For the transformation of tigers into neighbors, see Eric Dinerstein et al., "Tigers as Neighbours: Efforts to Promote Local Guardianship of Endangered Species in Lowland Nepal," in Seidensticker, Christie, and Jackson, *Riding the Tiger*, 316–332.

65. According to forest officers "Brij Mohan" and "Daulat," the tiger's death followed a hit-and-run accident; the vehicle's license number was taken down by some bystanders, and the police eventually arrested the driver for speeding. The Jaipur-Alwar highway was subsequently closed at night to traffic.

66. See, for example, Tushar K. Niyogi, *Tiger Cult of the Sundarvans* (Calcutta: Anthropological Survey of India, 1996); Greenough, "*Naturae Ferae*"; Jeffrey A. McNeely and Paul Spencer, *Soul*

of the Tiger: Searching for Nature's Answers in Southeast Asia (Honolulu: University of Hawaii Press, 1995); and Peter Boomgaard, *Frontiers of Fear: Tigers and People in the Malay World, 1600–1950* (New Haven: Yale University Press, 2001).

67. See Richard Ives, *Of Tigers and Men: Entering the Age of Extinction* (New York: Talese/Doubleday, 1996).

68. See note 10 above. During the year 2000, tigers finally disappeared from the entire Indian state of Gujarat.

69. The protective relationship between so-called top predators and multitudes of sheltering forest species, once a firm argument for megafaunal conservation, is now much more uncertain. While some biologists still maintain that "in the absence of the ecological functions performed by top predators, the whole ecosystem slides into imbalance and begins to spiral down into imbalance in a cascade of species losses" (John Terborgh, *Requiem for Nature* [Washington: Island Press, 1999], 16), others refute this idea, arguing that "the depletion of large-animal populations does not threaten the majority of the other species that comprise these forests; [for example,] forests that have been impoverished through hunting still have enormous conservation value" (Stephan Schwartzman, Adriana Moreira, and Daniel Nepstad, "Rethinking Tropical Forest Conservation: Perils in Parks," *Conservation Biology* 14, no. 5 [October 2000]: 1353; also, in the same journal issue by the same authors, "Arguing Tropical Forest Conservation: People versus Parks," 1370–1374).

70. Schwartzman, Moreira, and Nepstad, "Rethinking Tropical Forest Conservation," 1355.

71. See Eric D. Wikramanayake, Eric Dinerstein, John G. Robinson, K. Ullas Karanth, David Olson, Thomas Mathew, Prashant Hedao, Melissa Connor, Ginette Hemley, and Dorene Bolze, "Where Can Tigers Live in the Future? A Framework for Identifying High-Priority Areas for the Conservation of Tigers in the Wild," in Seidensticker, Christie, and Jackson, *Riding the Tiger*, 255–272.

72. See Kathy MacKinnon, Hemanta Mishra, and Jessica Mott, "Reconciling the Needs of Conservation and Local Communities: Global Environment Facility Support for Tiger Conservation in India," in Seidensticker, Christie, and Jackson, *Riding the Tiger*, 307–315.

73. See John Terborgh, "The Fate of Tropical Forests: A Matter of Stewardship," *Conservation Biology* 14, no. 5 (October 2000): 1358–1361. Those who try to reconcile the opposed positions (protected areas versus human uses) find themselves seesawing around opposed theoretical axes—that is, building bio-ironies. See Kent H. Redford and Steven E. Sanderson, "Extracting Humans from Nature," *Conservation Biology* 14, no. 5 (October 2000): 1362–1364.

74. Government of India, registrar general and census commissioner, at http://www.censusindia.net.

NANCY LEE PELUSO

WEAPONS OF THE WILD: STRATEGIC USES
OF VIOLENCE AND WILDNESS IN THE RAIN
FORESTS OF INDONESIAN BORNEO

The clean-cut young man dressed in stonewashed jeans and a white T-shirt hardly looked like a soldier, let alone a headhunter. "Robi" and I were chatting in a friend's home in a forest village in West Kalimantan, a province on the Indonesian side of Borneo.[1] He dragged his cigarette down to its very last bit of white paper; I enjoyed the all-too-brief cool of the dawning morning.

Robi described how he and many others in that village had prepared themselves for war on a similarly cool morning in February 1997. Two events of the past few days had caused them to gather. The "red bowl"—a traditional local symbol of Dayak solidarity in war, containing bird's feathers, a match, a piece of iron, and the blood from which it took its name—had arrived in the village, compelling them to formally join the sporadic fighting against another ethnic group, known as Madurese. Indeed, many villagers, including Robi, had already been involved in burning Madurese houses and even some killings in January. They were more immediately incited by news of an attack on a Dayak friend's child in the marketplace by a Madurese youth the day before. That day they would attack a Madurese village close to the marketplace, but they first needed to take some precautions.

As the cool of that February morning receded, they climbed to the forested site of the warriors' ancestral burial ground on the hillside behind the village. Groups of seven trekked up at a time; one of the local shamans solemnly met them. Each man drank water from bamboo containers filled from one of the closed jars marking these special graves. The shaman sacrificed a chicken and murmured prayers to make them brave, even immune to bullets. I asked Robi if he felt any different after drinking the water, but he said he felt "just normal."

When everyone had been protected in this way, the shaman killed a pig, spoke some other words, and let out a bloodcurdling yell—his voice was calling the *tariu*, the ancestor spirits of war.[2] Pausing for a moment, Robi looked at me as if realizing something for the first time. "*That's* when we first felt it [the effects of being possessed]. We were no longer afraid. We [the village men] followed its sound," he recounted, "The tariu preceded us into battle."

Three of their departing number never returned to the village. Two of them had not been through the protective ceremony; Robi speculated that the third had not drunk enough of the water. A different shaman, a woman, had recalled the souls of those returning, sending the tariu back to where they lived. Robi felt "normal" again then, he said, but was a bit shaken—not brave as when they had left.

Later, when I asked who still knew about these headhunting rituals, another friend told me it was only the old people.

"But this wasn't head-hunting [*ngayau*]," he added, "it was war [*perang*]."

An ethnic war had broken out in Sambas, a district of West Kalimantan, in December 1996. This province on the Indonesian side of the forested island of Borneo reeled from one set of horrific incidents to another through the first several months of 1997.[3] By the end of May, hundreds of people had been killed and thousands more left homeless, burned out of their homes and villages.[4]

Because the violence took place on Borneo and involved indigenous peoples collectively referred to as Dayaks, much of the reporting asserted or assumed "the return of the Borneo headhunters." To all the Dayaks I spoke with, however, these events constituted war, not headhunting.[5]

Journalists' lurid labels were belied by their own descriptions of the war-invoking scenes they encountered—accounts of military vehicles, curfews, and elite troops. Groups of young men patrolled in various types of uniforms, some wearing the battle drabs and insignias of various national military units, others in jeans and the symbolic red headbands that signaled Dayak identity.[6] Some writers casually juxtaposed their references to headhunting images from other times with discussions of human rights violations, hinting that something else was going on here. International journalists, in particular, freely conflated headhunting images with violent scenes from Southeast Asian wars more familiar to Western audiences. "The Killing Fields" and "Apocalypse Now" headlines, for example, evoked recent Hollywood movies "documenting" the horrors of forest-based wars in Cambodia and Vietnam, respectively. At the same time, the political and cultural dimensions of local histories of war were not more carefully analyzed; the images seemed intended to speak for themselves. Moreover, those talking about the violence depicted it as part of a forested landscape, or rather a "jungle," even though many of the incidents occurred in towns, settlements, or along paved roads.

The academic definition of headhunting as "an organized, coherent form of violence in which the severed head is given a specific ritual meaning and the act of head-taking is consecrated and commemorated in some form," fits many Dayaks' representations of these practices as something from the past rather than the present.[7] Headhunting had particular ritual/religious purposes; varied widely in form, context, and meaning across various groups of Dayaks; and had not been a part of religious practice for several generations. Missionaries and governments had seen to that.[8]

Nevertheless, Dayaks were performing some of the rituals associated with headhunting and "traditional" war, such as the passing of the red bowl to call allies to war, the performance of invulnerability rites in the sacred wooded places where warrior ancestors were buried, and the calling of the tariu to lead them while inspiring terror and fear in their enemies. Yet they used these rites and rituals in ways that could not exactly be called traditional.[9] The contemporary practices differed radically because the rituals were combined with tactics, movements, and representations of more modern forms of warfare. They were reminiscent of the guerrilla warfare tactics used in Southeast Asian forest theaters—including Borneo's rain forests—between the 1950s and 1970s.

Many elements of the 1997 violence between Dayaks and Madurese—the sites of evictions and killings, the people involved, the mechanisms for involving multiple villages, the terms used by people describing it, and even the tactics for killing—vividly echo those of the Borneo chapter in what Westerners euphemistically call the Cold War—the intense social and political upheavals that wracked the rain forests of this region in the 1960s.[10] In 1967–1968, in the aftermath of an undeclared war on Malaysia, the Indonesian army had incited West Kalimantan Dayaks to violently evict rural Chinese from the same districts where the ethnic war between Dayaks and Madurese took place thirty years later. In the 1960s, a calculated military decision led to the deployment of "The Borneo Headhunter" as both a mobilizing and cloaking tactic. The reemergence of The Borneo Headhunter in 1997, however, also helped obscure the parallels of this violence with the counterinsurgency operations of the 1960s. Thus, although the Indonesian military may not have explicitly mobilized Dayaks against Madurese in 1997, its long shadow of past practices extended the military's actual activities—patrolling the streets, keeping curfews, and housing refugees.

On one visit to my friends' village, I was met at a roadside shop by an old man who had lost his son in the violent events of 1997. Somewhat ironically, I thought, we sat in the stall of a middle-aged Chinese shopkeeper who had been evicted with his whole family thirty years earlier. The old man, a former field officer with the Indonesian police, implored me to learn and tell the world "the real truth" about what had happened in 1997. Other friends and acquaintances came and sat with us, and over the next few hours they raised issues that I heard repeatedly during the next several weeks. These Dayaks saw themselves as victims of circumstance, subjects of a cultural politics of aggression—a very different perspective from that of news reports depicting them as violent aggressors.[11] Neither the man who had lost his son nor any of the others referred to this violence as ngayau.

Both warfare and headhunting have played important parts in the histories and historical memories of the Dayak people involved in the ethnic violence of the 1990s. Both have shaped and been shaped by the rain forest landscapes of West Kalimantan. Yet for many observers in the press, government, and general public, that ever fascinating—and highly simplified—icon of a more exotic Dayak past has overshadowed the strategic tactics and implications of war.

The Borneo Headhunter has been used to explain all kinds of complex actions. Government actors have used the image to justify "civilizing" or "development" projects of the colonial or contemporary eras.[12] Some Dayaks' association of The Borneo Headhunter with strategic and honored service to colonial or postcolonial states leads them to proudly "remember" as well as justify contemporary violence by referring to periods and places unrelated to their own specific histories, thereby constituting another level of translation and simplification.[13] "Jungle" violence during the twentieth century thus constitutes a curious sort of ethnic/political violence: it is haunted by ancient likenesses of the participants themselves.

Rain forests and their people are an apparently endless source of romantic, literary inspiration as well as an idealized setting for ecological research. The kinds of violence described in this essay, however, are usually left out of both fictional and scientific reporting on the allegedly pristine rain forests, though they do make their way into accounts of so-called jungles.[14] Violence has played a role in the construction of virtually all the world's rain forest landscapes, not just those stalked by The Borneo Headhunter. But why should such an apparently sinister view be important either to aspiring novelists, journalists, or "scientific" forest managers—as well as political historians? I show here that the fear created by images and stories of wild and violent people in violent places (jungles) lives in historical, collective memory and can continue to affect the ways people behave generations later. Snippets of history and the lived experiences of all kinds of storytellers are combined into tales that emphasize some forms and memories of violence while suppressing others. Rain forests valued for their commodity and aesthetic value—however and by whomever they have been acquired—are invariably landscapes of conquest. In this essay, I explore how and why different actors use violent stories to make some violence visible even as they hide other forms.

VIOLENCE AND WILDNESS

The contrasting yet interwoven images of headhunting and war in Borneo correspond to parallel, mutually constitutive concepts of "wildness" and "violence," particularly as these more general terms are associated with rain forests or jungles. I use the term "violence" primarily to describe the construction of what humans do: it explicitly involves consciousness and choice. "Wildness" is a distinctive modality of violence because at the same time it can be used to label humans, it describes a characteristic of animals or even "Nature" itself. The term "wildness" implies behaviors based on primal instincts, as if they lack the peculiarly human aspect of self-conscious action. When notions of *both* violence and wildness are used to denote and differentiate human actions, it is important to determine who labels certain actions—and certain actors—wild or violent, and why.

From state authorities' disciplining frames of reference, headhunting appears uncontrollable, making it wild by definition.[15] Headhunting appears to be unmediated direct action by individuals or groups that springs from primitive practice rather than logic.[16] Violent actors thus have too much decision-making power, directly challenging the power of the ruler, commander, or another person of authority in an organized hierarchy of violence like an army. Headhunting also derives from a spiritual world associated with peoples living on the fringes of colonial power centers in the region. This world was never really understood by most nineteenth-century European rulers, missionaries, and traders, though many travelers and ethnographers wrote about Dayak "religion." Clearly, that now iconic, still spectacular past and its reappearances in present struggles have remained mysterious to postcolonial Indonesian regimes as well.

According to some of the Dayaks I interviewed, traditional war, like headhunting, is premeditated violence, which entails calling spirits to possess earthly bodies. After possession, enemies appear to the possessed as animals, and the spirit then uses their human bodies to hunt these enemies. Whether everyone who states this believes it is another question altogether. Nevertheless, as a spatial practice, headhunting resembles both hunting for animals and guerrilla warfare; that is, its form involves surprise, stealth, agility, and attacks by individuals or small groups hidden by the cloak of the forest. The locus of command and control is intentionally dispersed.

War is more understandable to states: it is organized violence. Attempting to maintain a monopoly on legitimate violence defines an important part of what states do.[17] War is premeditated, directed and carried out by government-trained professionals. War involves systematic human choices to kill certain people (combatants) and damage certain property (strategic sites). An army is like a violent machine. Conventional war has plans, strategies, and tactics meant to pressure the enemy. It involves moving people across the landscape in ways quite different than would traditional head-hunting. War zones have command posts at strategic points in the field, central places where on-the-spot decisions can be made or where communications with superior officers can be maintained.[18] Nevertheless, the contrast between war and headhunting isn't absolute: modern forms of guerrilla warfare combine the rationality of strategy and graduated force with headhunting-like forms such as stealth, dispersion, and surprise.

Three key historical periods illustrate the connections and contradictions between headhunting and war, wildness and violence, in West Kalimantan. First, in the middle and late nineteenth century, The Borneo Headhunter emerged in the West as a stereotypical image of the Oriental savage, influenced in its specific iconic forms by early travelers' notions of the Wild Man of Borneo.[19] Second, during the long decade between 1963 and 1975, the aftermath of a "low-intensity conflict" heated up the Borneo border between Malaysia and Indonesia.[20] Indonesian troops and their allies purposefully used headhunting images to incite people to violence and cover up the military's roles in this Cold War theater. The third period is the late 1990s, when to fight Madurese settlers, Dayaks themselves enacted violent practices associated with The Borneo Headhunter while simultaneously reenacting some of the guerrilla warfare movements used in the 1960s. Through their actions, and by using images from both these periods, West Kalimantan Dayaks and others have reconstructed a significant aspect of their public image as a timeless "wild tribe."[21]

These stories of violence and wildness—of organized guerrilla warfare and headhunters—are not simply accidental products of complex times. Rather, stories, labels, and images are plucked from the past to become powerful tools in the present. As scholar Anna Tsing has shown, for marginalized "indigenous" peoples, "going tribal" or "becoming a tribal elder" can be a green development fantasy as well as a strategy to use identity for (re)claiming lost resources.[22] By the same measure, "going wild" and "becoming a headhunter" would amount to strategically mobilizing a stereotypical Dayak identity to accomplish the goals of war. Put differently, equating Dayak identity with wildness may constitute the dark side of such a green Bornean dream.

The Dayak system . . . recognizes the efficacy of human sacrifices, a belief which, under various shapes and disguises, has tended more than any other cause to check the growth of population, to perpetuate barbarism, and to keep alive all those odious passions by which savages are distinguished from civilized men.—SPENCER ST. JOHN, *The Wild Tribes*

The Borneo Headhunter as a trope—a generic image or metaphor that comes to stand for a whole range of assumed characteristics, practices, and beliefs—is a product of the nineteenth century. This is not to say that head-hunting as a practice was "invented" during that time, though some claim it increased dramatically with the growing involvement of Europeans in the region.[23] Rather, the image and the things that it stood for took on a static, generalized, and highly stereotyped form.

For Victorian social theorists, savages were the nadir in a social evolutionary hierarchy that led to the pinnacle of European civilization.[24] Spencer St. John, while traveling through northern and western Borneo in the late eighteenth century for the British government, specifically refers to the "barbarism" (read: wildness) of "Dayaks" or "the Dayak system," as illustrated by the epigraph quote.[25] For St. John, Dayaks were characterized by their "odious passions" (read: uncontrollable instincts), which resulted in their failure to increase the population and improve the land—to transform it, in other words, from what St. John saw as a wild, allegedly unproductive state.[26]

Headhunting and even more so cannibalism created haunting images of wildness, as illustrated by descriptions such as the following, taken from a book by Charles Hose, a colonial officer in Sarawak's government and a famous chronicler of the Dayaks of the late nineteenth century—people he professed to love and greatly respect.

When the attacking party has quietly surrounded the house or houses, the bundles of shavings are ignited, and their bearers run in and throw them under the house among the timbers on which it is supported. Then ensues a scene of wild confusion. The calm stillness of the tropical dawn is broken by the deep war-chorus of the attacking party, by the shouts and screams of the people of the house suddenly roused from sleep, by the cries and squeals of the frightened animals beneath the house, and the beating of the alarm signal on the *tawak*. If the house is ignited, the encircling assailants strive to intercept the fleeing inhabitants. These, if the flames do not drive them out before they have time to take any concerted measures, will hurl their javelins and discharge their firearms (if they have any) at their assailants; then they will descend, bringing the women and children with them, and make a desperate attempt to cut their way through and escape to the jungle or, sometimes, to their boats. . . . It usually happens that the greater part of the fugitives escape into the jungle; and they are not pursued far, if the victors have secured a few heads and a few prisoners. The head is hacked off at once from the body of any one of the foes who falls in the fight; the trunk is left lying where it fell.[27]

Dayak headhunters rarely took—or sought—European heads, though there were an unfortunate few whose murders became legendary.[28] Interior groups generally directed their attacks against other Dayak (and sometimes "Malay") groups. Moreover, the reasons for and frequency of engaging in headhunting and traditional warfare varied widely across the island.[29]

This focused violence among other so-called wild tribes of Borneo did not allow Europeans to sleep more soundly. Hugh Low, secretary to James Brooke, the first "White Rajah" of Sarawak, spent thirty months in Sarawak and western Borneo in the 1840s. In his famous memoir, the first to treat in detail the lives and practices of the interior peoples of Borneo, Low claimed that headhunting was *the* specific characteristic binding the different aboriginal peoples and separating them from other Bornean groups such as Malays. In his words, "independent of their similarity of language and many customs, the great and distinguishing feature of their character—the barbarous custom of taking and prizing as objects of pride and triumph the heads of their enemies—is equally common to all the ramifications of [Dayak] tribes."[30] Thus, the generic Borneo Headhunter was born.

The suppression of headhunting as the epitome of lawlessness and irrational violence became a primary objective of colonial rule, just as law and order were obligatory everywhere in this part of the colonial world.[31] While the sheer terror invoked by the practice may have contributed to the formulation of this goal, more important for colonial policymakers were its dampening effects on trade.[32] As technology improved, and oil and coal were discovered (earlier, antimony and gold discoveries had been more important), colonial rulers hoped to extract profit from the interior districts and trade easily along the coasts. But the threat of headhunting constrained them. What European would want to engage in trade or extraction in wild places where so-called wild people might fall on them or their laborers? Colonial administrators also felt that "head-hunting is . . . the most difficult feature in the relationship of the subject races to their white masters, and the most delicate problem which civilization has to solve in the future administration of the as-yet independent tribes of the interior of Borneo."[33]

Victorian era Europeans, however, were fascinated by the savages of Borneo, devouring accounts of the exotic island.[34] European writings came in three flavors, and some combined them all. One was the adventurous traveler's account, which generally included somewhat fictionalized anecdotes of the author's personal experiences and imaginings. A second was the memoir a colonial official—such as Lowe—or his wife would write after a longtime posting in Borneo.[35] A third was that which anthropologists and geographers today squeamishly recognize as ancestral texts: the science-in-the-service-of-the-colonial-state sort of ethnographic survey or natural history. These studies generally involved the "scientific" cataloging of plants, animals, and peoples, and included enticing tidbits or long descriptive chapters about tribal practices, customs, languages, and dress. Pictures of skulls and data on head measurements were generally thrown in for good measure (figure 1). Whatever the genre, in European accounts, "wild" Dayaks were always deeply embedded in the "lush" forest or the "jungle." It's hard to say whether overtly literary or ostensibly ethnographic and political accounts contributed more to the resulting generalized association.

Early ethnographers and travelers also collected skulls along with other sorts of rain forest artifacts.[36] As "trophies" of their travels in Borneo, and for exhibits in the museums of Europe, skulls came to symbolize the primitive passions of, and for, wild people (figure 2).[37]

The Borneo Headhunter was also born of the slippages between *physical* evolutionary links connecting apes and humans. What better place to imagine these associations than the Borneo home of the "man of the forest," the orangutan? The Malay word *orangutan* literally translates as "person of the forest," and early European explorers translated person into "man."

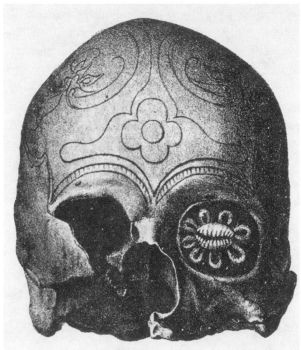

▲ 1. Dyak mode of
drying heads. From
Henry Ling Roth,
*The Natives of
Sarawak and British
North Borneo* (1896;
reprint, Kuala
Lumpur: University
of Malaya Press,
1980), 146.

◄ 2. Dyak skull in
Stockholm Museum,
front view. From
Henry Ling Roth,
*The Natives of
Sarawak and British
North Borneo* (1896;
reprint, Kuala
Lumpur: University
of Malaya Press,
1980), 152.

The legend of the Wild Man of Borneo developed from this confusion.[38] The Wild Man of Borneo is probably one of the earliest, most persistent, and most ambivalent links between wildness and violence in Borneo forests, and as such, has much to do with perceptions of headhunters and headhunting. Yet the average American, scholar or not, would be generally hard-pressed to say whether the phrase refers to man or beast.[39] A few explorations on the Internet—the first source for many a researcher or journalist today—convinced me it had become both.[40]

The confusion started early. Scholar Victor King traces the Wild Man legacy to an early-eighteenth-century visitor to Borneo, Captain Daniel Beeckman, who reported in 1718 that "the Natives do really believe that these were formerly Men, but Metamorphosed into Beasts for their Blasphemy."[41] He published a picture that looked more like a man than an ape, in that it was devoid of body hair, but it nonetheless had strikingly animalistic features (figure 3). Another account, an adventure story written by another sea captain, continued the notion that the orangutan was half man and half animal, though his descriptions of the red-haired ape brought it closer to Greek images of satyrs than to men.[42]

Low's memoir of his service to James Brooke also conflates the European-style evolutionary genealogies of links between animals and humans and Dayak genealogies linking the spirit world to the human world. The conflation of these two sets of worldviews takes place in Low's explanation of native violence. Talking of "Dayak spirits of the woods and mountains," he differentiates the appearance of the "Triu," who allegedly resemble the Dayaks in person, from Kamang, "who are as disgustingly ugly as they are barbarous and cruel in their dispositions: their bodies are covered, like those of the orangutan, with long and shaggy red hair: they are misshapen and contorted, and their favorite food is the blood of the human race."[43] When Low inquires about these spirits at various ritual invocations he attends, the participants tell him that these Wild Man–like spirits "[take] joy . . . in the misery of mankind, and . . . delight in war and bloodshed, and all the other afflictions of the human race. They mix personally in the battles of their votaries . . . that the carnage may be increased, for they are said to inspire desperate valour."[44] The Wild Man metaphor thus gets inserted into European imaginings of violent situations by conflating Dayak views of the spirit world with European anxieties about human evolution.

3. Beeckman's Wild Man of Borneo. From Victor T. King, *The Peoples of Borneo* (London: Blackwell, 1993), 12. First published in Daniel Beeckman, *A Voyage to and from Borneo in the Southeastern Indies* (London: T. Warner and J. Batley, 1718), 36.

The more widely read book today of this genre written in the mid-nineteenth century and published in 1869, is Alfred Russel Wallace, *The Malay Archipelago*. Tellingly, its subtitle is *The Land of the Orang-utan and The Bird of Paradise, A Narrative of Travel with Studies of Man and Nature*. One of the first and subsequently most famous illustrations shows an orangutan biting the arm and taking the spear of a nearly naked man in the jungle, while the pair is surrounded by other nearly naked, spear-carrying men (figure 4). The caption to this famous picture, "Orang-utan attacked by Dayaks," continues to confuse the issue of which is the "man of the forest" and which is the animal.[45]

Other graphic images and drawings of the time, and eventually photographs, contributed to the common associations of headhunters with Borneo forests. Some twelve years after Wallace's book, Carl Bock traveled

4. Orang-utan attacked by Dayaks. From Alfred Russel Wallace, *The Malay Archipelago* (1869; reprint, Singapore: Graham Brash, 1989), 31.

through southeastern Borneo and published a wildly successful and sensationalist account. *The head-hunters of Borneo* hit the streets of London and Amsterdam in 1881.[46] The book was accompanied by an "atlas," a series of Bock's own illustrations of scenes he encountered and people he met. As author and artist, Bock had the power to write his captions, including, "Wild People at Home" (figure 5), "The Chief of the Cannibals," "Tring Dyak's War Dance," "A Long Wahou Warrior," "A Chief of the Forest People," and "On a Headhunting Tour."

In their texts, Bock and other writers consistently label Dayak men, particularly young men, "warriors," or "cannibals," "headhunters," often stating outright that these terms describe the individuals' character.[47] Bock's exaggerations reinforced myths about the Wild Man of Borneo, evolutionary questions about the orangutan, and of course, the trope of The Borneo

5. Bock's drawing of "Wild People at Home." From Carl Bock, *The head-hunters of Borneo: A narrative of travel up the Mahakkam and down in Barito* (1881; reprint, Singapore: Oxford University Press, 1985).

Headhunter.[48] Foreshadowing the intentions of journalists today who are intent on finding "evidence of headhunting," Bock was fascinated also by reports of the purported "missing link," a race of people with tails.[49]

Colonial pacification efforts included colonial rulers in Sarawak using some headhunting groups against others, while the Dutch used colonial military force in their loosely held western Borneo territories. At this time, a few sympathetic Europeans tried to undo the stereotyped images of Borneo peoples as "wild" and "savage." Charles Hose, mentioned above, published several accounts of interior life. Though Hose's stated intent was to "attack the predominant idea that these people were contemptible savages," he included detailed chapters on war making and the critical role of war in agriculture and everyday life.[50] Nevertheless, by the mid-twentieth century, the image of The Borneo Headhunter had undergone subtle changes, paralleling shifts in colonial power in Borneo. The image evolved from that of a fierce, wild, bloodthirsty, semihuman being to a tamable, pacified, and ever-more-romantic figure.[51]

Europeans imagined the forests where headhunters lived as being as wild as their residents. Yet unlike images today, they considered Borneo's forest most impenetrable where the island was most heavily inhabited. In those places, mature (or in Wallace's words, "virgin") forest had been cut for agriculture and secondary growth dominated the landscape. Even where dominated by large trees, the vines and other obstructing vegetation in these inhabited areas made movement through the forest difficult even for adventurous travelers and naturalists.

Where to Europeans these forests were wild and impenetrable, they were highly social landscapes to the local population. These forests were sites of remembered events, places they had long manipulated, if not managed, which contained nodes of connection with the spirit world. Both long-term manipulations of these environments and the doggedness of great warrior ancestors' spirits kept Dayaks in the forest, close to those who aided them in headhunting and war (and agriculture, social relations, and the rest of life in general). This social forest, like the forms of violence used by the people who lived there, was frequently misunderstood by Europeans awed by its complexity and diversity. If knowledge is power, then, contemporary Dayak knowledge of these forests *localized* their control of the rain forest in this period. As respect and recognition of this local knowledge gave way to a preeminence of colonial-style scientific knowledge of the "rain forest," local peoples' power to control its representation slowly waned. The subtleties and nuances also disappeared from The Borneo Headhunter's public and popular appearances, especially after independent Indonesian and Malaysian nation-states replaced the Dutch and British colonial regimes of Borneo.

FAST FORWARD TO THE BORNEO JUNGLES OF THE 1960S

No leaf stirred, although leaves and the stems that bore them were the whole environment. . . . A hornbill shrieked indeed, but from a safe distance. Even further away, a family of long-armed gibbons, high in the trees, hooted with wild intensity and volume enough to echo eerily from the mountain behind, which marked the border with Sarawak in Malaysia. Thus do gibbons proclaim their territory and menace intruders; but when man plays the territory game, he does not hoot, he shoots, and since Lillico and his men were purposely intruding into Indonesia, they were very, very quiet.—PETER DICKENS, *SAS: The Jungle Frontier*

In forest war, the law of the jungle prevails.—Comment of a village head about the 1960s

Just about a hundred years after Wallace journeyed to Borneo, a neocolonial cold war drama began in these forests. Young nation-states—Malaysia and Indonesia—and their international allies engaged in military-led guerrilla warfare along and across their shared and heavily forested international boundary on Borneo. On the Indonesian side, military operations that started out in 1963–66 as an undeclared war by Indonesia's first president, Sukarno, against the Federation of Malaysia, were turned inward between 1966 and 1974, when his successor, Suharto, launched a counterinsurgency operation that depended heavily on psychological warfare tactics deployed within West Kalimantan borders. At the same time, Suharto embarked on an aggressive national development program that benefited greatly from the revenues of Kalimantan's rich resources.[52] The intent of these two parallel programs was to establish the central Indonesian state's control of Kalimantan forests and people.[53]

For the people living in border districts of West Kalimantan, militarization charged every day with terror. Thousands of Indonesian soldiers were posted there, many of them from the Indonesians' "Special Forces"—elite troops trained on Java. They built barracks and bases in villages and towns considered part of the border area.

In practice, "the border" encompassed a far larger geographical area than that immediately surrounding the boundary line itself.[54] One map shows how the military viewed the western Kalimantan international border area extending at least as far south as the town of Singkawang and as far east (interior) as Sanggau (map 1). No roads through the forests crossed the formal boundaries; instead, there were only slippery footpaths and rivers inhabited by snakes, crocodiles, and mosquitoes. Jungle camps and helicopter landing pads—and of course some isolated villages—dotted the interior terrain. Soldiers traveled on foot, by small canoe (*prau*), or helicopter.

War defined this landscape. As mentioned above, between 1963 and 1966, Indonesian president Sukarno started the military buildup to resist the for-

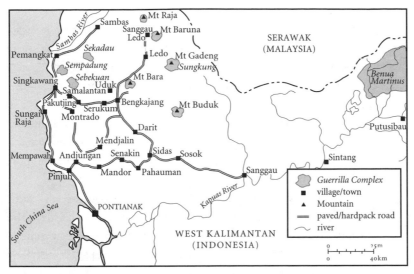

Map 1. Overlapping sites of ethnic violence, 1967–68 and 1997–98, western districts of West Kalimantan, Indonesia. Map by cartographer Darin Jensen.

mation of the Federation of Malaysia—a conflict known as Confrontation. West Kalimantan was a critical site because of its significant Chinese population and its extensive border with Sarawak, one of the former British Borneo colonies that had become part of the Federation of Malaysia. Violently opposing formation of this new nation-state, Sukarno's troops allied with and trained members of the Sarawak People's Guerrilla Force (known generally as PGRS) in guerrilla warfare tactics.[55] The military set up guerrilla training camps in West Kalimantan's border regions.

Suharto had different goals and different allies. One of his first acts on becoming Indonesia's second president in 1966 was to declare an end to Confrontation. He also outlawed the Communist Party (CP) of Indonesia, then the largest CP (legal or illegal) outside the Eastern Bloc.[56] Suharto kept military forces in West Kalimantan, both to help clear it of the communists, who had been his predecessor's allies, and eventually to enforce implementation of his new economic programs. He thus directed military force "inward," into the province of West Kalimantan. As a result, many people who had lived there for generations were now violently redefined as soldiers.

Counterinsurgency operations served to establish a strong federal "Indonesian" presence in West Kalimantan. Turning a Maoist notion on its head, one military officer described the general strategy as "draining the sea in which the communist fish swims."[57] To a large extent, communists were conflated with Chinese, who made up 11 percent of the population.[58] Chinese guerrillas could be routed out of the forest by cutting off their supplies, previously provided by Chinese families and shopkeepers living throughout the western rural districts of the province. This Chinese population was unusual in Indonesia, because so many were rurally based, unlike the more urban Chinese population elsewhere.

The way some local people tell it, however, the army consisted of "wild" outsiders. Most troops came from other Indonesian islands such as Java and Sumatra, and as such, could not easily distinguish "Chinese" from some of the other groups in West Kalimantan. Not only had Dayaks and Chinese long intermarried in the region, ethnic boundaries were blurred in other ways too: many Hakka-speaking Chinese were small farmers and poor laborers (unlike many Chinese in Java and stereotypes of them and elsewhere).[59] As they do today, many Dayaks and Chinese spoke multiple local languages fluently. Furthermore, while some Chinese and Dayak farmers consumed some different foods and grew different crops, these distinctions required close observation, and therefore more time than many soldiers had.

One strategy intended to pull apart the two groups was the purposeful creation of an atmosphere of terror. The military constantly threatened violence. Torture was not uncommon. Villagers I interviewed in the 1990s told about soldiers hanging friends and acquaintances upside down and poking them with sabers or rifle butts until the names of supposed communists or at least of people they could label Chinese spilled out of their mouths. Soldiers also held informants' heads under water until they spit out names, and put captives, male and female, in cages alongside village paths to make them into public examples.

Dayak trackers leading Indonesian soldiers faced different kinds of danger: the elite troops made forest trackers walk in front, where they were at the highest risk of being shot, either by oncoming enemies or the Indonesian soldiers from behind. Villagers recalled how groups of soldiers and trackers might return from jungle patrols with a headless body dressed in a soldier's uniform. Although the soldiers would report the loss to the military command as one of their own, allegedly killed by Chinese guerrillas, the trackers knew when one of their friends did not return. They also knew that sometimes soldiers decapitated the Dayak trackers in what could be seen as a twisted modern form of headhunting: the army was "wildly" raiding the ranks of their allies instead of those of the enemy. Local Dayaks had little choice but to join a tracking party when asked to "volunteer." To refuse meant acquiring a communist label. Once that affiliation was illegal, it meant instant execution or public torture.[60]

Using other psychological warfare techniques, Indonesian soldiers spread racially tinged rumors and choreographed racially charged crimes to make Dayaks believe the Chinese were turning on them.[61] Here, the image of The Borneo Headhunter served them well. These techniques were meant to create believable stories that appeared to emerge "logically" from the local cultural milieu.[62] Interviewing people some thirty years later, I heard several headhunting sorts of stories that were used as part of this terror campaign.[63] One story, for example, had all the elements of trickery, stealth, and savagery:

A Chinese who employed a Dayak husband and wife sent the wife out to collect firewood one day when the husband was slightly ill. When she came back, he gave her some meat to cook, and asked her to prepare a meal. It turned out later that the meat was the flesh of her husband. The Chinese had killed and chopped him into little pieces while she was gone. He even gave it to her for her dinner and she ate it.

Such tales had to be backed up by other strategic stories and experiences. Several villagers remembered Indonesian soldiers killing Dayak trackers and framing Chinese as the killers, thereby mobilizing them to "declare war" on the Chinese. As shown by the following story a Dayak friend told me years later, the Indonesian army was capable of crude ethnography—it had learned that sustained acts of violence by an individual or a "group" could incite Dayaks to blood revenge.[64]

A Dayak was killed one day and was found on the front steps of a Chinese shop. Dayaks from his village, on hearing of this, came down from their hillside farms en masse and demanded retribution. [But, the narrator asked, who could have killed him? And would the killer have left his body on the front steps of his shop?] So [he continued], even though he was not guilty, the Chinese owner paid the customary fines, fearing retribution from the Dayaks. Once the fines were paid, relations returned to normal, as is the custom. The army killed another Dayak and laid his body on the steps of the same Chinese shop. Because it now appeared that the Chinese owner had not learned from his first mistake, the Dayaks were no longer willing to forgive him, and the red bowl was eventually passed, calling all Dayaks to make war on the Chinese.

The above story was told in response to my question about what could have caused so many people to turn against their neighbors of hundreds of years. Why believe these soldiers from Java and elsewhere? Moreover, the economic and political differences among the so-called Chinese themselves had to be downplayed in order to create them as a single category of "the common enemy."[65] Thus Chinese youths hiding in the forest—whether they had in fact been guerrilla sympathizers or were just running scared—had to be connected to entrepreneurs and long-resident farmers. These connections were complicated because of mixed allegiances and identities: some of those "guerrillas" were the products of mixed marriages, some self-identified as overseas Chinese nationals, and some were Indonesian citizens.[66] A masterstroke of propaganda, then, was to convincingly spread the story that all the Chinese wanted to establish a province of China in Borneo. Violence and violent imagery helped convince people of this. The Borneo Headhunter acquired a newly vital role.

The son of a former Dayak subdistrict officer posted at the border (in Seluas) during the post-Confrontation period recounted the following story based on his own experience in 1967:

We [my father and siblings] were traveling somewhere by army truck when we were ambushed by someone; the vehicle was shot at. [He shows me the scar of an old wound on his leg—a two-inch-long reddish strip where a bullet had once lodged.] We were injured, but no one was killed. We were told it was "Chinese guerrillas" and the army took us to a "safe house," where we stayed for a month. During that time, word passed that we had been killed. People became very upset because my father was very beloved.

Who actually shot the vehicle and who spread the rumors that the Chinese were responsible for the "killing" remained unclear. But the story of the incident had a major effect on the mood of the Dayak population.

On 18 October 1967, the former governor, a Kayan Dayak of West Kalimantan, sent out the red bowl. Although he had lost his governor's seat the year before, he was still a powerful local leader. As a Kayan, passing the red bowl amounted to borrowing the ritual practice that signaled a call to war for Salako and Kenayatn Dayaks in these western districts; the red bowl was not part of his Kayan heritage. This bowl, a small Chinese rice bowl containing chicken and dogs' blood, a feather, and a piece of flammable tree resin, was carried from village to village across this part of the province and beyond.[67] Every local Dayak recognized it as the call to war, and its far-reaching path indicated that this was indeed a serious struggle.

What happened next became known subsequently as the Demonstrasi Cina (Chinese Demonstration).[68] Within three months of the fateful night when the red bowl circulated, Dayaks all over West Kalimantan had rousted more than 58,000 rural Chinese from their homes. The Chinese fled or were evicted to the cities.[69] Many lived in resettlement camps for months. Although those who left quietly by road in the first wave were often taken to safety, those who ran into the forest, refused to go, or had to be forcibly evicted were more likely to be slain.

Newspapers described heads taken from the bodies and left strewn along the few roads, the rivers, or the forest.[70] But it remains unclear whether Dayaks cut or took the heads, or whether they saved these cut heads and treated them ritually. No Dayak I have spoken with ever called these events ngayau.[71] Rather, the event is called Demonstrasi, using a borrowed English term with great political significance worldwide in the late 1960s. The term also gives a sort of performative quality to the whole spectacle of the violent evictions and their material consequences: a racialization of the Borneo landscape accomplished by removing the Chinese from rural areas.

Why did both the Dayak governor and Dayak public go along with the army's campaign against the Chinese? Shaken by the military buildup and actual fighting, the populace was further terrorized by stories depicting violence against Dayak bodies. The long existing fear of the European or Chinese headhunter in local lore turned these Dayaks into a generalized group of victims and laid the groundwork for "revenge" killings of Chinese.

The so-called law of the jungle cited in the epigraph to this section was to kill or be killed, to do or be done to—a familiar notion in colonial evolutionary thinking, policy, and practice.

Although rural dwellers, Chinese had not been inhabitants of the forest per se. Military propaganda transformed all of them into the guerrillas that the government needed to rout out of the jungle. Importantly, the Indonesian and international publics were reintroduced to The Borneo Headhunter. This trope was deployed to both stimulate and explain the conflict. By inserting this time-honored image into a so-called wild forest or jungle landscape occupied by wild forest people, the military could conceal its own role in the bloodshed.

Government intelligence reports from this period provide clear evidence that violence was choreographed specifically using familiar, culturally potent images.[72] While military strategists conceived of these tactics and local actors followed through wittingly or unwittingly, journalists played a role in "legitimizing" stories via, for one, a series of articles in the national newspaper, *Kompas*, by a journalist named Widodo, who disguised himself as a soldier in order to go to West Kalimantan.[73] In his ensuing reports from the field, Widodo re-created the generic Borneo Headhunter for his Indonesian audience, going so far as to transform this figure into a cannibal. In one 1967 *Kompas* story, "The Background of the Dayaks' World View," Widodo described a conversation he allegedly had with a "typical Dayak" in Bengkayang, West Kalimantan. He reported asking, "Have you ever drunk blood?" "How does it taste?" and "What about a human liver, how does that taste?" His Dayak subject reportedly answered these questions—the only ones actually asked or reported about his "worldview"—in grisly terms.[74]

All this was for the benefit of Widodo's audience on Java, for most of whom Dayaks were exotic, forest-dwelling primitives. The article, which contained little more than this interview and some comments on various rituals, implied that the "Dayaks' World View" was all about blood, and that across Borneo all Dayak customs were the same. The accompanying photograph suggested that Dayaks had barely emerged from their wild past.[75]

So, why was Widodo "allowed" to don a soldier's garb and observe this intensely violent scene? One reason may have been to disseminate this "explanatory" image of the fierce Dayak headhunter-cannibals—a tactic that strategically accompanied the torture and military-inspired terror destined to continue another seven years. The image therefore served an immediate strategic purpose in the army's efforts against the Chinese and the internal communist "threat." Whether or not the interview was accurate didn't really matter because the trope of the wild Borneo Headhunter already existed; it could easily be resuscitated and adapted to the context of national emergency in a new guise. The spectacle of mass violence also took on the mythic character of Borneo's romanticized jungle histories, conversely hiding—from those who did not really want to see—the even more horrific realities.

Thus, the reinforcement of the Borneo mystique of wild people (headhunters) and places (jungles) took several forms on the Indonesian side of the border in the 1960s and 1970s. In an unusually innovative twist, the government and military used the notion of wildness both to demonize Chinese as communists and to mobilize Dayaks against them during this period. Some Dayaks took violent action against Chinese in part because of their fear of the wildness of the Indonesian soldiers. Moreover, the military used elements of Dayak history and culture to reconstruct the political violence as the untamed acts of wild, primitive people.

In the end, the wildness would be turned against the Dayak actors, even as Suharto's New Order government celebrated them as "indigenous Indonesians (*pribumi* or "sons of the soil"). Because of their alleged primitivity and wildness, Dayaks needed "help" from their "older brothers" in Java.[76] Hence the historic irony of the picture used to illustrate Widodo's article: the disguised soldier's arm thrown casually around the shoulder of the scantily clad "tribesman." Although the gesture seemed intended to imply "we are friends," the parties in question were by no means equals.

The Dayaks' political and cultural marginalization facilitated their exclusion from many of the rain forest riches that Suharto would soon appropriate. Local Dayak forests were converted into national (and individual) treasure chests, not sources of funds for regional development. In 1967—the same year of Demonstrasi—the Indonesian government passed the Foreign Investment Act and the Forestry Act No. 5, enabling new foreign companies to set up extractive timber operations in Kalimantan's forest. The forest was divided up into hundreds of (often overlapping) timber concessions, many of which were allocated to the military.[77] Informal institutional links between the military and the forest reinforced these formal arrangements.

All foreign investment had to be joint ventures and most of the local partners were generals.[78] Lower-ranking military officers and even soldiers were heavily involved in extracting timber and non-timber products from these forests as well.

Forest wars, the law of the jungle, and The Borneo Headhunter: it is sadly fitting that in modernizing Borneo, Suharto's tactics would prove to be the wildest of all. While professing the Indonesian motto of "unity and diversity," Sukarno's successor used cultural politics to divide and conquer, employing the bioscript of a chaotic jungle to turn the supreme violence of the state against the ostensibly wild forest population.

CONTEMPORARY ITERATIONS OF THE "ANCIENT" BORNEO HEADHUNTER ICON

Both in its physical form and the ways the landscape was imagined, the Suharto regime eventually succeeded in reorganizing Borneo as a political terrain. Once a peopled "jungle" associated with terror and wildness, the newly subjugated "rain forest" could be described as a new territorial domain for industry and science. A suddenly powerful forestry establishment facilitated the entry of territorial institutions for timber extraction, plantation development, and conservation. These new rain forests were mapped, zoned, allocated, and industrialized or conserved. By mapping few villages and no local forest claims, the very peopled histories of these rain forests were wiped off the government's historical and contemporary record.

Following the lead of previous military operations that had made possible their occupation, foresters constructed local Dayaks as symbols of wild nature. They called them (and their counterparts on other islands) wild, shifting cultivators.[79] In part by blurring the lines between formative national politics and local violence in the form of headhunting, New Order authorities attempted to justify the taming of this landscape through development and resource extraction.

Development programs since the 1960s led to more rain forest conversion than ever before.[80] The military and central government had built roads into the forest to facilitate counterinsurgency activities as well as increasing surveillance and control of the populace (under the rubric of development); timber and plantation companies built roads for logging and estate crops. Roads replaced rivers as the primary means of transport, thereby changing the way visitors and locals would imagine and experience the forest environment.

Along these roads, migrants from within and outside West Kalimantan built houses and shops, further transforming the outer edges of the forested landscape. Resettlement, or what Indonesians call *transmigrasi*, exploded. Two types of government-sponsored transmigrasi are of importance here. First, the organized resettlement of people from densely populated areas of Indonesia (Java, Madura, Bali, and Lombok) to "less populated" areas such as Kalimantan. Second, a program to reward former soldiers and police for their state service, granting them a few hectares of land in special trans-migration tracts—respectively, called TransAD and TransPol). Both sorts of government-sponsored transmigration have entailed widespread forest con-version for oil palm and rubber plantations where the new settlers work. West Kalimantan's Sambas District has historically accommodated a major share of the settlers.[81]

Even more than forest conversion, these settlement areas constitute sa-lient symbols of state power inserted directly into once forested landscapes that were at one time more locally controlled. At least one TransPol and one TransAD site were located in former "forests" where key PGRS encampments had been located in the vicinity of villages whose histories I compiled. In addition to—and separate from—sponsored resettlement projects, a huge number of "spontaneous migrants" called *transmigran spontan* settled in West Kalimantan. Most of the Madurese who had come to West Kalimantan since the 1960s arrived in this fashion, through their own networks of friends and family.

Although the vastness of the West Kalimantan forest rendered replace-ment a long-term task, significant tracts had been converted by the late 1980s. By 1990, a traveler could get just about anywhere in West Kalimantan by road. By then, as well, small-holder agriculture had accounted for addi-tional clearance of forest for rice fields, pepper gardens, mixed rubber and fruit gardens, and managed forests. The government and international orga-nizations had established conservation areas and nature reserves. The rain forest was a patchwork of different institutional arrangements and often conflicting claims.

Few Chinese dared to return to the interior areas; those who did re-mained close to towns or market centers. By the early 1990s, a mix of Dayak,

Madurese, and some Javanese farmers were cultivating the wet rice fields left behind by Chinese farmers whose ancestors had converted the swamps decades or centuries ago. Upland farming and agroforestry remained largely the purview of Dayak farmers.

As described in the introduction, The Borneo Headhunter ripped into this superficially tranquil picture in 1997. He came under circumstances that had little to do with an earlier cultural politics of traditional headhunting. Even the rituals and symbols of headhunting that reappeared were distorted. For example, the red bowl was passed, but it was not clear who initiated its circulation or even what symbolic work it was meant to accomplish. Thus, a West Kalimantan Dayak who had married and settled in Sarawak, Malaysia, scoffed when I asked him whether the red bowl had signaled him to return to his family's village when the fighting started in 1996–97. "No way," he said, "the red bowl is too slow. My brother sent me a letter—by post."[82] If the red bowl was sending a message, it wasn't working only through some sort of "traditional" symbolism calling disparate Dayak villages and individuals to battle.

Similarly, village shamans summoned the tariu spirits to possess and protect the bodies of strong men and boys. Many of these "warriors," particularly those in the dominant demographic of male youths aged sixteen to thirty-five, now made their livings as drivers, plantation workers, loggers, or commercial farmers; few were forest-dependent hunters. Moreover, the geography of their movements was not reminiscent of headhunting. In other words, small bands of men were not slipping through the forest in the predawn hours to startle and surprise their intended victims. Rather, people involved described hundreds of "troops" (*pasukan*) traveling in trucks and vans and traversing the now-extensive network of roads.[83] *POSKO* (command posts) and *PosDesa* (village posts) were quickly set up to mobilize and organize these troops at key intersections. It is important to recognize that terms such as "pasukan" come from war, not headhunting, and are of Indonesian rather than Dayak (language) origin. Nevertheless, some local Dayak terms for "general" (*pangaladokn*) and a few magical words were used to describe what was transpiring.

The scenes, like the language, were highly militarized, and more reminiscent of war than headhunting. Many of the same sites of the 1960s' Confrontation (Sanggauledo, Bengkayang, Singkawang, Montrado, Sosok, and

Tayan) reemerged as targets (map 2). However, this time the Dayaks attacked the shops and homes of Madurese, not Chinese. More significantly, the attacks had not been incited by the military, at least not unambiguously.[84]

Most Dayaks blamed the new war on a series of rapes and murders committed by Madurese against Dayaks during the previous twenty years or so. Because many of these individual violent crimes (by Madurese against Dayaks) had never been settled according to customary practice, many Dayaks viewed them as repeat offenses, claiming that this indicated a lack of "community remorse" or responsibility.[85] In any case, anxiety about Madurese had long been high. In the aftermath of a stabbing incident at a rock concert in late 1996, these anxieties erupted into massive vigilantism.[86] Dayak mobs demanding some sort of justice spread out from where the two young Dayaks had been stabbed into other districts where Madurese lived in clusters. They burned Madurese houses, killing men, women, and children. The police were unable (or some say, unwilling) to stop them. Some Dayaks killed Madurese and took their heads, cut out their livers, or tasted their blood. Conscious of the effects, many spoke about these actions freely. As one man said, "We did these things so everyone would be really afraid."

Madurese retaliated forcefully, setting up roadblocks, murdering Dayaks pulled from their cars and motorbikes, and burning down Dayak houses in the region's major city, Singkawang. A siegelike atmosphere took over urban neighborhoods and villages. The Madurese ultimately "lost" after several months of fighting. Thousands of Madurese fled to refugee camps in urban areas or left the island.

The images, rituals, and practices of both headhunting and war were in active circulation from the outset of the violence, but as implied above, their audiences and the intentions of their circulators were not clear. From a multiplicity of directions, both actors and commentators resurrected and deployed the spectacle of The Borneo Headhunter in simultaneously new and timeworn ways. One BBC Internet reporter made a typical association for Western readers: "We had decided to travel out of Pontianak to investigate sketchy reports of a violent conflict between the Madurese settlers and indigenous Dayaks, *the original headhunters of the Borneo rainforest*" (my emphasis). The reporter's next sentence, "Much of the rain forest has long been cut down, but the Dayaks continue to live as subsistence farmers, alongside the incomers," implies that Dayaks live in a timeless vacuum, just

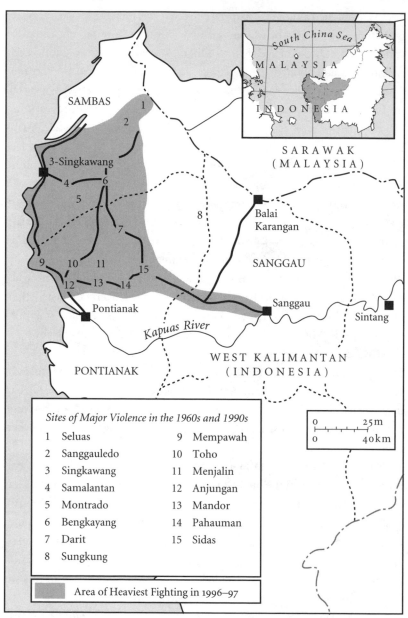

Map 2. Military operations in West Kalimantan border areas. Modified by cartographer Darin Jensen from original in Indonesian newspaper *Kompas*, 13 December 1967.

waiting for timeless, savage urges to stimulate them.[87] The reporter, perhaps unwittingly, used journalistic authority as a way to assert that Dayaks are "only" subsistence farmers (which has not been true for centuries). He rendered them more native, natural, primitive, and wild.[88]

Further on in the BBC report, the author switches to images from the 1960s. "Without any warning, we had stumbled into what looked like a war zone. Soldiers were everywhere, walking alongside the ricefields [sic] or driving east in trucks or on motorbikes. . . . They were all carrying automatic weapons, and all wore the uniform of Indonesia's elite combat regiment, which has been drafted in from other areas of the country to help deal with the fighting."[89] These images of weapons, the Indonesian Special Forces, military vehicles everywhere, and the comment on "a war zone" actually transports us from a forested scene with The Borneo Headhunter into the rice paddies and jungles of a generic Southeast Asian *Apocalypse Now*. Once the war zone was discovered, why did reporters and other observers continue to foreground the headhunting angle? Intentionally or not, quotes such as the following contributed to the public's view of Dayaks as wild "tribes" rather than modern violent actors or angry citizens.[90] Notes CNN, "A man was reportedly decapitated and his head paraded through a village *by screaming tribesmen* at the height of ethnic clashes in Indonesian Borneo that have claimed at least 59 lives."[91] Clearly, such decapitation was one of many charged symbolic acts, a purposeful evocation of The Borneo Headhunter. The ethnic violence had become a caricature of ethnic identities, a modern spectacle that used violent icons as a tactic in a strategic war.

Richard Parry, an English journalist covering the violence, was obsessed with the reemergence of this icon and what to him seemed to be "real live headhunters."[92] In an article reprinted in the *Utne Reader* but first published in *Granta*, he illustrated his story with an image of a young Dayak man. The man seemed to be dressed in some late-nineteenth-century or early-twentieth-century costume. He had a shaved forehead and a long, queue-like ponytail running down his back; he wore a loincloth, and held a homemade rifle; and his long bush knife was tied to a string around his waist (figure 6). Indeed, this was not a new picture: I first discovered an early print of it in a 1930's colonial science journal and later a copy in the library of the Colonial Institute of the Tropics in Amsterdam. Perhaps the real garb of a 1990s Dayak soldier-warrior—jeans, sneakers, and a red bandana—would not have

6. "Dayak Headhunter."
From Richard Parry,
"The Possessed," *Utne
Reader*, October 1998.
© Hulton-Deutsch
Collection/CORBIS.

conveyed an image wild enough for Parry's story. Moreover, a story about plantation labor, land expropriation, and the militarization of everyday life in the "developed" landscape of West Kalimantan would have evoked a very different kind of jungle tropicality than the more wild view Parry intended.

Curiously, the images and words used in Parry's 1996–97 stories echoed the 1960s' images of Dayaks conveyed by Widodo, the journalist disguised as a soldier described above. Parry's search for real "evidence of headhunting" also recalled Carl Bock's fierce search for "evidence of the men with tails" in the nineteenth century, resulting in an unconscious parody of himself as both an intrepid traveler and a field reporter on exotic peoples and places.

Writers far more familiar with Dayak life in the 1990s—and sympathetic with their political and economic disenfranchisement—also helped exoticize the public image of the Dayaks, even though their intent was to create a more

DAJAK VAN LANDAK.

7. Image of a "traditional" Dayak carrying a severed head. From the cover of a 1997 pamphlet by Pastor Yeremias explaining ethnic violence, *Sebuah Permenungan dan Refleksi Kerusuhan Etnis di Kabupaten Pontianak*, special issue of *Batakki* (January–March 1997). Original image comes from an 1854 monograph by P. J. Veth, *Borneo's Westerafdeeling.*

nuanced picture. A Dutch priest living in the heart of the violent districts published an issue of a periodic pamphlet, *A Reflection on the Ethnic Unrest in Pontianak District*, explaining what he claimed were Dayak beliefs and customary practices. On the pamphlet's cover is a line drawing of a man wearing a loincloth, carrying a machete-like bush knife in one hand and a severed head in the other (figure 7). Neither this image nor much of the text could have depicted the priest's actual experience with the long-converted Catholic Dayaks with whom he interacted on a daily basis. Indeed, the image comes from a historical-geographic four-volume account published in Dutch in 1849.[93]

Although the journalists did not analyze the comment, "It isn't head-hunting, it's war," it seemed that people stated it to me everywhere. The difference was important to local people, particularly rural Dayak leaders and urban NGOs. Dayak activists were among those writing to "explain" the histories of these ideas and their connections to the bloody events, trying to justify their relevance on the eve of the twenty-first century.[94] Whatever their motives, however, the effect was to amplify the representation of Dayaks as a unitary and perpetually timeless, wild tribe.[95]

The Borneo Headhunter trope was reproduced from many different symbols of "Dayakness" used throughout the violence: red head cloths, the red leaves of a specific plant, and, of course, the red bowl. Dayak villagers told stories about places and people who were the fiercest, most courageous, and most "primitif." The power implied by their use of the word "primitif" derived from times and places distant from—but still connected to—their own contemporary lives through stories told about their ancestry. Nevertheless, because of the layered histories of the trope, this contemporary icon represented an amalgamation of ideas brought by colonial and postcolonial ethnographers, natural historians, and other storytellers' tales from across this huge island.

As I have argued here, the 1960s' war and its aftermath taught local people that images of wildness can be effective tools in warfare, and the deployment and explanation of headhunting imagery is part of their contemporary repertory of "weapons of the weak."[96] Unfortunately, however, these wild stories live on, after the violence has subsided, haunting the key claimants to the Borneo rain forest. Sadly, they provide the means for preventing Dayaks from developing beyond what they can violently demand.

VIOLENCE, WILDNESS, PEOPLE, AND RAIN FORESTS

I have tried in this essay to examine the ways wildness and violence inform a common view of forests and "forest people" in western Borneo. Glancing at only a few snapshots of this region's history—recent or colonial—reveals the layers of an icon of wildness embedded in the sociological and forested landscape.

How a rain forest landscape and its inhabitants are described, depicted, and analyzed constitutes a critical part of the politics of their control, and a way of allocating or denying access to the valuable resources they contain. Wildness and violence, thus, are complements within a larger scheme of iconic simplification in which both nature and natives are reduced to caricatures for political and economic ends.

Rooted in a particularly human rain forest icon, created and transformed over time, the shifting meanings of The Borneo Headhunter have derived in part from the specific networks within which it has been deployed and the spectacles within which it has appeared. I expect that The Headhunter will continue to emerge in future contexts as a weapon of both the weak and powerful, and continue to have political implications for the forests and people of West Kalimantan.

NOTES

Thanks to Charles Zerner for suggesting the first part of this essay's title. This essay could not have been written without the time and space provided by a generous fellowship from HRI in the winter and spring 2000 quarters, or the support provided by my fellow fellows: Candace Slater, Paul Greenough, Alex Greene, Suzana Sawyer, Scott Fedick, and Charles Zerner. I alone am responsible for any remaining errors of judgment and fact.

1. Robi is a pseudonym.

2. Pastor Yeremias, *Sebuah Permenungan dan Refleksi Kerusuhan Etnis di Kabupaten Pontianak* special issue of *Batakki* (January–March 1997): 10 (quote translated by the author), describes the *tariu* or *triu* as one of the most powerful kinds of spirits in these Dayaks' historical cosmologies. He claims the tariu are called from outside a village by three piercing shrieks issued while decapitating a dog and a chicken, both auburn-colored. "When someone is one with the *tariu* and other *Kamang* spirits, they acquire extraordinary strength, unflappable courage, and can become invisible, turn into a dog, resist bullets, and so on" (ibid.). As mentioned later in the essay, Hugh Low, secretary to James Brooke of Sarawak, who wrote a memoir of his service to the White Rajah, *Sarawak: Its Inhabitants and Productions* (London: Richard Bentley, 1848), described the *triu* and *kamang* of the "Land Dayaks" as "spirits of the woods and mountains" and "the martial genii of these people," who "accompany [the Dayaks] on their expeditions against their enemies" (250).

3. Another bout of violence broke out in West Kalimantan in 1999, this time between Malays and Madurese, which later involved local Dayaks and some Chinese. The 1999 violence, described by many observers as "*yang paling ganas*" ("the most savage of all") is not treated in this essay. And in March 2000, Central Kalimantan Dayaks violently evicted thousands of Madurese for allegedly similar (but not exactly the same) reasons.

4. Human Rights Watch, "Communal Violence in West Kalimantan" (New York: Human Rights Watch, 1997).

5. A paper put out by the Dayak NGO, the Institute for Dayakology Research and Development (IDRD), also claims that headhunting constituted a specific and different practice from war. See IDRD, "The Role of *Adat* in the Dayak and Madurese War" (paper circulated at INFID con-

ference, Berlin, 1997). Conversations with NGO members and my own observations in the villages confirmed my position on this.

6. Low, *Sarawak,* also talked about red kerchiefs adorned with feathers and shells that Dayak men wore during war in the nineteenth century (179).

7. See Janet Hoskins, introduction to *Headhunting and the Social Imagination in Southeast Asia,* ed. Janet Hoskins (Stanford: Stanford University Press, 1996), 10.

8. Various "subgroups" of Dayaks practiced headhunting and traditional war—also called ngayau—but neither the practices nor their significance were homogeneously across Borneo. See, for example, Allen Maxwell, "Headtaking and the Consolidation of Political Power in the Early Brunei State," in Hoskins, *Headhunting,* 90–126; Robert Pringle, *Rajahs and Rebels: The Iban of Sarawak under Brooke Rule, 1841–1941* (Ithaca: Cornell University Press, 1970); Peter Metcalf, "Images of Headhunting," in Hoskins, *Headhunting,* 249–290; Anna Tsing, *In the Realm of the Diamond Queen: Marginality in an Out-of-the-Way Place* (Princeton: Princeton University Press, 1993); Henry Ling Roth, *The Natives of Sarawak and British North Borneo* (1896; reprint of vols. 1 and 2, Kuala Lumpur: University of Malaya Press, 1919), IDRD, "The Role of *Adat.*"

9. See Hoskins, *Headhunting,* esp. chap. 1. My first inkling of important discrepancies in the interpretation and representation of this violence came in the course of conversations I had in the region in May 1997.

10. For a detailed discussion of these, see Nancy Lee Peluso, "Passing the Red Bowl," in *Violent Conflict in Indonesia,* ed. Charles Coppel (London: Curzon Press, forthcoming).

11. Without exception, the people I spoke with expressed an unambiguous conviction that they had been under siege by Madurese during and before the violence discussed here.

12. See, for instance, Pringle, *Rajahs and Rebels;* and Emily Harwell, "The Un-Natural History of Culture: Ethnicity, Tradition, and Territorial Conflicts in West Kalimantan, Indonesia, 1800–1997" (Ph.D. diss., Yale School of Forestry and Environmental Studies, 2001).

13. For example, the famous forest operation against the Japanese during the occupation, in which Sarawak Dayaks were led by Tom Harrison. See Judith M. Heimann, *The Most Offending Soul Alive: Tom Harrison and His Remarkable Life* (Honolulu: University of Hawaii Press, 1997).

14. On the differential work that tropes of jungles and rain forests do, see Candace Slater, "Amazonia as Edenic Narrative," in *Uncommon Ground: Rethinking the Human Place in Nature,* ed. William Cronon (New York: W. W. Norton, 1996).

15. Except when the state mobilizes or authorizes people to headhunt in situations of extreme loss of control. This occurred at several times in recent Borneo history, including under the Brooke regimes in the nineteenth and early twentieth centuries. All three Brooke regimes authorized headhunting at certain times—for the state's purposes—but criminalized all other forms (see Pringle, *Rajahs and Rebels;* and Harwell, "The Un-Natural History of Culture"). And as mentioned in note 13, during the Japanese occupation of Borneo during World War II, British and Australian troops parachuted into Borneo jungles beyond direct Japanese control and authorized Iban Dayaks to hunt Japanese heads.

16. This European "fear" or distrust of the wild derives from ancient Greek and Romans' fears of the "wild people" and "monsters" on the edges of their empires. See the excellent chapter on

this by Merryl Wyn Davies, Ashis Nandy, and Ziauddin Sardar, "Them," in *Barbaric Others: A Manifesto on Western Racism* (London: Pluto Press, 1993).

17. As defined by Max Weber, *Economy and Society*, ed. Guenther Roth and Claus Wittich (Berkeley: University of California Press, 1978). On the disciplinary dimensions of state power, see Michel Foucault, *Discipline and Punish: The Birth of the Prison* (New York: Vintage, 1979).

18. See Nancy Lee Peluso and Michael Watts, introduction to *Violent Environments* (Ithaca: Cornell University Press, 2001), 3–39. Evolutionary theorists associate highly organized—and increasingly mechanized—warfare with "high civilization."

19. For a brief history of the Wild Man's imagery, see Victor King, *The Peoples of Borneo* (London: Blackwell, 1993). The hybridization of men and beasts was a common practice of Greek and Roman geographers: with satyrs, centaurs, minotaurs, and other creatures being "documented" by travelers venturing to the edges of the empires—in what is now India or Sub-Saharan Africa. See Davies, Nandy, and Sardar, "Them."

20. The low-intensity conflict referred to here was called "Confrontation" or "*Konfrontasi.*" The term refers to a conflict between Indonesia and Malaysia over the inclusion of the Borneo states of Sarawak and Sabah in the newly formed Federation of Malaysia. In the aftermath of Confrontation, as discussed below in the text, the Indonesian Communist Party was criminalized and the new Indonesian army under the second president, Suharto, embarked on eight years of counterinsurgency operations to rout out remaining "communists" from their forest hiding places along the Kalimantan-Sarawak border.

21. See Spencer St. John, "The Wild Tribes of the Northwest Coast of Borneo," *Journal of the Malayan Branch of the Royal Asiatic Society* (1862): 232–243, and discussion below.

22. See Anna Tsing, "Becoming a Tribal Elder and Other Green Development Fantasies," in *Transforming the Indonesian Uplands*, ed. Tania Li (Singapore: Harrowed Academic Publishers, 1999), 159–202.

23. See, for example, Pringle, *Rajahs and Rebels*; and Maxwell, "Headtaking."

24. The literature on savages and development is huge. See, for one review, Michel-Rolph Truillot, "Anthropology and the Savage Slot: The Poetics and Politics of Otherness," in *Recapturing Anthropology: Working in the Present*, ed. Richard Fox (Santa Fe: School of American Research, 1991). On wildness, an equally large literature, the classic piece is Hayden White, "The Forms of Wildness: Archaeology of an Idea," in *The Wild Man Within: An Image in Western Thought from the Renaissance to Romanticism*, ed. Edward Dudley and Maximillian E. Novak (Pittsburgh: University of Pittsburgh Press, 1972), 3–38.

25. The word barbarian comes from the Greek *barbaroi*, which means "babbler" and was used in reference to someone who could not speak Greek. See Davies, Nandy, and Sardar, "Them."

26. "Improvement" was the catchphrase of British and Dutch colonialism. See, for instance, John Drayton, *Nature's Government* (New Haven: Yale University Press, 2000).

27. Charles Hose and William MacDougall, *The Pagan Tribes of Borneo* (1912; reprint, London: Cass, 1966), 172–173.

28. See, for example, John Dalton, "On the Present State of Piracy amongst These Islands and the Best Methods of Its Suppression," in *Notices of the Indian Archipelago and Adjunct Countries*, comp. J. H. Moor (Singapore, 1837), 15–29; and King, *Peoples of Borneo*.

29. For example, some groups used violence more routinely than others; some engaged in revenge attacks, but only if they were attacked first; and still others responded to headhunting by fleeing. On Sea Dayaks/Iban, see Benedict Sandin, *The Sea Dayaks of Borneo before White Rajah Rule* (East Lansing: Michigan State University Press, 1967); Vinson Sutlive, *Change and Development in Borneo: Selected Papers from the Conference of the Borneo Research Council* (Kuching, Malaysia: Council, 1990); and Pringle, *Rajahs and Rebels*. On Land Dayaks, see W. R. Geddes, *Nine Dayak Nights: The Story of a Dayak Folk Hero* (Singapore: Oxford University Press, 1957). On Meratus, see Anna Tsing, "Telling Violence in the Meratus Mountains," in Hoskins, *Headhunting*, 184–215. On Berawan, see Metcalf, "Images of Headhunting," in Hoskins, *Headhunting*. On Brunei groups, see Maxwell, "Headtaking." On Selako, see William Schneider, "The Social Organization of the Selaka" (Ph.D. diss., University of North Carolina, 1973).

30. Low, *Sarawak*, 98.

31. In addition, Borneo's value was politically symbolic. Borneo became something of a pawn in the competition for global supremacy between the British and Dutch, despite the fact that the state of nineteenth-century technology for resource extraction and production was not conducive to the kinds of profitable colonial exploitation of land-based resources possible in Java and on the Malay Peninsula. See Graham Irwin, *Nineteenth-Century Borneo: A Study in Diplomatic Rivalry* (Singapore: Donald Moore Books, 1955). Pacification (of headhunting and tribal warfare) ceremonies were held in Sarawak in 1914. In Dutch Borneo in what is now central Kalimantan, they were held in 1894.

32. See Low, *Sarawak*; and Sylvia, *Queen of the Headhunters: The Autobiography of Sylvia, Lady Brooke, the Ranee of Sarawak* (1885; reprint, Singapore: Oxford University Press, 1990).

33. Carl Bock, *The Headhunters of Borneo* (London: S. Low, Marston, Searle, and Rivington, 1882), 215.

34. See King, *Peoples of Borneo*. Frank S. Marryat, an early-nineteenth-century explorer and midshipman on the H.M.S. *Senarang*, in his *Borneo and the Indian Archipelago with drawings of costumes and scenery* (London: Longman, 1848), made many references to the American Indians in his classifications and comments. On this fascination with New World savages, see, for example, Robert F. Berkhofer, *The White Man's Indian: Images of the American Indian, From Columbus to the Present* (New York: Vintage Books, 1979). On wildness in the Peruvian Amazon, see Michael Taussig, *Shamanism, Colonialism, and the Wild Man: A Study in Terror and Healing* (Chicago: University of Chicago Press, 1986).

35. Low, *Sarawak*.

36. Sandra Pannell's excellent article "Travelling" (1992) in the journal *Oceania* compares the headhunting practices of Alfred Russel Wallace (who actually only *measured* heads) and other natural historians such as Henry Forbes to those of Kodinese headhunters of the island of Sumba. See also Hoskins, *Headhunting*, chap. 1. Anna Tsing, in *Diamond Queen*, compares local people's views of government officials as headhunters.

37. The cranial collectionism of these first *European* headhunters of Borneo contributed to long-term *Dayak* terrors as well. Whenever Europeans began the construction of bridges, dams, roads, and other major infrastructure, stories would circulate among Dayaks that their

success required the burial of a head—a human sacrifice—at the site. See, for instance, Hoskins, *Headhunting*; Tsing, "Telling Violence in the Meratus Mountains," and *Diamond Queen*; Metcalf, "Images of Headhunting"; and Maxwell, "Headtaking." All these writers provide stories from different parts of Borneo of local peoples' fears of European or government (Indonesian and Malaysian) headhunters. They draw on a literature that documents similar fears in the Pacific Islands (for one, Pannell, "Travelling"). In my West Kalimantan research and much earlier in East Kalimantan, different versions of this story would surface from time to time.

38. King, *Peoples of Borneo*.

39. A reader of a draft of this essay—a U.S. historian—knew nothing at all about Borneo, including its geographic location, but he *had* heard of "the Wild Man of Borneo."

40. Some of my findings included references to orangutans as "the original Wild Man of Borneo" and a rock group that has claimed the title for itself. See http://www.guitar9.com/wildmanofborneo.html.

41. Daniel Beeckman, *A Voyage to and from Borneo in the Southeastern Indies* (London: T. Warner and J. Batley, 1718); cited in King, *Peoples of Borneo*, 11.

42. Captain Mayne Reid, *The Castaways: A Story of Adventure in the Wilds of Borneo* (1870); cited in King, *Peoples of Borneo*, 11. Compare Davies, Nandy, and Sardar, "Them."

43. Low, *Sarawak*, 250. Triu is now spelled tariu, as in my introduction.

44. Ibid.

45. The separation between "Orang" and "utan" also adds to the confusion in translation. The text tells us what is left out of the picture and the actual origins of the caption: the orangutan had proved too strong for the native hunters ordered by Wallace to capture it. Wallace eventually shot and killed it himself. Alfred Russel Wallace, *The Malay Archipelago* (1869; reprint, Singapore: Graham Brash, 1989), 31.

46. The book was a popular version written after his report to the Dutch governor-general of the Netherlands East Indies, by whom Bock was commissioned. See R. H. W. Reece, introduction to *The head-hunters of Borneo: a narrative of travel up the Mahakkam and down the Barito*, by Carl Bock (1881; reprint, New York: Oxford University Press, 1985).

47. Bock practically cites verbatim Low's comment excerpted above, but gives him no credit for it: "The barbarous practice of Head-Hunting, *as carried on by all the Dyak [sic] tribes*, not only in the independent territories, but also in some parts of the tributary states, is part and parcel of their religious rites" (*The head-hunters of Borneo*; emphasis mine).

48. Bock's book had a huge impact across Europe. It was published in 1881 in English, translated to German the next year, and then translated into his native Norwegian.

49. See also Reece, introduction. King, *Peoples of Borneo* (11), writes that this may be a notion that Bock picked up from a story published in 1875 by T. Skipwith, "Men with Tails," in *Waiting for the Tide on Scraps and Scrawls from Sarawak*, ed. W. M. Crocker (Kuching, Malaysia: Crocker and Chapman, 1875). See my comments later in this essay on 1960s' journalist Y. Widodo and 1990s' journalist Richard Parry. Unlike Bock, Parry spoke no Indonesian or Malay, and spent a short time "researching" his story.

50. Reece, introduction, xiii.

51. Similarly, scientific surveys of the peoples of Borneo, such as the two-volume study by Henry

Ling Roth, contributed to the conflation of all Dayaks as wild headhunters through both an overload of information and an obsession with Dayak forms of violence. Three out of twelve chapters in his second volume are catalogs of violence: "War and Weapons," "Headhunting," and "Peace, Slaves, and Captives" (Roth, *The Natives of Sarawak*).

52. In the 1980s, for example, half of Indonesia's timber concessions were its Borneo (Kalimantan) provinces. See Lesley Potter, "Forest Degradation, Deforestation, and Reforestation in Kalimantan: Towards a Sustainable Land Use?" in *Borneo in Transition: People, Forests, Conservation, and Development*, ed. Nancy Lee Peluso and Christine Padoch (Oxford: Oxford University Press, 1996), 13–40.

53. The local peoples' progressive loss of land and resources and their change in legal status are detailed in Nancy Lee Peluso and Emily Harwell, "Territory, Custom, and the Cultural Politics of Ethnic War in West Kalimantan Indonesia," in Peluso and Watts, *Violent Environments*.

54. I define "boundary" as the *line* demarcated in the field and marked on the map, and "border" as a broader area surrounding that line, following Thomas M. Wilson and Hastings Donnan, eds., *Border Approaches: Anthropological Perspectives on Frontiers* (Lanham, Md.: University Press of America, 1994), 24.

55. PGRS is an acronym for Pasukan Gerilya Rakyat Sarawak.

56. For more on Confrontation, see J. A. C. Mackie, *Konfrontasi: The Indonesia-Malaysia Dispute, 1963–1966* (Kuala Lumpur: Australian Institute of International Affairs, 1974; and Greg Poulgrain, *The Genesis of Konfrontasi: Malaysia Brunei Indonesia, 1945–1965* (London: Crawford House Publishing, 1998).

57. Cited in Herbert Feith, "Dayak Legacy," *Far Eastern Economic Review*, 21, 1968. The phrase originally comes from Mao Tse-tung, promoting the creation of a favorable milieu for communism.

58. Chinese made up only 3 percent of Indonesia's overall population. These figures come from Charles A. Coppel, *Indonesian Chinese in Crisis* (Kuala Lumpur: Oxford University Press, 1983).

59. On different patterns of migration and marriage in the early years of settlement, see ibid. See also Mary Somers-Heidhue, *The Chinese of West Kalimantan* (Ithaca: Cornell University Press, 2003).

60. Only Dayaks served as trackers, stereotyped as they were as forest people. Malays, according to held-over colonial subjectivities, were stereotyped as fishers and bureaucrats; Chinese were, by now, "the Communist enemy" or "the sea."

61. See Michael Klare, *War Without End* (New York: Vintage Books, 1972), 324–355; Audrey R. Kahin and George McT, *Subversion as Foreign Policy* (New York: New Press, 1995); SCUT report (Special Bureau of Chinese Affairs 1967); and Charles Coppel, personal communication with author, July 2000.

62. Klare describes these and other strategies of guerrilla warfare in *War Without End*.

63. The following stories are from author interviews in Sambas, October 1998.

64. A tactic taught by both Mao and Indonesian general Nasution under Sukarno. See Klare, *War Without End*, 39; and H. Nasution, *A Handbook of Guerrilla Warfare* (Jakarta: Semana Masa Djaki, 1970).

65. It is also telling that these stories drew as much on the Javanese soldiers' constructions of Indonesian Chinese based on their experiences with Chinese living outside of West Kalimantan. In Java, most Chinese were shopkeepers or managers of various enterprises; in West Kalimantan, many were subsistence farmers and laborers.

66. See Coppel, *Indonesian Chinese.*

67. Note that in different times and places, different combinations and types of blood and illuminants (that is, stick matches or tree resin) were used or preferred. Illustrating the historical messiness of tradition mixing, today's red bowl is actually a Chinese bowl, and was used in payment of customary fines in the past.

68. In some contemporary reports, it was called the Demonstrasi Dayak. See, for example, Feith, "Dayak Legacy."

69. Many of those forced out, mostly Hakka, were so local that they had never been to the capital city of the province or only spoke dialects of Chinese and local Dayak languages—no Indonesian or even trade Malay, the lingua franca of the area. Although how many were killed or died remains unclear, estimates range around three thousand in the eviction process, more in the refugee camps.

70. See *Kompas,* October 1967–February 1968.

71. Jerome Rousseau, *In Central Borneo,* says the same term, *ngayau,* is used among the Kayan as well to indicate both ritual war and head-hunting. See also Yeremias, *Sebuah Permenungan.*

72. See SCUT report.

73. He tells us in the first of the series of articles that this disguise was the only way he could travel through West Kalimantan at the time.

74. Widodo, "Lata belakang pandangan hidup Suku Daya," *Kompas* 20 December 1967; translated from Indonesian by this author. The Indonesian text (in old spelling) reads as follows: "*Teman saja jang tidak takut-takut bitjara itu mentjeritakan pengalaman2nja waktu ia ikut menggerebek sebuah sarang* PGRS. *"Pernah djuga minum dara?" tanja saja. "O, pernah pak." "Bagaimana rasanja?" tak terkendalikan rasa pingin tahu saja. "Darah manusia jang sudah dituang kedalam mangkuk rasanja tidak enak. Betul-betul tidak enak. Anjir. Tapi darah jang segar, rasanja hangat pak, dan gurih. Tjaranja minum, begitu kepala dipantjung, djangan tunggu darahnja mengembur ketempat lain. Minum darah itu langsung dari gembungnja, segar pak." "Kalau hati manusia, bagaimana rasanja?" saja bertanja terus. "Hati manusia mentah saja belum pernah makan pak. Kalau saja, hati itu saja bakar dulu. Dan kita ambil hatinja. Lalu kita panggang diatas api. Merah pak warnanja. Rasanja gurih dan halus. Njaman pak, lebih njaman dari hati ajam."*

75. The photo shows the reporter with his arms around the shoulders of the Dayak speaker, dressed not in a loincloth but scant clothing. The representation is a far cry from Hose's and others' depictions of headhunting regalia of the nineteenth century, but an eerie premonition of the way the military later claimed a closeness to the Dayaks. See Hose and MacDougall, *Pagan Tribes*; and Roth, *Natives of Sarawak.* A detailed account of the military reports of the 1960s violence, written at the same time as this piece, is Jamie Davidson and Douglas Kammen, "Indonesia's Forgotten War," *Indonesia,* April 2002.

76. Even at late as 1995, I was shocked to hear a prominent Javanese sociologist (Omar Khayyam) giving a lecture at Yale make mention of the importance of "helping our younger brothers" from Kalimantan "come down out of the trees and join the developing nation-state."

77. Prior to this, government companies had done some extracting, as former president Sukarno's policy had discouraged foreign investment. His involvement in Konfrontasi and other military operations generally superseded any immediate concerns with forest management.

78. See Richard Robison, *Indonesia: The Rise of Capital* (North Sydney, Australia: Allen and Unwin, 1986); and Chris Barr, "The Rise of Bob Hasan and Apkindo," *Indonesia*, no. 65 (April 1998): 1–30.

79. *Peladang liar* (literally, "wild farmers").

80. For a few examples of the impacts, see Joan Hardjono, *Resources, Ecology, and Environment in Indonesia* (New York: Oxford University Press, 1991); Potter, "Forest Degradation, Deforestation, and Reforestation"; Harold Brookfield, Lesley Potter, and Yvonne Byron, *In Place of the Forest: Environmental and Socio-Economic Transformation in Borneo and the Eastern Malay Peninsula* (Tokyo: United Nations University Press, 1995); and "Indonesian Environment," World Bank report, 2000.

81. See Jamie Davidson, "State Instigated Violence in the Outer Islands of Indonesia: West Kalimantan," *Journal of Select Asian Studies Papers* (1998).

82. Interviews, Jangkar village, Sarawak, October 1999.

83. But toward the end of this conflict, some men did "hunt" for Madurese who had run into the forests and plantations to hide.

84. Some debate as to whether military or police had planned their noninvolvement has ensued since these events, but no conclusive evidence one way or the other has been produced. See, for example, Human Rights Watch, "Communal Violence"; and Benny Subianto, "Kerusuhan di Pontianak" (manuscript in author's possession).

85. For an extensive analysis of this grievance, see Peluso and Harwell, "Ethnic War."

86. For further development of this argument, see ibid. Some journalists conflated what they interpreted as general anxiety about increasing numbers of transmigrants—of which Javanese are the largest number—with this directed targeting of Madurese migrants.

87. "From Tapol: BBC Stumbles on 'War Zone' in W. Kalimantan," http//:2/13/97.

88. Dayaks all over Kalimantan have been involved in the international trade in forest products—selling goods for cash or trading them for other goods—since at least the third century B.C. See, for example, Kenneth R. Hall, *Maritime Trade and State Development in Early Southeast Asia* (Honolulu: University of Hawaii Press, 1985); and Nancy Lee Peluso, "Markets and Merchants: The Forest Products Trade of East Kalimantan in Historical Perspective" (master's thesis, Cornell University, 1983).

89. "BBC Stumbles on 'War Zone.' "

90. Many ethnographers who use the term "tribe" claim it is a strategic move to empower the "indigenous" in development politics. See the work of Michael Dove; and Anna Tsing, personal communication. See also Tania Li, "Articulating Indigenous Identity in Indonesia: Resource Politics and the Tribal Slot," *Comparative Studies in Society and History*, 42, no. 1 (2000): 149–179.

91. CNN, 1999, http://www.cnn.com/WORLD/asiapcf/9903/19/indonesia.01.

92. Parry spoke no Indonesian and his translator seems to have been a Malay, whose conveyances of nuances would have been filtered through his own fears.

93. Pieter Johannes Veth, *Borneo's Westerafdeeling, Geographisch, Statistisch, Historisch, Voorafgegaan Door Eene Algemeene Schets Des Ganschen Eilands* (Zaltbommel: J. Norman en zoon, 1854–56), 2 vols.

94. Some of these accounts include IDRD, "The Role of *Adat*"; and Yeremias, *Sebuah Peremungan.*

95. On "timelessness" and the study of "primitive headhunters," see the classic piece by Renato Rosaldo, *Ilongot Headhunters, 1883–1974: A Study in Society and History* (Stanford: Stanford University Press, 1980).

96. Emily Harwell also argues this point in her dissertation ("The Un-Natural History of Culture") on Dayaks and Dayak identity in the Danau Meninjau lakes region of West Kalimantan.

CHARLES ZERNER

THE VIRAL FOREST IN MOTION:

EBOLA, AFRICAN FORESTS, AND

EMERGING CARTOGRAPHIES OF

ENVIRONMENTAL DANGER

Some geographic regions of the globe, in Western representations, have been branded with the mark of particular cultural imaginings and meanings.[1] This representational burden, although not fixed and certainly permitting historical variations in contour and emphasis, is also carried by the world's rain forests.[2] In Amazonia, for example, Candace Slater documents a variety of intertwined representations that she calls the Edenic rain forest.[3] This forest is distinguished by its vulnerability, fragility, and radiance, in addition to a sense of biological equilibrium within coherent ecosystem borders.[4]

Representations of African forests, like Africa itself, have also carried a heavy burden of meanings in the Western imagination. Containing in their recesses virulent pathogens, warrior tribes, and lurking predators, they possess a potential for violence, the capacity to explode or implode, and the possibility that their "endemic" virulence would seep across borders and saturate adjacent territories.

This essay explores recent changes in rain forest representations in popular media in the United States, specifically representations of viral biodiversity in African equatorial forests. At the height of rain forest romanticism in the United States, a period that began roughly in the 1970s and lasted through the mid-1990s, a certain kind of rain forest imagery, associated particularly with Amazonia and to a lesser extent Southeast Asia, prevailed. Whether described in popular magazines, the literature of international environmental campaigns, or in a more scientific discourse as "biological diversity," this tropical rain forest was in danger and in need of protection. Like the unicorn encircled by the king's hunters, it was surrounded by threats on all sides. Images of encroachment and conversion, of intrusion, penetration, and invasion, by loggers, settlers, and a multitude of pernicious forms of resource extraction were widespread.

The image of a fragile, bounded, ethereal Amazonian rain forest, of course, was only one of a multitude of representations of rain forests that circulated during the 1970s and 1980s. Among others were images of rain forests as: targets for the modernization project, sites for clearance and development in the form of vast plantations or utopian urban settlements; "nature's pharmacy," sites of salvation for the modern world; and raw resources, sites made for massive extraction of timber as well as precious and strategic minerals.

Within the panoply of overlapping rain forest representations produced by conservation organizations during the mid- and late 1990s, I argue that a particular representation of African forests began to become prominent. Within this representation, the directionality and significance of forest penetrations were reversed: the channels and processes that had been conceptualized by conservationists as penetrations into the heart of the rain forest—roads, timber trucks, agricultural conversion, and development— began to be described as exits from the forest to the world beyond. They became the first connections to the global highway of commerce and spread of tropical diseases. I call this lethal aspect of representations of African equatorial forests "the viral forest."

The picture of the viral forest that is just coming into focus, however, is complicated by representations of globalizing forces and flows of media, commerce, capital, ideas, and technologies.[5] When the African tropical forest is linked to regional and global circuits of movement and exchange by roads, jet transport, sexual congress, the trade in monkeys, blood transfusions and organ transplants, global tourism, and local warfare, it becomes a traveling threat to humankind: the viral forest in motion.[6] Representations of a dangerous, outward-spiraling African viral forest emerging during the late 1990s appear to invert the image of the delicately balanced, benevolent Amazonian rain forest. What sense are we to make of the growing salience of this representation of the African rain forest as a "viral forest," and more particularly, a representation of an outwardly spiraling "viral forest in motion"?[7]

If the image of a lethal virus escaping from an African tropical rain forest and traveling across the globe, across national boundaries and bodily defenses, is the icon at the center of this chapter, the spectacle is the story of how this icon—Ebola virus on the loose—has been, and continues to be, inserted into international networks in global health, national security, movements against genomic experimentation, and regional, national, and transnational networks mobilizing against globalization. How has Ebola been seized, rendered, and rearticulated as a figure conjuring up fears, anxieties, and meanings? How are we to make sense of the appearance, in 2001, of a *National Geographic Magazine* cover story titled "The Green Abyss Megatransect," featuring an account of "choking vegetation, impassable swamps, and rumors that virulent Ebola virus has struck gorillas in Gabon"?[8] How have varied social and environmental movements deployed these figurations of Ebola in information flows and networks ramifying across the globe?

While rain forest representations may reflect realities that already exist as "facts on the ground," they are also cultural interventions in social life and thought, shaping the possibilities of the social imagination, policy, and everyday conduct.[9] My principal concerns here are to investigate the character and implications of the image of a "viral forest in motion" through a detailed examination of Ebola on the move.[10] I ask these questions through a close reading of a popular nonfiction book, *The Hot Zone*, and a Hollywood movie, *Outbreak*, loosely based on it. I am interested primarily in articulating how this representation is constructed. I am also concerned with understanding the range of social and political implications of this environmental representation for U.S. perceptions of civil life within the United States, as well as the potential for shaping U.S. perceptions of and policies toward the tropical developing world, especially Africa.[11] At the end of this chapter, I speculate on possible links between themes elaborated here and larger anxieties about the consequences of globalization in other domains.[12]

Richard Preston's book about the travels and depredations of a rain forest microorganism, *The Hot Zone*, was received with great enthusiasm by a general reading public on its publication in 1994.[13] Beginning with an account of the death of a French engineer in Kenya in 1980, *The Hot Zone* provides accounts of previous Ebola outbreaks in southern Sudan and Zaire. From those scattered villages and towns within or on the periphery of African rain forests, Preston tracks Ebola's movement into the world he knows the best, the West.

At the center of *The Hot Zone* is an account of a 1989 Ebola outbreak among a population of monkeys housed in the Reston Primate Quarantine Unit in Reston, Virginia. Many of the afflicted monkeys, captured in coastal rain forests of the southern Philippines and destined for use as laboratory animals in the United States, were decimated by a variety of the virus known as Ebola Zaire. Ebola Zaire was not fatal to human beings. Although four people associated with the Reston outbreak were infected, Ebola Zaire did not make them ill.

As Preston shapes the story of the Reston incident, the drama revolves around the horrors made possible by globalization and its ramifying networks—the possibilities opened by jet transport along with webs of trade and commerce in exotic primates—to carry these diseases across the world, and penetrate national borders and the bodies of everyday U.S. citizens. The pivotal question for the Reston monkey house drama is: Will Ebola get out into the world? Will it create a fatal epidemic in "a prosperous community just 10 miles west of Washington, D.C.?"[14] Preston devotes almost half of *The Hot Zone* to an account of the Reston incident, focusing on efforts to contain the virus within the confines of the primate unit. His account emphasizes the unconventional military involvement of the United States Army Medical Research Institute of Infectious Diseases (USAMIRID) rather than the civilian-controlled Centers for Disease Control (CDC), which would normally have handled the outbreak.[15]

Within a year, *The Hot Zone* was offered at least four options for production as a film. *Outbreak*, a Hollywood-made bioterror movie about an African virus and the defense establishment, was loosely based on the book. Publication of *The Hot Zone* was also followed by an efflorescence of fictional and nonfictional accounts of frightening microorganisms, and their destructive capacities and travels.[16] The commercial success of *The Hot Zone*, the number of similar nonfictional accounts of viruses and their dangers in a moment of globalization, and the popularity of a website called *Outbreak* all suggest the U.S. public's apparently boundless fascination with the potentials of viral travel, invasion, and infection.[17]

The Hot Zone was packaged as a "revenge of the rain forest" story, explicitly linked to the terror of AIDS and its spread across the planet.[18] It is a given that these lethal viruses are frightening. Several questions piqued my curiosity after seeing Ebola images on full-page *New York Times* advertisements linked to protests concerning the World Trade Organization and the dangers of genetic engineering, as well as printed in red letters on envelopes for a Physicians for Social Responsibility fund-raising letter trying to mobilize constituencies against global warming.[19] How and why has a rain forest virus from Africa become an icon? What range of meanings, fears, and anxieties does Ebola trigger at a new millennium? Because Preston's *Hot Zone* is a key popular work in a proliferating genre of viral nonfiction and fiction, I wanted to know more about this author's techniques for stimulating concern and fear.[20]

Fear has a literary and social history. North Americans and Europeans can no longer be seriously scared by stories about werewolves, for example. Particular ages produce figures of fear that resonate widely with different publics and specific historical concerns. But what kinds of fears did representations of Ebola and the viral forest in motion generate? If environmental representations of Ebola and the African viral forest are interventions, in what ways might they affect social attitudes and policy within as well as outside the borders of the United States? How might these recent representations suggest the shape and trajectory of emerging environmental threat and danger?

How has Preston fashioned an anxiety-saturated environment through the skillful selection of facts, contexts, tone, and imagery? If rhetoric is the study of persuasive speech, then what are the means by which Preston conjures up fear and terror about African rain forests and their links to globalization?[21] How do Preston's images resonate with particular social and historical contexts? How does his selection of particular representations constitute intervention in the social sphere? I conclude, in part, that Preston's language yokes Ebola to centuries of Western imagery about African environments, peoples, and disease while coupling these fears with deep, less accessed regions of primal bodily anxieties. Preston joins these representational legacies with a family of unsettling representations of circulation: "reverse flows" of microorganisms moving from the South and penetrating the global North.

Although Preston's arsenal of poetic technique is formidable, it is only part of the story. Preston's Ebola terror story is so successful because it amplifies prevalent, intense anxieties about the spread of HIV, a virus believed to originate in Africa and to be genealogically related to primates inhabiting equatorial Africa. It also plays on a legacy of negative images of Africa as primitive, chaotic, disease ridden, and violent: the place of monkeys, the jungle, tribal excesses, and unexplained deaths.[22]

The Hot Zone also resonates with contemporary concerns about networked societies—their dangers, technological links, and unpredictable consequences. A standard environmental narrative underlies widespread fears about the consequences of globalization: an exotic threat from a foreign location somehow becomes dislodged from its natural moorings/place of origin, travels in global circulation networks, crosses borders stealthily/undetected, and explodes/proliferates within the nation/body, body somatic, and body cultural.[23] This standard environmental narrative is easily nationalized. In the United States, arguably a center of economic globalization, fears are proliferating that our borders will no longer hold against alien movements—whether the aliens are tropical viruses, nonnative plants and animals, or refugees or illegal immigrants.

FASHIONING AN AFRICAN RAIN FOREST

To be an effective trope, the Ebola virus has to come from somewhere, a place that is itself suggestive of terror. Preston fashions an image of a dangerous rain forest and its viruses that is distinctively African. The story of Ebola in *The Hot Zone* begins with the journey of Charles Monet, an expatriate Frenchman tending the water supplies of the Nzoia Sugar Factory in western Kenya. We know from the first page that Monet, on an excursion to Mount Elgon and Kitum Cave, is going to be a victim of a "Biosafety Level 4 hot agent" (1994: 3). But we do not know where or how he will contract the virus.

As we follow Monet and his girlfriend up the slopes of Mount Elgon, Preston configures the landscape in which the Ebola virus reservoir is located.

At the center of central Africa, Mount Elgon is described as an embattled icon of African rain forest biodiversity, a place of extinction of species. Preston quickly fashions a Manichaean split between wild nature and human culture. The mountain is besieged by human activities.[24] As Preston describes it, "The villages form a ring of human settlement around the volcano, and the ring is steadily closing around the forest on its slopes, a noose that is strangling the wild habitat of the mountain. The forest is being cleared away, the trees are being cut down for firewood or to make room for grazing land, and the elephants are vanishing" (1994: 6).

Preston selects a cave on Mount Elgon as a probable site of the Ebola reservoir, indulging heavily in what might be called the African gothic. He enlists a macabre cast of animals and animal organs—"hundreds of bats' eyes, like red jewels" (1994: 9)—and Dracula-like acoustics—"echoed back and forth, a dry, squeaky sound, like many doors being opened on dry hinges" (9)—in the service of fashioning the cave as a site of the uncanny.[25] The interior of the cave appears as an ancient, ur–rain forest: "Then they saw the most wonderful thing about Kitum Cave. The cave is a petrified rainforest. Mineralized logs stuck out of the walls and ceiling. They were trunks of rain-forest trees turned to stone—teaks, podo trees, evergreens" (9). Although Preston occasionally describes African forests as rain forests, evoking soft-focus, romantic Amazonian imagery, the emphasis throughout *The Hot Zone* is on the African tropical forest as a jungle: a site of a violent, ancient, and threatening nature. In his description of Kitum Cave, the petrified rain forest becomes a lethal natural landscape: "An eruption of Mount Elgon about seven million years ago had buried the rain forest in ash, and the logs had been transformed into opal and chert. The logs were surrounded by crystals, white needles of minerals that had grown out of the rock. The crystals were as sharp as hypodermic syringes, and they glittered in the beams of the flashlights" (9). Beginning with his lurid account, Preston's preoccupation with the integrity of the surface of the human skin and the possibilities of rupture becomes obsessive. He muses on how Monet may have contracted the virus: "Did he run his hands over the stone trees and prick his finger on a crystal. . . . There were spiders hanging in webs among the logs. The spiders were eating moths and insects" (9). Did Monet touch these spiders? Did the spiders touch him without Monet knowing it?

The ecological web, object of countless rhapsodic paeans in popular and scientific literature, is transformed into a series of dangerous connections. Every surface and creature is suspect—as potential sites of viral reservoirs, hosts, and transmission—the objects of an anxious gaze. Yet much of this evocation is fanciful, unrelated to actual sites and surfaces that virologists consider likely for human contact with Ebola virus.

EBOLA, AFRICA, AND AIDS

Ebola, Africa, and AIDS, each powerful enough as singular presences, are combined in Preston's story to augment anxiety and apprehension in a rapidly ascending spiral. As I began to track the way the tropical forest and Ebola virus were fashioned, I started to see that Preston's images of the virulent, African microbial world were stand-ins for the African rain forest—the jungle. With Tom Geisbart, a young microscopist attempting to identify a virus extracted from a monkey house in Reston, we take a safari through the lens of an electron microscope. Preston shows us what Geisbart sees in a cell: "He could see forms and shapes that resembled rivers and streams and oxbow lakes, and he could see specks that might be towns, and he could see belts of forest. It was an aerial view of rain forest. The cell was a world down there, and somewhere in that jungle hid a virus" (1994: 135). The Ebola virus, a malevolent presence, rules over African tropical nature. Images of the virus morph from buckets of rope, to worms, to snakes, until they become the Medusa, the face of nature herself: "He saw virus particles shaped like snakes, in negative images. They were white cobras tangled among themselves, like the hair of Medusa. They were the face of nature herself, the obscene goddess revealed naked. This thing was breathtakingly beautiful. As he stared at it, he found himself being pulled out of the human world into a world where moral boundaries blur and finally dissolve completely" (1994: 137).

Preston's menagerie of African viruses—Ebola Zaire, Ebola Sudan, and their filovirus sibling, Marburg—are characterized by an astonishing array of constantly mutating metaphors and similes.[26] These viruses are at one moment lethal, mechanical, and metallic—likened to "exocet missiles" (1994: 32). Then they are described as deadly as nuclear "radiation" (1994: 37) or as magnificent and terrifying as predators like "sharks" (1994: 59). They are disguised as the Grim Reaper in viral drag—a "slate wiper" (27)—animate, "breathtakingly beautiful" (1994:), charged with predatory intentions, and primitive promiscuity. Ebola and Marburg, in Preston's hands, call forth the power of something resembling the "romantic sublime" and put it into play: they spellbind, ravish, terrify, and transfix their victims in a beguiling spectacle of beauty.[27]

Filoviruses are not just generic viruses; they are described by Preston as African viruses, as ancient African viral predators:

> The more one contemplates the hot viruses, the less they look like parasites and the more they begin to look like predators. It is a characteristic of a predator to become invisible to its prey during the quiet and sometimes lengthy stalk that precedes an explosive attack. The savanna grass ripples on the plains, and the only sound in the air is the sound of African doves calling from acacia trees. . . . In the distance, in the flickering heat, in the immense distance, a herd of zebras graze. Suddenly from the grass comes a streak of movement, and a lion is among them and lands on a zebra's throat. . . . Some of the predators that feed on humans have lived on the earth for a long time, far longer than the human race, and their origins go back, it seems, almost to the formation of the planet. (1994: 93–94)[28]

Ebola and the ancient African nature it signifies are vested with malevolent agency, charged with the sinuosity of animal movement. As Preston puts it, Ebola "savaged patients and snaked like chained lightning out from the hospital through patients [visiting] families," and it "jumped quickly through the hospital via the needles" (1994: 68). Ebola, like a wild African predator, retreats after a kill and prepares to strike again. Preston observes that it "retreated to the heart of the bush, where undoubtedly it lives to this day, cycling and cycling in some unknown host, able to shift its shape, able to mutate and become a new thing, with the potential to enter the human species in a new form" (1994: 69).

His impressive arsenal of poetic technique cannot fully explain how these images of African landscapes and viruses generate such intense anxieties. In part, these images intersect and resonate with a centuries-old legacy of European images of the African landscape as a dark, chaotic, pestilential, savage world, a distinctly threatening configuration of an African other.[29] A key template for these images can be found in the inward spiral of Joseph Conrad's *Heart of Darkness*, where a heavy atmosphere of anxiety and foreboding permeates representations of the African rain forest.[30] "Going up that river was like traveling back to the earliest beginnings of the world," remarks Conrad, "when vegetation rioted on the earth and big trees were kings. An empty stream, a great silence, and impenetrable forest. The air was warm, thick, heavy and sluggish. . . . And this stillness of life did not in the least resemble a peace. It was the stillness of an implacable force brooding over an inscrutable intention. It looked at you with a vengeful aspect."[31] Conradian images of the African forest seem to pervade Preston's representation of the Ebola virus itself in *The Hot Zone*: "It seemed to emerge out of the stillness of an implacable force brooding on an inscrutable intention" (1994: 69–70).[32]

Although older as well as more recent representations of the African rain forest possess a sense of lurking threat, they differ in their direction of flow. In *The Heart of Darkness* and other older representations, the danger (especially to white westerners) is one of "being sucked into the forest and absorbed by it" ("going native").[33] In the more recent representations, including Preston's, the danger resides more in the forest's capacity to spread rapidly outward, beyond its limits within the forest and into the world beyond. The trajectory of Conrad's narrative and the cartography of danger he maps is an inward-turning spiral, moving toward a center where a meltdown occurs. The trajectory of the African viral forest in motion at the beginning of the twenty-first century is, in contrast, that of a spiral moving out into the world from a lethal African center.

Underlying and reverberating below the narrative of *The Hot Zone*, like a bass obbligato—barely audible, but sounding nevertheless—are dark fears of AIDS, the iconic plague of the late twentieth century, emanating out of Africa. Even the copy on the book jacket is structured to invoke these fears (map 1).[34] Fears of AIDS are fused in *The Hot Zone*, which entangles stories about Ebola, Marburg, and other hemorrhagic viruses in ways suggesting common geographic origins, common modes of jet travel, and horrific outcomes. Although *The Hot Zone* is about Ebola, the map at the front of the

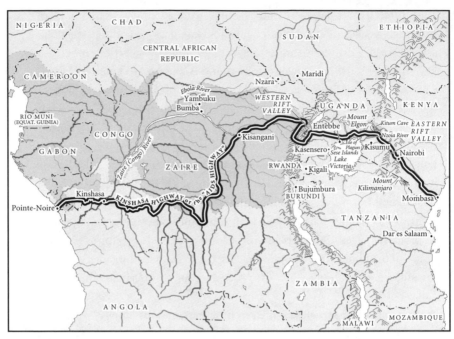

Map 1. The AIDS highway. Map by David Lindroth, from Richard Preston, *The Hot Zone* (New York: Random House, 1994). © 1994 by David Lindroth. Reprinted by permission of Random House, Inc.

book tells a different story: the largest, double-black-line road leading from Mombasa on the east coast to Pointe-Noire on the west coast of Gabon is denominated "Kinshasa Highway" or the "AIDS HIGHWAY." Central Africa, the Congolese rain forest, and the landscape of AIDS are as significant to this story and its effectiveness as the bold snaking line of the Kinshasa Highway is to the story of the dispersion of AIDS. Fears of a new viral plague from Africa move and merge with currents of anxiety about a known virus, HIV.

Like Conrad in his time, Preston is not alone in deploying negative images of African landscapes, peoples, and social institutions on the screen of contemporary policy analysis and prognostication. Preston's sense of horror about African viruses, for example, and the inevitability of their transmission to developed countries resembles policy analyst Robert Kaplan's determinist views of the state and future of African peoples, environments, and polities.[35]

But African environments are not the inevitably degraded, deforested sites of violence and social chaos evoked by Kaplan.[36] Indeed, Kaplan's dire views of African environments, driven by Malthusian visions of population and resource limits, have been soundly disputed by environmental and political analysts. Many of these analysts argue that hunger and resource scarcity are driven, even created, by inequitable and unwise state policies and practices creating differential access to resources, rather than by deficiencies in environmental attributes or population density. There is nothing inevitable, these authors argue, about famines or resource degradation, including rain forest deforestation, in Africa.[37] Preston's vision of a virulent rain forest coming here and breaking out in the suburbs of Washington, D.C., echoes Kaplanesque visions of Africa as a site of inevitable environmental degradation, disease, and social chaos.[38]

EMERGING CARTOGRAPHIES OF DANGER

Preston stimulates anxiety through collapsing a sense of distance and time, a narrative technique that suggests the African rain forest's apparently growing capacity to relocate itself in sites and networks formerly thought to be secure. I focus first on the ways in which formerly safe places—arrayed in a descending spatial scale from nations and national borders, cities, affluent suburbs, laboratories, and hospitals, to the living, individual human body— are shown to be vulnerable to the entry of the African rain forest. I then show how Preston stimulates anxiety about the safety and viability of networks—transport, communications, air-conditioning—on which middle- and upper-middle-class Americans increasingly depend.

An emerging cartography of environmental danger is embodied in these travels.[39] The African jungle in the form of Ebola and other viruses does not merely travel across the globe or across national boundaries in all directions. Preston focuses on a trajectory of viral travel from the South—the zone of poorer, developing countries—in the bodies of African green monkeys captured in equatorial forests, to the developed North. For example, the emergence of Marburg virus in a factory located in central Germany emphasizes the threat to northern civilized life within an ancient European city. In *The Hot Zone*, Preston fashions a contrast between the well-ordered, peaceful northern European landscape and its sedate city, "surrounded by forests and meadows, where factories nestle in green valleys" (1994: 25) with the lurid, bloody havoc wrecked by Marburg, "an African organism, but it has a German name" (25).⁴⁰

It is not within Europe, however, that the African jungle makes its most disturbing relocation, but rather within the borders of the United States. Preston devotes more than half of *The Hot Zone* to describing and dramatizing the discovery and suppression of an Ebola virus infection among monkeys housed just a few miles from Washington, D.C. Preston's descriptions of Ebola among monkeys at the Reston facility suggest the transformation of a portion of "one of the first planned suburbs in America, a visible symbol of the American belief in rational design and suburban prosperity . . . where disorder and chaos were given no sign of acknowledgement and no places to hide" (1994: 119) into a wild jungle.⁴¹ To provoke a heightened sense of anxiety, of things out of place in a newly dangerous neighborhood, Preston first conjures up Reston as a site of built and natural U.S. order: "Reston was surrounded by farmland, and it still contains meadows. In the spring, the meadows burst into galaxies of yellow-mustard flowers, and robins and thrashers sing in stands of tulip trees and white ash. The town offers handsome residential neighborhoods, good schools, parks, golf courses, excellent day care for children" (1995: 109). Preston's account of pandemonium in the Reston monkey house begins with his description of "a hundred wild monkeys from the Philippines" (1995: 111). He emphasizes presumably primitive, repulsive features of their primate bodies, food-eating preferences, and social behavior, establishing an atmosphere and a dramatis personae characterized by crudity and violence: the monkey's heads had a "protrusive, doglike snout with flaring nostrils and exceedingly sharp canine teeth, able to rip flesh as easily as a honed knife" (1995: 122). Their shrieking, aggressive, crab-eating habits are described in ways that suggest savagery: "A crab comes out of its hole, and the monkey snatches it out of the water. . . . He grabs the crab from behind as it emerges from its hole and rips off the claws and throws them away and then devours the rest of the crab" (1995: 123).

When the monkeys show symptoms of an Ebola infection, Preston creates a forest full of paranoid, depressed, and violent primates—an outbreak of African jungle in the midst of one of the more placid suburban towns in the United States. "Seventy pairs of monkey eyes fixed on a pair of human eyes in a space suit," writes Preston, "and the animals went nuts.

They were hungry and hoping to be fed. They had trashed their room" (1995: 243). As a miliary suppression of the outbreak proceeds, the primate unit resembles a cross between a wild African jungle and a madhouse "filled with hysterical, screaming, leaping, bar-rattling monkeys" (1995: 257).

Between the nation and the suburb lies the city. In the primate unit incident, the nearby city happens to be the capital of the United States. Preston is clear about raising the anxiety level of the reading public and focusing it on the possibilities of the African equatorial jungle appearing in Washington, D.C. He stages his descriptions of the primate unit infection in ways that emphasize this proximity: "On a fall day, when a western wind clears the air, from the upper floors of the office buildings in Reston you can see the creamy spike of the Washington Monument, sitting in the middle of the Mall, and beyond it the Capitol Dome" (1994: 109). Through Preston's Reston-centered optic, the reader sees and senses the possibility of the monkey-house horrors spreading just a few miles within the Beltway, and infecting the bodies of our nations' political leaders: an epidemiological meltdown of the nation's capital, its political and symbolic center, and by extension, the nation itself.

THE FOREST IN THE BODY

The human body presents astonishing possibilities for the stimulation of anxiety. Preston fully explores these possibilities in creating scenes of invasion and viral proliferation. Of all sites in *The Hot Zone*, Preston is often at his lurid best in fashioning somatic scenes that are bloody and disturbing. Rather than focusing on the gore, I would emphasize the ways in which Preston's imagery of the living body is articulated in governmental, military, and architectonic metaphors: "As Ebola sweeps through you, your immune system fails and you seem to lose your ability to respond to viral attack. Your body becomes a city under siege, with its gates thrown open and hostile armies pouring in, making camp in the public squares and setting everything on fire" (1994: 46–47).

Homologies between besieged cities and infected human bodies abound, reinforcing the central images of failing border defenses, the vulnerability of boundaries and openings, and the lack of a capacity to respond to an enemy attack. As if fashioning a series of Chinese toy boxes, Preston has constructed a series of nested homologies. The African rain forest appears within national borders, upper-middle-class communities, and that most intimate of sites, the individual human body.

FAILURES IN THE WEB, INFESTATIONS OF THE NETWORKS

In *The Hot Zone*, the viral African forest also appears within the networks and webs of communication, transport, and connectivity that are increasingly interpreted as the emerging technologies of globalization. Preston suggests the growing capacity of the African viral forest to invade and populate boundary-spanning networks on which the managerial and professional classes increasingly depend.[42] Preston suggests the potential horror of these connections.[43] "A hot virus from the rainforest lives within a twenty-four-hour plane flight from every city on earth. All of the earth's cities are connected by a web of airline routes. The web is a network," explains Preston, adding that "once a virus hits the net, it can shoot anywhere in a day—Paris, Tokyo, New York, Los Angeles, wherever planes fly. Charles Monet and the life form inside of him had entered the net" (1994: 11–12). Within the bodies of monkeys who are themselves within the bodies of jets, the viral forest is in a constant state of motion and relocation. Preston amplifies anxiety about the lethal potential of global transport networks and other forms of border-crossing connectivity: "Why is the Reston virus so much like Ebola Zaire, when Reston supposedly comes from Asia? If the strains were from different continents, they should be quite different from each other. One possibility is that the Reston strain originated in Africa and flew to the Philippines on an airplane not long ago. In other words, Ebola has already entered the net and has been traveling lately" (1994: 261). Marburg virus is portrayed as a globetrotter by Preston: "The Marburg virus was a traveler: it could jump species; it could break through the lines that separate one species from another, and when it jumped into another species, it could devastate the species. It did not know boundaries" (1994: 5).

Boundary-spanning forms of connectivity and their risks are not only large-scale phenomena such as jet flight paths, transcontinental highways, or interstate highway systems. Even air-conditioning systems are potential paths for traveling viruses. Rain forest enemies, Preston asserts, are only hours away from us. Preston's story of Ebola depends on images of almost instantaneous travel, from the South to North, on a global scale.[44]

Globalization processes themselves, especially jet travel, have indeed become the objects of intense anxiety about time and space compression.[45] Legitimate concerns about the risks of disease spread are recognized by a large group of eminent virologists and public health specialists as well as medical journalists including Laurie Garrett, the author of *The Coming Plague: Newly Emerging Diseases in a World out of Balance*.[46] While accurately portraying the horrors of the kinds of symptomatology that Ebola produces in monkeys as well as humans, Preston's accounts of Ebola infection exaggerate the ease with which the virus can disseminate itself in space, the ease with which it can be transmitted to and among humans, and the likelihood that those strains that are transmitted across species barriers are fatal for human beings.[47]

Questions about the risk of jet transport spreading Ebola and the possibilities of a major epidemic as a result are far from settled.[48] Preston's fevered risk assessments of globally spread Ebola epidemics are parodied within the scientific community. As Dr. Brian Jelle, a researcher at the University of New Mexico, writes: "In Preston-speak?: Bricks of bad information and fear-mongering set up a highly-efficient, deadly cycle of hysteria replication in the populace. The public hemorrhages, spilling hysteria to the next unwitting victim. Fear gushes from every media orifice. No one is safe from the hype."[49]

FIGHTING THE FOREST, ENTER THE MILITARY

The ubiquitous presence of the U.S. military in the biocontainment and cleanup operation at the primate unit is one of the more remarkable aspects of the staging of *The Hot Zone*. Preston fashions a narrative in which military control over the Reston incident seems natural. The branch of the federal government charged with jurisdiction over emerging diseases is the CDC. As the story of the Ebola infection in the primate unit unfolds, the question of whether the CDC or the USAMIRID staff has legal and operational authority over the situation is treated in a brief, cavalier fashion. General Philip Russell, M.D., the commander of the United States Army Medical Research and Development Command, is fully aware that the primate unit infection, as a matter of law and public policy, lies within the authority and powers of the CDC. As he says in *The Hot Zone*, "The Army doesn't have the statutory responsibility to take care of this situation, but the Army has the capability. The CDC doesn't have the capability. We have the muscle not the authority" (1994: 158–159). Despite the CDC's congressionally mandated authority, Russell opines that this was a job for soldiers operating under a chain of command. At the time of the primate unit incident, Russell explains to his colleagues: "We [the military] are going to do the needful, and the lawyers are going to tell us why it's legal" (159).

Hollywood-style, military-action chapter titles reinforce the apparent reasonableness of the militarization of public health: "Chain of Command," "Shoot-Out," "Reconnaissance," "Insertion," and "A Man Down." Military metaphors and idioms merge with the languages of biomedical research institutions: "It looks like we're going to have to go down and take those monkeys out and we're going to do it in Biosafety Level 4 conditions," Colonel C. J. Peters says to Colonel Jerry Jaax, a USAMIRID veterinarian designated as commanding officer of the operation (1994: 187). Attempts are made to conceal the involvement of military personnel in the search-and-destroy operation at the Reston primate unit. Military teams wear civilian clothes to avoid attracting attention and setting off a panic triggered by the spectacle of "soldiers in uniforms and camouflage putting on space suits" (1994: 201). Unmarked vans as well as unmarked military vehicles transport the troops and commanding officers to the unit.

In a visit to the primate unit after the infection was successfully suppressed, Preston warns the reader, in a moment of Wordsworthian melancholy laced with a sense of threat: "The time of year being autumn, the spider had left egg cases in its web, preparing for its own cycle of replication. . . . Ebola had risen in these rooms, flashed its colors, fed, and subsided into the forest. It will be back" (1994: 291).

EBOLA AT THE MOVIES: SCENES FROM *OUTBREAK*

In 1995, one year after *The Hot Zone* was published, the film *Outbreak*, starring Dustin Hoffman as an army medical expert, Renee Russo as his physician wife, and Donald Sutherland as a military biowarfare villain, appeared in local movie theaters around the country. One morning in fall 1995, as I rode the subway downtown to my job at the Rainforest Alliance office in Greenwich Village, I noticed wall-high posters in the subway station entrance at West Fourth Street announcing *Outbreak*'s release. I recall Hoffman in a biocontainment space suit in the foreground, a glamorous Russo near him, several nasty-looking monkeys, and a fringe of tropical forest. Palm trees may have waved in the distance.

Outbreak is the story of a fictional virus, Motaba, and its discovery in the Congo in 1967 and reappearance in the 1990s in roughly the same area.[50] The drama focuses on military attempts to contain the spread of the virus in a fictitious town, Cedar Creek, located somewhere in northern California, and efforts to keep knowledge of a Motaba-based bioweapon and its vaccine secret.

Like its portrayal in the posters, the jungle plays only a peripheral role in *Outbreak*. The film, like *The Hot Zone*, is a morality tale about what a disrupted, invaded, ravished African nature has in store for the nations and peoples of the global North.[51] *Outbreak* begins in equatorial Africa in 1967. A village ravaged by local warfare, surrounded by forest, and afflicted with Motaba virus is destroyed when U.S. military personnel descend on it. After drawing blood from a dying combatant, a U.S. mercenary, the U.S. military firebombs the entire compound, presumably destroying every living organism, virus, and human being in a burst of orange flame. Secretly, samples of the virus are collected, carried back to the United States, and used to develop a biological weapon and its antidote, both based on Motaba. In the 1990s, another outbreak occurs. This time, the U.S. military sends Hoffman, the irreverent hero working for the military's medical division, back to the Congo to identify and gather samples of the virus for the U.S. military.

In *Outbreak*, a series of monkey scenes set up the story line and show us how Motaba moved from central Africa to the United States. A young capuchin monkey is shown watching combat around the village that is eventually destroyed by the U.S. military in 1967. Another monkey observes Hoffman and his colleagues fly away from a newly infected village in the 1990s. Yet another monkey is shown captured in a hunter's net. And a fourth, later to be named Betsy, becomes a victim of the international trade in primates for scientific experimentation and pets. Betsy is an ambiguous symbol for Africa in the late twentieth century and the multivalent equatorial African rain forest—its viral dangers and fate in an era of globalization, militarization, an international trade in tropical animals, and the manipulations of biomolecular science in the service of the U.S. national security establishment.

We first watch Betsy in captivity onboard a mysterious Korean vessel as she is shipped to the United States. On shore, Betsy is warehoused in an animal holding facility in San Jose, California. She is then stolen by an unscrupulous young worker who wants to sell her, as he has done with other tropical primates, to the private pet trade. Provoked by the man's disrespectful manner, Betsy squirts water from her mouth onto his face, thus beginning a chain of Motaba infection that spreads via her bloody scratch on the arm of a northern Californian pet store owner, followed by its transmission through a remarkably animated cough in a provincial movie theater in a small working-class town and a kiss planted firmly on the mouth of the thief's girlfriend. Finally, prior to a cross-country jet flight, the thief releases Betsy into the great and majestic silence of the Pacific Northwest forest, where she wanders, a disoriented, potentially lethal carrier of Motaba. As the thief flies east, Motaba swiftly and silently proliferates within Cedar Creek.[52]

As in *The Hot Zone*, *Outbreak* generates and heightens fears of a viral forest in motion, and its capacity to penetrate borders and formerly secure spaces. The placid, working-class, rural town of Cedar Creek, a community nestled near the Pacific Ocean and flanked by the darkly majestic trees of a northern Californian forest, is reduced to a mass of people in panic, while others, weakened or dying, are laid on stretchers. Cedar Creek's probable closeness to San Francisco, as Reston's proximity to Washington, D.C., in *The Hot Zone*, suggests the possibility of an epidemic in a major metropolitan center.

Outbreak, like *The Hot Zone*, focuses on the potential of webs and networks to facilitate the dispersion of infectious diseases. The virus is carried to U.S. shores on an ocean freighter. It travels, via the U.S. interstate highway system, from a probable location in San Jose to sites along the road in northern California. Motaba is sucked up into the vortex of air-conditioning ducts in an improvised field hospital in Cedar Creek, only to be expelled into another room, where it finds new victims. From California to Boston, Motaba flies in a transcontinental jet.

Fears of disease dispersion in an era of jet travel are graphically displayed when Sutherland, in an effort to convince the president and his advisers to permit the firebombing of Cedar Creek and its civilian population, presses a button to reveal a map of the entire United States. The map displays relatively small clusters of red dots signifying the geographic distribution of isolated Ebola cases. Within moments, the entire map is saturated with red. Using cartographic imagery as a spectacle and emotional prod, this scene makes the phenomenon of epidemic dispersion and its networked connections a bloody red reality. And like *The Hot Zone*, *Outbreak* creates anxiety by obsessively focusing on the nodes and portals within webs: airports, shipping facilities, the air-lock doors of the temporary military-medical center of operations airlifted into Cedar Creek, and the levels and locks of the biosafety labs.[53]

Outbreak also emphasizes the militarization of public health interventions. The enormous, awesome, fear-inspiring resources of the U.S. military and its technologies are deployed to mark and police the perimeters of the afflicted Cedar Creek. Its disenfranchised citizens are herded from their homes to fenced-in containment areas, escorted by armed, intimidating military personnel. Hovering over this working-class community are dragonlike helicopters that fire their weapons against a few families attempting to escape the concentration camp–like containment zone. Insulated physically from the normal citizenry and air by their impressively sealed, high-tech biosafety operations center, these techno-science troops are as frightening as the Motaba virus. Exempt from the normal restraints and obligations of citizenship, these troops can kill civilians in the service of the state's proprietary interest in safeguarding the weapons of bioterror/national defense.

Unlike *The Hot Zone*, which seems to render reasonable the use of the U.S. military force within spheres of civilian life, particularly in connection with public health operations, *Outbreak* appears to condemn the militarization of public health and everyday life. Sutherland, as the chief of the secret bioterror division, is an amoral figure. He wants to keep his stock of Motaba, collected during the 1967 initial outbreak, as a secret weapon of the U.S. military. His decision to exterminate the remaining population of Cedar Creek is based on a desire to retain this U.S. monopoly on a bioweapon and its remedy.

But *Outbreak* actually makes the interventions of a certain kind of military—the kind of military embodied by familiar antics of spunky, clever, and well-intentioned Hoffman—seem attractive and reasonable. By setting up a boldly drawn contrast between the appealing Hoffman and Sutherland as the gaunt, cold, military bioweapons commander, *Outbreak* makes the decision easy. It is just a question of what kind of military we should welcome into our civic sphere, not whether there is a legitimate role for the armed forces in civilian life.

EBOLA IN THE NEWSPAPERS: THE TURNING POINT PROJECT GENETIC ENGINEERING ADVERTISEMENTS

Recently, a series of newspaper advertisements used images of African Ebola virus to provoke fears of genetic recombinant research and xenotransplantation.[54] The series was sponsored by the Turning Point Project, an alliance of remarkably diverse social, environmental, and economic advocacy groups.[55] It is useful to briefly look at the ways in which Ebola, African nature, and images of border-crossing in nature are used by the producers of these advertisements.

On 1 November 1999, the text of a full-page advertisement in the *New York Times* asked, in large, bold-faced type, "Where will the next plague come from?" The headline was located directly above a greatly enlarged photograph of an Ebola virus in the center of the page (figure 1).[56] The text below the headline answers by making a misleading analogy between the processes of disease transmission among primates and human beings in African forests, and interspecies viral transmission resulting from laboratory-based genetic engineering and xenotransplantation practices in North America and elsewhere in the developed world: "HIV and the Ebola virus . . . crossed from primates to humans. The 1918 influenza virus started in pigs. Now, biotechnologists who shuffle live organs and genes between humans and animals could end up unleashing deadly new diseases without hope of a cure." In a section called "Crossover Diseases," the text shifts from fears of Ebola to anxieties about AIDS: "The history of the twentieth century provides ample evidence of the catastrophes that can result when disease agents cross species boundaries. . . . A version of HIV, the retrovirus that leads to AIDS, probably resided harmlessly in a wild forest primate before people invaded its habitat and contracted it. Some forty million people worldwide will soon be carrying that deadly disease."[57]

In these intentionally disturbing advertisements, African viruses and

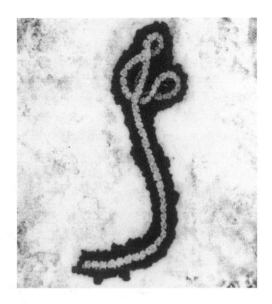

1. Ebola virus. From genetic engineering advertisement no. 4 by the Turning Point Project. © 1998 by Turning Point Project.

their movements from equatorial forests are used as figures of fear. "From day one," a knowledgeable participant in the Turning Point Project commented, "the goal [of the ad series] was to articulate a description of a problem that would send you over the edge." Using the names of Ebola and AIDS and linking these terrifying viruses to contemporary experiments in genetic recombinance amounts to a form of word magic: the transfer, by means of analogy and association, of characteristics from one empirical domain to another.[58] Coupling these names with photographs of the Ebola creates a wide spectrum of associations and anxieties: epidemics, uncontrolled dispersions, unholy border crossings, and deaths. These ads use the movement of "African" viruses—AIDS and Ebola—across species boundaries and geographic space as a metaphoric means of inciting anxiety about the risks of another domain, xenotransplantation.[59] Among the key words in this series are species barrier, integrity, and identity.[60] In language that refers to the science fiction movie about a mad scientist, *The Cabinet of Dr. Caligari*, Turning Point genetic engineering advertisement no. 1 conjures up a monster escape scenario that echoes Preston's overwrought descriptions of Ebola on the loose: "These things are alive. And they don't want to go back into their test tubes."[61]

By choosing to place ads only in the major print media of the Northeast, the creators of the Turning Point series sought to provoke concern among elite opinion and policymakers in that corridor of the United States.[62] By linking accounts of naturally occurring cross-species movements of genetic material with scientifically generated genetic change, by linking these processes to the feared AIDS and Ebola, and by conjuring up the representation of movement as a source of danger, the Turning Point genetic engineering ads create a vortex of invasion imagery as well as impending catastrophe across species, ecosystems, and national boundaries: "Biotech creatures and microbes are unpredictable. They can reproduce, cross-pollinate, mutate and migrate. They can jump across species using virus vectors. They can hitch rides in cars, boats and planes or in your socks. The can show up in other ecosystems. Like the Gypsy Moth, Dutch elm disease, and Kudzu vines, 'exotic organisms' can run amok and cause unparalleled environmental destruction."[63]

RHETORICAL INTERVENTIONS:
CARTOGRAPHIES OF DANGER AND MILITARIZATION

African rain forests, like Africa itself as imagined by Euroamericans, have long been the object of intense ambivalence in the West. A sense of potential threat, whether of predators or plagues, has permeated many representations of equatorial African rain forests. As I have argued, in the late twentieth century, Edenic images of the Amazonian rain forest made popular during the 1970s and 1980s were rejoined by earlier representations of the African forest as a lethal environmental zone. The trajectory of danger, however, has been reversed. While earlier European travel narratives to Africa followed a spiral inward and downward to a fevered and chaotic center, the trajectory of African biodanger is now represented as an outward movement, a spiral originating within the heart of the equatorial forest, arcing across the globe, and penetrating the core nations of the North.

I have called this emerging representation "the viral forest in motion." Once regarded as stable, coherent, bounded associations of endemic species, representations of tropical African forests increasingly emerge in popular media descriptions as ruptured, ancient rain forests from which a plaguelike explosion of multiple viruses is liberated, exiting on the same rough roads that only recently forced an entry.[64] Rather than an inward journey into the African forest, we observe the dangers of "reverse flows."[65]

The image of a virus escaping from the forest and relocating in the North, as I have also argued, has come to represent the African forest and its potential to travel, spread, and relocate within a variety of social spaces. This representation has coalesced at a moment of ramifying anxieties about the consequences of globalization, border-crossing flows, and possibly U.S. military establishment in search of new missions and enemies.

In analyzing two popular texts, *The Hot Zone* and *Outbreak*, I have asserted that these representations are interventions that shape perceptions of phenomena as well as particular ways of imagining reasonable policies and practices that deserve public support. Both texts, and the Turning Point advertisements, intentionally use the image of a mobile, virulent African biological diversity as an instrument to achieve certain aims. *The Hot Zone* and *Outbreak* use this representation to produce anxiety about a foreign other arriving within U.S. (or European) borders, passing through a variety of boundaries, and proliferating within and taking over sites that were formerly considered to be safe. These representations are instruments for the production of fear about security and survival, within the North or developed world, most broadly, and within U.S. and European sites—nations, cities, suburban enclaves, middle- and upper-middle-class homes and bodies— more specifically. The trajectory of this movement out of Africa, across oceans, and into the North suggests the outlines of an emerging cartography of environmental danger.

This chapter has sought to provoke discussion about the possibility of an emerging environmental cartography of danger that crosses empirical and conceptual borders. Some striking, if partial parallels exist between recent talk about the travels and threats of rain forest viruses, discussion about the cross-border movements and character of exotic plants and animal species, and discourses on the movements of illegal immigrants into the United States.[66] Many of these discourses focus on surveillance, capture, and control of the flow of aliens.[67] These narratives suggest a turn toward militancy at the border, even militarization within the United States and abroad, to the extent they propose early warning systems, surveillance technologies and border interdiction strategies, and the infrastructures for monitoring and control.[68]

Is there a standard environmental narrative about endemic ecological and cultural systems, whose vulnerable borders are to be fortified for protection from foreign threats?[69] How are we to account for the formal repetition of invasion analogies across several natural domains?[70] Why, at this historical moment, are the movements of rain forest microorganisms, nonindigenous plant and animal species, and immigrants apparently conceptualized in similar ways?[71]

What consequences might these representations have? If they are effective, then it is quite possible that they contain implications for social attitudes and practices, for policy and politics. A heightened, diffuse fear of alien organisms is one of the possible consequences of these representations. This fear or anxiety could embrace immigrants as well as microscopic viruses, nonnative plants, and exotic animals. The gamut of living creatures set in motion and moving from the global South to the global North might well become the objects of an increasingly anxious gaze.

It is also possible that these representations and the diffuse anxieties they engender could lead to increasing scrutiny and intensified surveillance around the borders of areas formerly considered as safe and secure. These areas might include portals of entry and exit, ranging in scale from national borders and border-crossing areas, to building security peripheries and entryways in institutional as well as domestic settings.

The emerging cartography of danger suggested in these pages could reinforce a generalized fortress mentality in which a politics and policies of hardened borders, clarified boundaries, and strengthened barriers began to be seen as desirable and justified. Is there a convergence of methods, models, and strategies leading toward a hardening of security and surveillance concerns in domains as disparate as domestic architectural design and urban planning, conservation policy and planning, border control and immigration policy in the U.S. West, and epidemiology and virology?

The Hot Zone and *Outbreak* both fashion a world so full of invasive, moving, border- and barrier-proof biological danger that military intervention, at home at least, seems not only natural but a godsend. Both texts, in quite different ways, make the insertion of military force into the public sphere and, in particular, public health contexts seem inevitable. With its opposition of plucky, spunky, heroic Hoffman as the representative of the "good" military in opposition to Sutherland as the cold-fish biogeneral,

Outbreak does not allow the viewer to imagine a world without a medical establishment merged with the marines or army. Preston, similarly, has fashioned medical-military heroes out of the U.S. army colonels and privates who illegally accomplished the monkey-killing and decontamination mission in the Reston Primate Quarantine Unit. These representations may constitute a new map, a new cartography, of permissible military intervention in public health and civilian life in the United States.

What are the potential social and political implications of this representation beyond U.S. national borders? Both texts along with the Turning Point advertisements, revitalize and ratify a view of Africa, its environments in general, and its equatorial forests and biological diversity in particular as dangerous. They are environments and organisms, the representations seem to imply, that need to be stabilized, contained, and probably avoided. These conceptions and the images that support them dovetail neatly with older Eurocentric images of Africa as the dark continent. They also resonate and reinforce recently fashionable doctrines of inevitable African environmental, cultural, and economic decline. This new-old cartography of the African continent as well as the African equatorial forest portrays Africa as the site of catastrophic events, including seemingly inevitable famine, plagues, and political terror. This simplified, totalizing vision constitutes what might be termed a racist cartography with the potential to become a self-fulfilling prophecy.

What implications might this representation have for the management and fate of the African equatorial rain forest? Given the volatile nature of this particular forest representation, its tendency to escape its bounds, to go traveling in places it doesn't belong and make mischief, perhaps the central tropical forest policy implication of these texts is: keep this forest in its place. A hard regime of strengthened borders, heightened surveillance, and military patrols may be imagined as the logical policy implication of this representation.

The virus has become one of the master metaphors of the early twenty-first century.[72] Tropical African nature and its plethora of microorganisms are being held hostage to the ravages of diffuse anxieties about border crossings, alien takeovers, and penetrations of and threats to the coherence of the self—individual, national, cultural, ecological, biological, and economic. The image of Ebola virus and the equatorial African rain forest—

representing Ebola's natural origin and ruptured container—have become points of articulation within a remarkable variety of social, economic, and environmental movements. Ebola has been invoked as a symbol of nature's irrepressible instability and its virulence—its tendency to morph, jump species, move across boundaries, invade individuals, and penetrate Western civilization.[73]

The image of African Ebola, like a gladiator's shield emblazoned with a terrifying countenance, has become a symbol and site—of articulation and linkage—through which a variety of movements speak on behalf of nature and humankind. Although the issues and anxieties animating these movements differ significantly—biomedical and epidemiological security, national defense, rain forest conservation, genetic recombinant research and xenotransplantation, world trade, and biological weapons development—they share common concerns with keeping endemic nature in its place, keeping the aliens outside, and policing the borders.[74] Ebola, uncoupled and released from its rain forest origins in equatorial Africa, continues to flash its colors—as a template, a moving target, a chameleonlike living vessel for proliferating anxieties in an era of ramifying networks and intensified globalization.

Now is the time for engaged scholars to interrogate the sources, effects, and sites of production of these fearsome rhetorics of environmental danger.

NOTES

I wish to express my thanks to all the HRI seminar members, who generously contributed their insights, energy, critique, and knowledge during the course of our work together in Irvine and Yucatán. Thanks are due to Candace Slater, who directed our seminar with grace, wit, and intelligence; Paul Greenough, who enthusiastically encouraged this journey into the world of global health issues; and the HRI staff, who graciously facilitated our conversations and research. Roger Rouse provided highly detailed and insightful comments on an earlier version of this chapter, contributing materially to its final form. Michael Watts, Ian Boal, and participants at the University of California Environmental Politics Colloquium, where a version of this chapter was presented in March 2001, as well as Toby Alice Volkman, read and provided helpful comments on this chapter. Versions of this essay were presented at the Health, Science, and Society lecture series at Sarah Lawrence College (2003), the Global Migrations Program Seminar at Hampshire College (2002), and the Crossing Borders Program and Global Health Studies Program of the University of Iowa (2001). At Sarah Lawrence College, I am indebted to Ginger Hagan and Alayna Baldanza, who provided research and editorial assistance. Thanks also to

Barbara and Bertram Cohn, generous supporters of Sarah Lawrence and the Environmental Studies Program. All the usual disclaimers apply.

1. Tropical forests and their representations have been regarded ambivalently by the West since the fifteenth century, alternately emphasizing paradisiacal or darker, negative aspects. On the history of the idea of the tropics and tropicality, and variations in relative emphasis along this axis from paradise to green hell, see David Arnold, *The Problem of Nature: Environment, Culture, and European Expansion* (Oxford: Blackwell, 1996). For a generalized account of shifting conceptions of "the tropics," I rely on Arnold's chapter "Inventing Tropicality," 141–168.

2. By the early eighteenth century, environmental historian David Arnold has shown, a "full fledged myth of tropical exuberance" had been elaborated by European observers. By the late eighteenth century, however, the darker aspect of representations of tropic environments, including forests, were accentuated: they were sources of fevers and disease; a torrid zone of humid climate as well as pestilential swamps and forests afflicting European colonizers and travelers. See Philip Curtin, *The Image of Africa: British Ideas and Action, 1780–1850* (Madison: University of Wisconsin Press, 1964), 55–60; cited in Arnold, *The Problem of Nature*, 145. On the destruction of the scrub and forests of Bengal as a means of improving the ventilation of miasmas, see Richard H. Grove, *Green Imperialism: Colonial Expansion, Tropical Island Edens, and the Origins of Environmentalism, 1600–1860* (Cambridge: Cambridge University Press, 1995).

3. I simplify Slater's more nuanced analysis of the Edenic rain forest and, of course, make some enormous historical leaps. The reader is encouraged to pursue the fine-grained histories of rain forest representations that can be found in Arnold, *The Problem of Nature*. See also Candace Slater, "Amazonia as Edenic Narrative," in *Uncommon Ground: Toward Reinventing Nature*, ed. William Cronon (New York: W. W. Norton, 1995), 114–131. Slater elegantly articulates the qualities of an Edenic forest representation characteristic of the 1970s and 1980s: "Beautiful in the extreme, the rain forest is also intensely vulnerable" (127). See also Curtin, *The Image of Africa*.

4. See the publications of biologist E. O. Wilson for a recasting of the romantic, sublime imagery of rain forests in terms of biological diversity. For an example of Wilson's rhapsodic, cosmologically benevolent image of biological diversity, see *The Diversity of Life* (Cambridge: Harvard University Press, 1992), 11–15.

5. For characterizations of globalization processes, see Arjun Appadurai, *Modernity at Large: Cultural Dimensions of Globalization* (Minneapolis: University of Minnesota Press, 1996). See also Zygmunt Bauman, *Globalization: The Human Consequences* (New York: Columbia University Press, 1998).

6. The tropical forests of equatorial Africa always lay below the shadow of a darker, racist, Western gaze, as demonstrated in Curtin, *The Image of Africa*; and Christopher Miller, *Blank Darkness: Africanist Discourse in French* (Chicago: University of Chicago Press, 1985).

7. I am grateful to Roger Rouse's detailed comments on the emergence and significance of the "viral forest" imagery.

8. David Quammen, "The Green Abyss Megatransect," part 2, *National Geographic* (2001) (March): 8–37.

9. Roger Rouse contributed to the wording of this passage.

10. My focus is on the rhetoric of African forests, specifically rhetorics in which discussions of Ebola virus are embedded. This chapter is not an assessment of policy risks or options but an exploration of persuasion, imagery, and their contexts. I understand rhetoric as embedded in and inseparable from all accounts of the world, following Stanley Fish, "Rhetoric," in *The Stanley Fish Reader*, ed. H. Aram Vesser (Malden, Mass.: Blackwell Publishers, 1999), 114–142. See also Richard Rorty, who asserts: There . . . are two ways of thinking about various things. . . . The first . . . thinks of truth as a vertical relationship between representations and what is represented. The second . . . thinks of truth horizontally—as the culminating reinterpretation of our predecessors' reinterpretation of their predecessors' reinterpretation. . . . It is the difference between regarding truth, goodness, and beauty as eternal objects which we try to locate and reveal, and regarding them as artifacts whose fundamental design we often have to alter (*The Consequences of Pragmatism*) (Minneapolis: University of Minnesota Press, 1982), 92.

11. Due to limitations in time and scope, I was not able to explore several other important questions that should properly form part of this project. What was the nature of the audience at which these representations were directed? What kinds of conditions may have made this image popular with certain U.S. audiences?

12. I am not an African specialist, nor am I a scholar of rhetoric and poetics. I am a lawyer and scholar who has been engaged in research on environmental justice, culture, and politics in Southeast Asia, with a keen interest in the ways in which environmental representations are shaped and strategically situated within different social movements.

13. Attention to Ebola has not faded. In fall 2000, an outbreak of Ebola in Uganda was widely covered in U.S. and European newspapers as well as on National Public Radio in the United States. Mistaken diagnoses or suspicions of Ebola in Germany, and more recently Canada, during fall 2001 were the objects of intense and hyperbolic media coverage. A description of the medical-military treatment of a patient, Ullman, recently returned from shooting a nature appreciation film in central Africa can be found in Allan Hall, "Scared to Death," *Scotsman*, 1 August 1999. Hall states that crowds were "kept back by the machine-gun toting police, security guards with rottweilers and at night by hastily erected searchlights which scanned the crowds. . . . Ullman was as much a prisoner as anything else." In fall 2000 and winter 2001, central African jungles and Ebola figured prominently in three issues of *National Geographic* as an eminent conservation writer, David Quammen, chronicled the travels of Dr. Fay, a conservation biologist intent on traversing by foot the jungles, swamps, and rivers of central Africa. See Quammen, "The Green Abyss Megatransect."

14. Richard Preston, *The Hot Zone* (New York: Random House, 1994), 107. Hereafter, page numbers will be cited parenthetically in the text from two editions: the 1994 hardcover edition and the 1995 paperback edition (*The Hot Zone*, New York: Anchor Books Doubleday, 1995).

15. I am indebted to Paul Greenough for this insight.

16. See, for example, Marc Lappe, *Breakout: The Evolving Threat of Drug-Resistant Disease* (San Francisco: Sierra Club Books, 1995); Richard Preston, *The Cobra Event* (New York: Random House, 1997); William Close, *Ebola: A Documentary Novel of Its First Explosion in Zaire* (New York: Ivy Paperbacks, 1995); Johan Marr and J. Baldwin, *The Eleventh Plague: A Novel of Medical*

Terror (New York: Cliff Street Books, 1998); Joseph McCormick, Susan Fisher-Hoch, and Leslie Alan Horvitz, *Level Four Virus Hunters of the* CDC (Atlanta: Turner Publishing, 1996); Ed Regis, *Virus Ground Zero: Stalking the Killer Viruses with the Centers for Disease Control* (New York: Simon and Schuster, 1996); and C. J. Peters, *Virus-Hunter: Thirty Years of Battling Hot Viruses around the World* (New York: Doubleday, 1997). The shift from a focus on the benign aspects of biodiversity to a malign vision of its potentialities can be seen perhaps in the shift from Euell Gibbon's account of his personal search for and delight in edible and useful wild plants, albeit northern biodiversity, in *Stalking the Wild Asparagus* (New York: D. McKay, 1970) to Regis's lurid subtitle, *Stalking the Killer Viruses.*

17. Pragmatica, Inc., *Outbreak*, 26 April 1998, www.outbreak.org/cgi-unreg/dynaserve.exe/index.html (accessed on 6 March 2000).

18. For a popular rain forest "revenge" narrative, see Gary Strieker, "The Rainforest Strikes Back," CNN *Interactive* (online), 12 June 1997, www.cnn.com. Strieker asks the question, "Under attack by growing populations of humans . . . can the tropical rainforest defend itself?" He suggests that outbreaks of Ebola and other viruses, and their movement from animals to humans through zoonosis, occurs because "humans are suffering punishment for disturbing the primeval rain forest."

19. The names of six "modern plagues" including Ebola, printed in red ink and large type, together with a question, "What terrifying Modern Plague will we face next?" were on the envelope I received in late April 2000 from the Physicians for Social Responsibility. This respected NGO's letter read, in part: "Deadly new diseases are cropping up all over. Global climate change commonly known as 'global warming,' is creating a breeding ground for diseases. And unless we take action now, more 'modern plagues' that sicken and kill people could be coming our way."

20. More than twenty-five nonfiction and fiction titles are listed on the *Outbreak* website.

21. For recent analyses of environmental rhetoric, see Rom Harre et al., *Greenspeak: A Study of Environmental Discourse* (London: Sage Publications, 1999). See also Kevin Michael DeLuca, *Image Politics: The New Rhetoric of Environmental Activism* (New York: Guilford Press, 1999); and M. J. Killingsworth, *Ecospeak: Rhetoric and Environmental Politics in America* (Carbondale: Southern Illinois University Press, 1992).

22. On racist characterizations of AIDS and its links to cultural and environmental topographies, see Susan Sontag, AIDS *and Its Metaphors* (New York: Farrar, Straus and Giroux, 1989): "Africans who detect racist stereotypes in much of the speculation about the geographical origin of AIDS are not wrong. . . . The subliminal connection made to notions about a primitive past and the many hypotheses that have been fielded about possible transmission from animals (a disease of green monkeys? African swine fever?) cannot help but activate a familiar set of stereotypes about animality, sexual license, and blacks" (52).

23. The phrase "standard environmental narrative" was first used by Paul Greenough in analyzing scholarship on South Asian environmental history. I use it to mean widespread narrative structures informing contemporary anxieties about aliens, cross-border flows, purity, and danger within a variety of bodies—national, cultural, ecological, and somatic. On the border-crossing, alien, invasive character of plague, see Susan Sontag, *Illness as Metaphor, and* AIDS *and*

Its Metaphors (New York: Doubleday, 1990): "One feature of the usual script for plague: the disease invariably comes from somewhere else . . . there is a link between imagining disease and imagining foreignness. It lies perhaps in the very concept of wrong, which is archaically [*sic*] identical with the non-us, the alien" (135–136).

24. Preston's phantasmagoric description of Mount Elgon stimulated touristic interest in the area. Mount Elgon and Lake Victoria were the objects of a travel writer's visit, resulting in an article that might be considered a new genre of travel writing, viral tourism, in Jim Keeble, "Travel: On the Wild West Kenya Trail," *Sunday Telegraph*, 11 January 1998 (cited 28 January 2000) (available at *Lexis Nexis: Academic Universe*; see Ref. Lexus Nexus, 18/82). In an unending cycle between representations, the desires they stimulate, and the material movements of people and microorganisms, viral tourism may result in an increased probability that African viruses will be carried to foreign tourists' countries of origin.

25. Preston's frightening image of Kitum Cave has apparently succeeded in elevating it to the status of an international tourist attraction. See, for example, Keeble's account of his visit to Kitum Cave in "Travel."

26. For example, the Marburg virus, one of Ebola's viral "sisters," is described by Preston as invasive, promiscuous, and obscene (1994: 93).

27. On the power of the romantic sublime to induce states of terror and awe, see William Cronon, "The Trouble with Wilderness; or, Getting Back to the Wrong Nature," in *Uncommon Ground: Toward Reinventing Nature*, ed. William Cronon (New York: W. W. Norton, 1995), 69–90.

28. Or in another example from *The Hot Zone*: The "virus subsided on the headwaters of the Ebola River and went back to its hiding place in the forest"; "viruses never go away, they only hide and Marburg continued to cycle in some reservoir of animals or insects in Africa" (89).

29. For an analysis of the historical interconnections between discourses on African climate, natural history, and the emerging science of biology, see Jean Comaroff and John Comaroff, "Africa Observed: Discourses of the Imperial Imagination," in *Perspectives on Africa*, ed. Roy Grinker and Christopher B. Steiner (Oxford: Blackwell Publishers, 1997). For more on images of and ideas about Africa from the late eighteenth century through the middle of the nineteenth century, see also Curtin, *The Image of Africa*. On Africanist discourse in French thought and letters, see Miller, *Blank Darkness*. For an analysis of Joseph Conrad's *Heart of Darkness*, including a vitriolic account of the production of the idea of Africa as the site of the other, see Chinua Achebe, "An Image of Africa," *Massachusetts Review* (winter 1977): 782–795. On the nature of intertextuality, see Julia Kristeva, *Desire in Language: A Semiotic Approach to Literature and Art* (New York: Columbia University Press, 1980). As Kristeva asserts there: "Any text is constructed as a mosaic of quotations: any text is an absorption and transformation of another" (66).

30. See Curtin, *The Image of Africa*, however, which demonstrates how Conrad's African imagery itself is built on early images of the continent as the "whiteman's grave."

31. Joseph Conrad, *The Heart of Darkness* (New York: Dover, 1990), 41–42.

32. Conradian images of Africa are variations on a far older and more general theme. As Achebe, the distinguished Nigerian novelist, writes in an acerbic passage from "An Image of

Africa," "Conrad did not originate the image of Africa which we find in this book [*Heart of Darkness*]. It was and is the dominant image of Africa in the Western imagination. . . . Africa is to Europe as the picture is to Dorian Gray—a carrier onto whom the master unloads his physical and moral deformities so that he may go forward, erect and immaculate" (792).

In Preston's Ebola images, Conradian imagery is not limited to African forests. Ebola is conjured up as a she-devil of a virus, murderous and triumphant in her red-hot chamber. She has an ancestress in Conrad's vision of Kurtz's companion in *The Heart of Darkness*: "She walked with measured steps, draped in striped and fringed cloths, treading the earth proudly, with a slight jingle and flash of barbarous ornaments. . . . Her hair was done in the shape of a helmet; she had brass leggings to the knees; brass wire gauntlets to the elbow, a crimson spot on her tawny cheek, innumerable necklaces of glass beads on her neck; bizarre things, charms, gifts of witch-men, that hung about her, glittered and trembled at every step, . . . She was savage and superb, wild-eyed and magnificent; there was something ominous and stately in her deliberate progress" (75–76).

33. Roger Rouse, personal communication, 6 April 2001.

34. On the rhetorical character of maps in general, see Denis Wood and J. Fels, *The Power of Maps* (New York: Guilford Press, 1992). See also David Harvey, *Justice, Nature, and the Geography of Difference* (Cambridge: Blackwell Publishers, 1996).

35. Robert Kaplan, "The Coming Anarchy," *Atlantic Monthly* 273 (1994): 44–65. Kaplan does not seem to be as concerned about the fate of Africa as much as he is anxious about the possibilities of the future: the *over there* coming *here*, to America and the other developed northern countries. Kaplan's dire prognostications about the fate of African environments and societies has an intellectual lineage. See Jessica Tuchman Mathews's earliest formulation of the environment as a security domain in "Redefining Security," *Foreign Affairs* (spring 1989): 162–177; and Thomas Homer-Dixon's analysis of the causal links between environmental scarcity and violent conflict in "Environmental Scarcities and Violent Conflict," *International Security* 19 (1994): 5–40. For a critical commentary on Kaplan's assumptions, analysis, and prognostications, see Paul Richards, "Out of the Wilderness: Escaping Robert Kaplan's Dystopia," *Anthropology Today* 15 (1999): 16–18.

36. A substantial body of literature on the anthropogenic management and generation of the African forest has been generated within the past decade. For empirical studies and theoretical accounts disputing colonial and postcolonial assumptions about the limits of local knowledge as well as the negative effects of local forest management practices, see Robin Mearns and Melissa Leach, *The Lie of the Land: Challenging Received Wisdom on the African Environment* (Oxford: International African Institute, 1996); and James Fairhead and Melissa Leach, *Misreading the African Landscape: Society and Ecology in a Forest Savannah Mosaic* (Cambridge: Cambridge University Press, 1996).

37. For a trenchant critique of Kaplan's anachronistic, environmentally determinist views of the African forest, see Elizabeth Hartmann, "Population, Environment, and Security: A New Trinity," in *Dangerous Intersections: Feminist Perspectives on Population, Environment, and Development*, ed. Jael Silliman and Ynestra King (Boston: South End Press, 1999), 1–23.

38. For powerful critiques of Kaplan's determinist scenario for Africa, see Elizabeth Hartmann,

"Will the Circle Be Unbroken? A Critique of the Project on Environment, Population, and Security," in *Violent Environments*, ed. Nancy Lee Peluso and Michael Watts (Ithaca: Cornell University Press, 2001), 39–62, and "Population, Environment, and Security"; Eric Ross, *The Malthus Factor* (London: Zed Books, 1998); and Nancy Lee Peluso and Michael Watts, "Violent Environments," in *Violent Environments*, ed. Nancy Lee Peluso and Michael Watts (Ithaca: Cornell University Press, 2001), 39–62.

39. The phrase "emerging cartography of environmental danger" is intended here to broadly refer to the spatial patterning of fears and emerging risks that are increasingly salient at the dawn of the twenty-first century. The phrase is also intended to include the particular spatial forms, geographic locations, and regions that are saturated with these ramifying anxieties. Sub-Sahelian Africa, for example, in the work of Robert Kaplan and Homer Dixon-Smith is a vast subcontinental region that has increasingly come to be associated with environmental risk, while global transportation and information networks including jet transport and the Web can be seen as techno-science sites of global circuitry. These circuits are increasingly seen as sites of risk and danger, subject to potential attack or breakdown.

40. On Southeast Asian environmental racism and its embodiment in racist environmental terminology in Thailand, see Larry Lohmann, "Forest Cleansing: Racial Oppression in Scientific Nature Conservation," *Corner House* (January 1999), briefing 13, 1–24. Lohmann analyzes the use, by Thai governmental officials and other lowland groups, of racist epithets for vegetables, including "Hmong cabbages," as ways of stigmatizing and deriding the entrepreneurial activities of upland ethnic groups such as the Hmong.

41. On the logic underlying how out-of-place things become dangerous, see Mary Douglas, *Purity and Danger: An Analysis of the Concepts of Pollution and Taboo* (London: Routledge, 1992).

42. Roger Rouse, personal communication, 25 May 2001.

43. In an op-ed article titled "Digital Defense" (*New York Times*, 29 July 2001, A19), Thomas L. Friedman criticizes President George W. Bush for emphasizing a strategy of impregnable walls rather than one focusing on the threat of attacks launched at networks and the Webs: "The Bush missile defense plan is geared to defending the country from a rogue who might fire a missile over our walls. But the more likely threat is from a cyberterrorist who tries to sabotage our webs. The more tightly we get woven together, the more we become dependent on networks, the more a single act of terrorism can unleash serious chaos."

44. For a striking reversal of the narrative I am describing, see Mike Anane, "Ghana Rejects U.S. Chimp Sanctuary," *Environmental News Service*, 14 February 2000 (cited 15 February 2000), at www.lycos.com. Anane recounts the rejection of a Friends of Animals' proposal to set up a sanctuary in the Volta region of Ghana for the rehabilitation of chimpanzees used in North American laboratories for experimentation with drugs, vaccines, and genetic alterations by the Ghanaian authorities, including President Jerry John Rawlings. A leading Ghanaian campaigner against the project argued that the animals would be a health threat to local communities near the area of the proposed sanctuary. A statement by the Volta Caucus in the Ghanaian Parliament stated that "the primates would be carrying some diseases . . . [w]hich in the tropics might explode into uncontrollable proportions among the people."

45. See Harvey, *Justice, Nature, and the Geography of Differences.*

46. See Stephen Morse, "Regulating Viral Traffic," *Issues in Science and Technology* 7 (1990): 81–84; and Joshua Lederberg, "The New Global Security: A Common Defense against the Microbe Hordes," *New Perspectives Quarterly* 15, no. 1 (winter 1998): 35–38. On the relationship between recent outbreaks of yellow fever and environmental disturbance by road building, see Richard Horton, "The Plagues Are Flying," *New York Review of Books* 48, no. 13 (9 August 2001): 53–56. Laurie Garrett, in "The Return of Infectious Diseases," *Foreign Affairs* 74 (1996) 66–79, articulates the possibilities of jet transport speeding the movement of pathogens around the globe: "Every day one million people cross an international border. One million a week travel between the industrial and developing worlds. And as people move, unwanted microbial hitchhikers tag along. . . . In the age of jet travel, however, a person incubating a disease such as Ebola can board a plane, travel 12,000 miles, pass unnoticed through customs and immigration, take a domestic carrier to a remote destination, and still will not develop symptoms for several days, infecting many other people before his condition is noticeable."

47. I am indebted to Roger Rouse for emphasizing this point; personal communication, 25 May 2001.

48. In response to a www.virology.net correspondent asserting that "Ebola is the mightiest threat mankind has faced yet," Dr. Ed Rybicki ("The Ebola Virus: The End of the Civilized World," 18 August 1995, www.uct.ac.za/microbiology/ebothrea.html [accessed 6 March 2000]) offers the following assessment from Dr. Margaretha Isaacson, a senior virologist in South Africa: "Ebola . . . is of absolutely no danger to the world at large. It is a dangerous virus, but it's relatively rare and quite easily contained. . . . The media is scaring the world out of its wits, and movies like "*Outbreak*" are doing people a great disservice." Likewise, David Frazier, of the Epidemic Intelligence Service and Department of Microbiology and Molecular Genetics at Harvard Medical School, comments that "while rampant speculation on the dangers of Ebola aerosols is amusing in a manner similar to telling ghost stories around a campfire, such speculation has little grounding in truth. . . . It seems unlikely that a non-airborne pathogen, such as Ebola, would be transmitted to other passengers on a commercial flight" (15 May 1995, at virology@net.bio.net).

49. See Brian Jelle, "Re: The Ebola Virus: The End of the Civilized World," citing David Orenstein on 24 August 1995, at virology@net.bio.net.

50. The name Motaba suggests a combination of two equatorial viruses, Ebola and Marburg, both of which have caused deaths—the former having caused the deaths of monkeys in Reston, Virginia; the latter having killed several German technicians.

51. On arriving in the village, decked out in his biosafety space suit, Hoffman points to a local medicine man, who is gesticulating and making obscure utterances on a cliff above him. "He is talking to the gods," an African in the employ of the U.S. military-bioterror establishment informs Hoffman. "The gods," he continues, were "woken up by white men cutting trees."

52. *Outbreak*, in both the film and the original film script, also reveals its kinship with centuries of prejudice about Africa, its environments, and its peoples. While African forests, monkeys, and medicine men may be innocent and interesting, they are not worth saving. In the 1967 firebombing of the African village in *Outbreak*, the killing of Africans and the incineration of

the village was not protested by anyone. In Hoffman's epic struggle to prevent the bombing of predominantly Caucasian Cedar Creek in northern California, he earnestly protests to his African American superior officer, "These people are Americans!"

53. Tropical forests are often conceptualized as the sources of cures as well as disease. The Periwinkle Project of the Rainforest Alliance, an international nonprofit environmental organization dedicated to rain forest conservation, focused on publicizing the great variety of natural sources of medicine found in tropical forests, suggesting the tropical rain forest was "nature's pharmacy." The lineage for this vein of thinking extends further back in time: A medieval doctrine asserting that tropical forests contain not only a panoply of specific disease agents but also their specific remedies was articulated in H. Fracastorius' *Contagion, Contagious Diseases, and Their Treatment*, trans. W. C. Wright (New York: G. P. Putnam's Sons, 1930). Paul Greenough brought this point to my attention. More recently, a phytochemist, Dr. Maurice Iwu, the descendant of a family of Nigerian healers and founder of the Bioresources and Conservation Programme, announced that some his studies of compounds obtained from the bitter kola tree have quashed some strains of flu virus. Iwu was quoted as saying, "The same forest that yields the dreaded Ebola virus could be a source of a cure" (cited in "Folk Remedy Zaps Ebola in a Lab Test," *Science News* 156 [14 August 1999]: 110).

54. In discussing these advertisements, I focus on the intentions of their producers rather than the reception of the ads by particular audiences. In this ad series xenotransplantation is explained as "the transplantation of animal organs and tissues, including those which have been genetically engineered into humans."

55. Even a partial listing of organizations for advertisement no. 2, the ad with the Ebola image in the center, suggests the remarkable span of issues and agendas united under the Turning Point Project: the Humane Society USA; Mothers for Natural Law; Institute for Agriculture and Trade Policy; Friends of the Earth; Organic Consumers Society; the Council of Canadians; Rainforest Action Network; and the Idaho Sporting Congress. While the advertisement emphasizes a common purpose, stating that "signers are all part of a coalition of more than 60 nonprofit organizations [the International Forum on Globalization] that favor democratic, localized, ecologically sound alternatives to current practices and policies," the Turning Point series and the forum itself are deeply complicated political projects. The signers of Turning Point series were not only interested in endorsing provocative advertisements. They were simultaneously engaged in cross-border, transnational alliance building across issues, ideologies, and political boundaries. For instance, while the Institute for Agriculture and Trade engages with genetic engineering and biotechnology from the perspective of family farms, the Humane Society USA has the interests of domesticated animals and their well-being in mind; the Rainforest Action Network creates linkages to this advertisement as an advocacy organization for the conservation of tropical rain forests and biological diversity.

56. Although space will permit only a brief glimpse at the institutional context and historical background to these advertisements, even a momentary view is illuminating. The Turning Point series, which began in September 1999, was conceptualized by Jerry Mander, an advertising professional formerly based in New York, who on moving to the San Francisco Bay Area, founded the Public Media Center. The goal of the Turning Point Project was to create a graphic

series of advertisements—preceding and leading up to the protests against the World Trade Organization meetings in Seattle in December 1999, and continuing through the following year—that were "flat out provocative." Mander, the chair of the International Forum on Globalization—an alliance of many organizations working on a remarkably diverse set of issues ranging from economic globalization, global governance, and intellectual property rights, to genetic xenotransplantation and organic agriculture—sought to fashion a "new genre" of public media advertising that would create an expectation of future messages and an "overall buzz."

Under the image of an Ebola virus from a tropical forest in Africa, an astonishing alliance was being formed, not only to protest globalization but also to articulate linkages and resistances across geographic space and issues: "You look for allies," says Mark Ritchie (personal communication, 7 July 2000), "in clean water, in migrant labor, in antibiotics. We have made some of what I call odd alliances. We are finding support from the religious Right, the Rural Life Southern Baptists, who would say things like, 'I can own cattle, but not the species.' Factory farms can be linked to other constituencies, labor issues, and public health. We are taking on the CEOs, economists, and the ruling elite. We are challenging the entire ruling class. Of course, there are members of this community which sponsored the ads that are so local they might even be a little xenophobic in their positions."

57. Turning Point Project, "Where will the next plague come from?" *New York Times*, 1 November 1999, A9.

58. On the semantic logic of spells, see Stanley Jeyaraja Tambiah, *Magic, Science, Religion, and the Scope of Rationality* (New York: Cambridge University Press, 1990). See also Kirk Endicott, *An Analysis of Malay Magic* (Oxford: Clarendon Press, 1970). On the power of metaphoric and analogical transfers of properties from one class of objects to another, see Michelle Zimbalist Rosaldo, "Its All Uphill: The Creative Metaphors of Ilongot Magic Spells," in *Sociocultural Dimensions of Language Use*, ed. Mary Sanchez and Ben G. Blount (New York: Academic Press), 177–203. On the role of metaphoric transposition of qualities from plants to boats in fishing magic and spells, see Charles Zerner, "Sounding the Makassar Strait: The Politics and Poetics of an Indonesian Marine Environment," in *Culture and the Question of Rights: Forests, Coasts, and Seas in Southeast Asia*, ed. Charles Zerner (Durham: Duke University Press, 2003).

59. For critical assessments of the representation of ecological systems as coherent, bounded, and ahistorical entities, see Cronon, "The Trouble with Wilderness." See also Daniel Botkin, *Discordant Harmonies: A New Ecology of the Twenty-First Century* (New York: Oxford University Press, 1990); and Karl S. Zimmerer, "Human Geography and the 'New Ecology': The Prospect and Promise of Integration," *Annals of the Association of American Geographers* 84 (1994): 108–125.

60. In the first advertisement of the Turning Point Project series, captioned "Who plays God in the twenty-first century?" underneath a genetically engineered mouse with a human ear on its back, the copy reads: "Breaking the species barrier. Whether you give credit to God, or to Nature, there is a *boundary between lifeforms* that gives each its integrity, and identity."

61. While Turning Point deploys the image of Ebola in a complex campaign against contemporary science and genetic research, the scourge of traveling diseases is also utilized in a publica-

tion by the series architect, Jerry Mander, to critique the globalization of the economy and rain forest destruction: "Horrible new disease outbreaks are very thoroughly reported with ghoulish relish in the Western press. The part that is omitted, however, is the connection between these outbreaks and the destruction of rain forest and other habitats. As economic expansionism proceeds, previously uncontacted organisms hitch rides on new vectors for new territory" (Jerry Mander, "Corporate Colonialism," *Resurgence* (online), September/October 1996 (cited 15 December 2000), at http://www.gn.apc.org/resurgence/articles/mander.htm.

62. At the time of this writing, the full-page Turning Point advertisements have appeared only in the *New York Times* and *Washington Post.*

63. Turning Point, "Who plays God in the twenty-first century?"

64. For a critical history of the idea of wild nature as a sacred precinct and indigenous ecologies, specifically as coherently bounded, ahistorical entities, see Cronon "The Trouble with Wilderness," 69–90. See also Michael Pollan, *Second Nature: A Gardener's Education* (New York: Delta, 1991). On the idea of wild nature reserves and regions of tropical biodiversity as coherently bounded entities targeted for protection by global funding, see Charles Zerner, "Telling Stories about Biological Diversity," in *Indigenous Peoples and Intellectual Property Rights,* ed. Steven Brush and Doreen Stabinsky (Washington: Island Press, 1994). For a contemporary critique of the notion of tropical ecological systems as coherent, stable entities, see Botkin, *Discordant Harmonies;* and Karl Zimmerer, *Nature's Geography: New Lessons for Conservation in Developing Countries* (Madison: University of Wisconsin Press, 1998). See also Zimmerer, "Human Geography and the 'New Ecology.' "

65. Roger Rouse, personal communication, 25 May 2001.

66. Anxieties about alien invasions, maintaining integrity in the face of border penetration and control, and racialized rhetorics of fear about exotic others penetrating national boundaries and subverting U.S. national identity are transparent in recent policy position papers and programs of the U.S. Immigration and Naturalization Service (INS). In 1993, the INS adopted and proceeded to implement an aggressive immigration control campaign on the U.S.-Mexico border seeking to develop "a host of new strategies and technologies designed to 'restore integrity and safety' to Southwestern borders" (Lisa Sanchez, manuscript, 2000, 1). These programs involved high-tech tracking and monitoring equipment deployed to restrict the flow of illegal aliens across a 2,000-mile stretch of the U.S. southwestern border. As Lisa Sanchez, a legal scholar exploring the intersection of discourses and representations of disease and criminality in justifying repressive immigration policies, states: "Construed as armed soldiers and dangerous criminals in a war against American community and economy, they [illegal immigrants] are as much represented as symbols of *disease,* seeping across the *fragile membrane* that separates 'first' and 'third' world, poised to infect the racial and cultural *integrity* of U.S. citizens" (manuscript, 2000).

67. On other parallels between xenophobia and nativist anti-immigration discourse, on the one hand, and anti-immigration, endemic species discourses in conservation biology, see Banu Subramaniam (*Chronicle of Higher Education,* November 2001, B9), a biologist and women's studies scholar, who asserts that discourses on alien plant and animal invasions constitutes

a "xenophobic" rhetoric. See also Mark Sagoff, "What's Wrong with Exotic Species," 1–13, http//www.puaf.umd.edu/IPPP/fall1999/exotic_species.htm.

68. On the production of narratives of fear, see R. D. Lipshutz, "Terror in the Suites: Narratives of Fear and the Political Economy of Danger," *Global Security* 13, no. 4 (1999): 411–439. Lipshutz articulates a state-generated security narrative that encompasses cosmic as well as calculated threats as "part and parcel of the production and reproduction of the sovereign, autonomous nation-state." Lipshutz attributes widespread national security anxieties and the proliferation of "threats" in the post–cold war era in the United States to "globalisation and the social uncertainty it has generated" (414). As well, Lipshutz focuses on the need of military and political leaders in the post–cold war era to articulate the locus of a new threat and, concomitantly, a new military mission.

On the use of metaphors of criminality, disease, or toxicity-based representations and rhetoric in analyzing the flow of nonindigenous species of plants and animals, see Bill N. McKnight, *Biological Pollution: The Control and Impact of Invasive Species* (Indianapolis: Indiana Academy of Science, 1993); Nature Conservancy, *America's Least Wanted: Alien Species Invasions of U.S. Ecosystems* (Nature Serve Publication, 1996), Robert Devine, *Alien Invasion: America's Battle with Non-Native Animals and Plants* (Washington: National Geographic Society, 1998); Chris Bright, "Bio-Invasions," in *The World Watch Reader on Global Environmental Issues*, ed. Lester Brown (New York: W. W. Norton, 1998), 115–134, and *Life out of Bounds: Bioinvasion in a Borderless World* (New York: W. W. Norton, 1998); David Quammen, "Planet of Weeds," *Harper's* 297 (1998): 57–69; Stanley Temple, "The Nasty Necessity: Eradicating Exotics," *Conservation Biology* (1990): 5; and Bruce Coblentz, "Judas Goats in the Tropics," *Aliens* (Species Survival Commission newsletter), March 1995. The literature in conservation biology journals and the news media using this family of metaphors within the alien-invasion narrative is enormous and continuing to grow. For accounts critical of the alien alarm against invading nonnative species, see Michael Pollan, "Nature Abhors a Garden, in *Second Nature*, 45–64, and "Weeds Are Us," in *Second Nature*, 116–138; and Mark Sagoff, "Why Exotic Species Are Not as Bad as We Fear," *Chronicle of Higher Education*, 23 June 2000, B7. See also the critical commentary of Banu Subramaniam (*Chronicle of Higher Education*, December 2001), who contends that the rhetoric of biological invasion is one more symptom of the displacing onto outsiders and foreigners of anxieties about the economic, social, political, and cultural changes associated with globalization.

69. This essay was drafted during my tenure as a fellow at HRI, well before the World Trade Center disaster of 11 September 2001. Clearly, the reflections on borders, surveillance, and the increasing normalization of a militarized U.S. society, at "home" and "abroad," are increasingly relevant in the post–11 September 2001 era.

70. I am indebted to Michael McKeon for this formulation.

71. There is a large and rapidly proliferating literature on "bio-invasion"—the movement of exotic species from other regions and countries, the effects of these species on native ecologies, and the war that is proposed—by many conservation biologists and nongovernmental conservation groups against them. The same kinds of metaphors of criminality, pathogens, the

indigent, hitchhikers, and stealth species that are deployed in constructing rhetorics on tropical viruses and their "invasions" as well as in anti-immigration discourses are used in the immense bio-invasion literature. For expressions of concern about the potential parallels between racist, anti-immigration thought and the bio-invasion literature, see Sagoff, "Why Exotic Species Are Not as Bad as We Fear." See also Sagoff, "What's Wrong with Exotic Species?" in which he states that biologists often attribute to exotic species the same "disreputable characteristics that xenophobes have attributed to immigrant groups. These undesirable characteristics include sexual robustness, uncontrolled fecundity, low parental involvement with the young, tolerance for 'degraded' or squalid conditions, aggressiveness, predatory behavior, and so on. This kind of pejorative stereotyping may be no more true in the ecological than in the social context" (http://www.puaf.und.edu/IPPP/fall1999/exotic-species.htm).

72. This was true by the end of the twentieth century, prior to the grotesque, terrifying spectacle of the destruction of the World Trade Center in New York City on 11 September 2001 and the terror-inducing anthrax deaths that followed. For post–11 September discourse on the virus as master metaphor, see the announcement for a cross-disciplinary conference titled "VIRUS!" circulated on the Web in early January 2002 at http://www.kah-bonn.de/fo/virus/Oe/htm.

73. On international news coverage of Ebola outbreaks, and the concept of a stable Western civilizational core and a periphery in constant turmoil, see Aldo Benini and Janet Bradford, "Ebola Strikes the Global Village: The Virus, the Media, the Organized Response," October 1995, at http://www.outbreak.org/cgireg/dynaserve.exe/Ebola/benini.htm. Benini and Bradford emphasize the way in which the post–Cold War has accelerated a map whose imagery of core and chaotic periphery echoes an ideological formation articulated during Roman times: "The Roman world was one of strict, even physical division between the reign of its law versus the peoples who lived outside its fortified borders, the limes."

74. Stuart Hall's concept of articulation, in the sense of establishing a linkage between two domains or networks previously unrelated, and in the sense of "speaking for," provides a useful way of understanding the uses of Ebola in a variety of political and environmental movements. See Stuart Hall, "On Postmodernism and Articulation: An Interview with Stuart Hall," in *Stuart Hall: Critical Dialogues in Cultural Studies*, ed. David Morley and Kuan-Hsing Chen (London: Routledge, 1996). For applications of Hall's concept of articulation in transnational political meanings and movements, see Anna Tsing, "Becoming a Tribal Elder, and Other Green Development Fantasies," in *Transforming Indonesian Uplands: Marginality, Power, and Production*, ed. Tanya Murray Li (Amsterdam: Harwood Academic, 1999), 159–202; Tanya Murray Li, "Articulating Indigenous Identity in Indonesia," *Comparative Studies in Society and History* 42 (1999), 149–179; and Charles Zerner, "Moving Translations: Poetics, Performance, and Property in Indonesia and Malaysia," in *Culture and the Question of Rights: Forests, Coasts, and Seas in Southeast Asia*, ed. Charles Zerner (Durham: Duke University Press, 2003), 1–23.

AFTERWORD: THE ONGOING SEARCH

CANDACE SLATER: As this book enters the production process, we find ourselves on the eve of the big Johannesburg United Nations Conference on the Environment, more commonly known as Rio Plus Ten. Although the present barrage of media reports on the conference will soon fade, Rio Plus Ten invites us to think in a more lasting way about some of the things that have—and haven't—happened in regard to environmental, and related social and political, issues over the past ten years. It's also particularly interesting to consider some of the changes that have taken place in rain forests and accompanying transformations in that generic Rain Forest that was almost certainly the single most compelling symbol for environmental destruction back in 1992.

Rio Plus Ten pushes us to think back on our shared concern with the relationship between environmental images and environmentally important actions. We've spoken directly or indirectly about this relationship at various points in our essays, and Rio Plus Ten gives us an incentive to recap our thoughts. This recap also offers us an opportunity to invite the participation of the three visitors to our seminar who were a significant part of the discussions that underlie this book. As conservation biologists who have worked in a variety of tropical forests, Francis Putz and Claudia Romero have their own perspectives on how these forests are depicted. Likewise, David Baron's work as an environmental journalist has given him a non-academic focus quite different in some ways from our own. How, then, would they—as well as the authors—revisit the essays in this book? What, in a nutshell, would appear to be the relationship between rain forest representations and the sorts of practical actions that affect today's rain forests? How would we sum up the possibilities as well as the limitations of environmental discourse analysis in light of ongoing events?

CHARLES ZERNER: I live in New York City. Perhaps it is no accident, then, that I often experience moments of intense environmental fantasy while walking through tunnels or beneath buildings, making connections. I began the research for my chapter after observing a picture of Dustin Hoffman on a monkey-populated, jungle poster on the wall of the West Fourth Street subway station in Greenwich Village. Two years later, just after the September 11 catastrophe, throughout the long, anxious winter of 2002, I continued to catch the North Hudson–Harlem line train to Bronxville. Each morning I walked past machine-gun-toting marines and heavily armed New York City police officers, restraining muzzled German shepherds and guarding the portals of Grand Central Station. The premonitions I once had in sunny southern California while working on "The Viral Forest" can no longer be dismissed. The U.S. public inhabits a world in which civil liberties, privacy, the right to travel, immigrants, and guarantees against profiling and unreasonable searches and seizures are increasingly restricted. Terror—including a Pandora's box of environmental terrors such as anthrax, smallpox, immigrants, nuclear reactors, and water reservoirs—has become the animating image in the hands of policymakers for uprooting, reorganizing, and rupturing regions of public policy and law that were until recently, considered fundamental, protected spheres of freedom and U.S. political culture.

In this climate of anxiety as well as diffuse paranoia about uncertain risks of aliens in the environment, including people as well as microorganisms and the channels through which they move—the networks—environment and security have become fused at the hip. According to some scientific and cultural observers, the virus is the master metaphor of the twenty-first century. When I drafted my chapter, I focused on potentially troublesome ways of visualizing and talking about geography, epidemiology, and ecology. I was also concerned with environmental rhetorics and metaphors about Africa, its peoples, rain forests, and microbes; discourses about illegal aliens (persons as well as nonhuman life-forms) crossing national borders, body boundaries, and ecological zones; and imagery and rhetorics linking non-endemic species with criminals, germs, and illegal immigrants. I suggested that these environmental representations needed to be examined in a critical light.

I hope to remind those of us in the environmental and civil rights communities that we have a responsibility as public intellectuals and scholars to hold the images, metaphors, and rhetorics of environmentalism up to scrutiny. Plato feared artists and rhetoricians for their power to move us, to shake and shape our beliefs and behaviors. Percy Bysshe Shelley boldly asserted, "Poets are the unacknowledged legislators of the world." Unlike Plato, Shelley acknowledged the impossibility of banning poets from our lives, or rhetoric from our scientific expressions. Nor would he consider that outcome desirable. In science and as in fiction, indeed in life, there is no apprehension of the world without the mediation of metaphor, model, and trope. We are all within the charmed and animate circle of language. In the complicated, threat-saturated world in which we walk and work, in matters of policy and personal conduct, the moment of choice is pivotal—the moment when we select, from a range of alternatives, our metaphors and models for human and environmental relations. Representations have an unnerving tendency to become realities.

These days, across staggeringly different domains, the emphasis in representations seems to be on war and the security state: militarized nature and armed troops defending it. In the world of global conservation discourse, more than a few respected figures are suggesting a return to a militarized "security approach" to conservation, surely a euphemism for state or multilateral-sanctioned violence, a policing of strictly enforced boundaries between the nonhuman and human, the natural and cultural. These ideas are not only scientifically flawed but morally and politically suspect.

I am hoping that in the wings, there are poet-legislators like Michael Pollan, who casts metaphors for nature as garden and humans as gardeners, and finds a voice that touches audiences beyond the world of science and policy. There are alternatives, as writers such as Pollan, Simon Schama, and

William Cronon remind us, to visions of nature as inviolate, pristine sanctuary divorced from history and protected from culture—visions that lay the groundwork for a world in which the "defense of nature" mirrors the practices of the militarized security state. As a concerned citizen and scholar, I cast my vote for those poet-candidates whose metaphors for nature and environmental practices gesture toward humane, democratic, just, and plural possibilities. I cast my vote for representations of a nonmilitarized nature, for metaphors that may mobilize citizens to detoxify an industrial culture and its relations to the natural world, while preserving a citizen-gardener's freedom to shape multiple designs.

DAVID BARON: Around the time of the first Earth Summit, I made a pilgrimage to the Amazon. My head filled with images from Greenpeace posters and *Time* magazine, I wanted to view the celebrity forest firsthand, to see the pristine, imperiled ecosystem before it vanished. Yet like a fan glimpsing a movie star on the street ("He looks so much *taller* on screen"), I found the experience initially disappointing and disillusioning. The Amazon was not what I expected.

Monstrous, crowded, industrial Manaus—smokestacks belching and gas flares burning—provided a bizarre welcome to the world's largest rain forest. Amazonian Indians greeted me with pet sloths and outstretched hands, offering to be photographed for a price. Litter floated down an oil-slicked Rio Negro. So much for the pristine Amazon rain forest.

I fled Manaus and ventured deep into the jungle, but it too failed to satisfy. The forest turned out to be, well, a *forest*. Sure, it sported exotic lianas and big-rooted kapok trees, but most of the vegetation appeared (to my untrained eye) remarkably similar to the woods of New England. Where were the spotted jaguars that stared out from Greenpeace posters? Where were the multihued parrots? Where was the much vaunted, mind-staggering biodiversity?

Years later, invited to join the rain forest discussion at HRI, I reflected back on my disappointment at meeting the Amazon. How had I been misled about the rain forest's on-the-ground reality? What images had I internalized? Were those images ultimately harmful? In the course of our conversations, I came to see my experience as an object lesson, a small example of the benefits and hazards of presenting ecosystems in dramatic, idealized terms.

I had been enticed to Brazil by a compelling tale, oft-repeated in the late 1980s and early 1990s, of evil corporations, innocent Indians, and a magnificent ecosystem on the verge of collapse. Though not wholly untrue, this story was exaggerated and oversimplified, like the Hollywood adaptation of a nonfiction book. The tale came about and prospered for a clear reason: it made good copy. Reporters love to tell a smashing yarn, a tale of crisis, of good guys and bad guys, of the earth in peril. Environmental activists, aware of how journalists work, fed the media this tale while conveniently ignoring the complexities.

The resulting image of an incredible, endangered Amazon—broadcast, printed, and direct mailed to millions a decade ago—produced a concrete result: it brought First World money and attention to a Third World environmental problem. It raised global consciousness. But it contained a snag. Just as I was disappointed to discover the Amazon's complex reality, environmentalists have been forced to come to a similar realization. A decade post-Rio, tropical deforestation has not appreciably slowed. Activists now admit that many of the solutions initially proposed were naive, unsophisticated. Struck by the difference between the bumper sticker idealism of a decade ago and the messy reality of today, environmentalists have begun to see that if they are to "Save the Rain Forest," they will need to adjust their image of what that rain forest is.

I went through such a process of adjustment during my brief stay in the Amazon. After my initial frustration at finding the rain forest unlike the images on television, I came to appreciate what I was actually seeing and returned home with new images in my mind: a peanut vendor in the streets of São Gabriel de Cachoeira; a highway of leaf-cutter ants; the grand opera house in Manaus; a canoe trip up one of the small tributaries known as *igarapés* with a local guide; a sip of *guaraná* (an Amazonian berry soft drink) from an aluminum can; a blue morpho; a thunderstorm; an evening of dominoes with a Swiss expat, a Brazilian businessman, and an Amazonian Indian. Though not as pristine nor as immediately dramatic as the forest I had expected, the Amazon still amazed. What made it fascinating was the mélange of people, forest, and water, the interaction of humans and their multiple environments.

With thoughtful guidance (including the essays in this volume), the general public can hopefully come to appreciate this more nuanced view of rain forests—a view that may lead to more effective methods of protecting the ecosystems and helping the people who live within them. By embracing the complexities, activists, governments, corporations, and native inhabitants we should be able to move beyond posturing and into problem solving. But a question remains.

The world would not have paid such rapt attention to rain forests, indeed I would not have traveled to the Amazon, without the romanticized image of a decade ago. If environmentalists now promote a more realistic, multi-faceted view, one with less glitz and drama, will the media remain interested? Will the public still care? The authors of this book believe the answer is yes, and I hope they are right. The planet may have no other choice.

SCOTT FEDICK: I'm writing these remarks after having just returned from another field season in the Maya forest of northern Quintana Roo, Mexico. As expected, I saw that development in the Cancún region is accelerating at a dizzying pace along the coastal zone of the Maya Riviera as well as the inland areas. The dirt road that takes one from the highway to the north and into our study region has a new billboard advertising lots for sale. Word is out that the road is about to be paved, and the whole region will be opened up to development. As I drove out to the field for the last time, my friend Ross and I were shocked at being stopped in a traffic jam on the dirt road, which rarely had any vehicles on it. It was election day, and it so happened that a political meeting was being held with the new landowners of the area to discuss the planned paving of the road. In fact, two bulldozers and several dump trucks just happened to be doing repairs on the section of road directly in front of where the meeting was being held . . . and nowhere else. All along the road were new ranchos that had recently been purchased and completely stripped of all vegetation by non-Maya Mexican immigrants who had cleared the land as a first step toward "improving" it. Interspersed between these newly cleared properties were other recently abandoned, overgrown plots with bracken fern or grass, with a hand-painted "For Sale" sign propped up by rocks in the soiless patch of destruction.

Why do I react so negatively to this situation, when in my chapter, I call for people to be included in the Maya forest? Because so much destruction could be avoided if new immigrants to the region could only recognize and incorporate the traditional knowledge of local Maya farmers and gardeners.

For about every ten new ranchos we saw along the road that looked like the surface of the moon, there was one where the new owner had selectively cleared some of the trees and other vegetation while planting an amazing variety of other productive trees. These new farmsteads were being established either by Yucatec Maya settlers from other parts of the peninsula or immigrants from other parts of Mexico who had learned from their Maya neighbors how to establish a homestead in a sustainable manner. In other words, they were resettling a garden gone to seed and returning it to a managed plot of the Maya forest.

There is hope. I know conservation biologists who are now studying traditional Maya farming as a means of preserving the forest biodiversity. I met a tour guide who wants to open a new business specializing in visits to ancient Maya sites that are located on the lands of small Maya communities—an ecotourism that features sustainable farming and forestry practices in many of those same traditional communities. The School of Ecological Agriculture, in the state of Yucatán, is both preserving and promoting traditional Maya agricultural practices while incorporating new scientific technical knowledge being developed in cooperation with the Autonomous University of Yucatán. Staff members of the Sian Ka'an Biosphere Reserve in Quintana Roo are working with local communities to develop wildlife corridors that connect existing reserves while assuring the economic viability of settlements within those corridors.

The Maya forest has proven itself over several millennia to be a resilient habitat that has always included humans. While working to preserve some zones of old-growth forest, we should at the same time recognize the value as well as ecological benefit of conservation through managed and sustainable use. And we should recognize both the power of images and the urgent need for their active reinterpretation. Now, more than ever, we need the ability to see that supposedly primordial rain forest above which Maya temples soar on a hundred different posters for what it is—an abandoned garden that with care, can bloom again.

SUZANA SAWYER: Without doubt, representations, politics, and policy are intimately entwined. The question is how. As I've tried to demonstrate in the case of Ecuador, constructing the rain forest as a purely bio-geophysical realm carries with it specific dangers. This focus not only obscures the fact that the rain forest is part and parcel of ongoing human historical relations but it also diverts attention away from the complex webs of power that the extractive industry creates and that are so divisive to the human lives in the region.

The fact that representations help shape policy became even more apparent to me last summer when Leonardo, an Ecuadorian indigenous leader visiting my home in California, told me of a meeting in which he had recently participated. The meeting was part of a program called Energy, Environment, and Population (EAP). In 1998, the World Bank in conjunction with the Latin American Energy Organization (OLADE) established EAP to create a forum in which representatives from petroleum companies, state governments, and indigenous organizations could come together to discuss the challenges facing hydrocarbon activity in the Upper Amazon—the area that includes parts of Venezuela, Colombia, Ecuador, Peru, Bolivia, and Brazil, and in which, according to OLADE, oil exploration and exploitation, a fragile biological environment, and millenarian indigenous communities coincide. The purpose of the EAP program was to allow oil companies and indigenous organizations interested in maintaining a fruitful dialogue to promote joint action for developing the region sustainably.

According to the indigenous leader, the EAP meeting was just as confining as the Technical Environmental Committee I discuss in my chapter. "We were good for their photo-ops," Leonardo noted. "La compañía loved to have their picture taken with us. But beyond that, we scarcely shared a thing." Those defining the meetings' framework and agenda largely saw and labeled the Amazonian region as a biophysical realm. Their concern was to develop the technologies and know-how to preserve this precious and rare biosphere. For the most part, indigenous voices were present to praise multinational corporate activity. Indians were not invited to challenge or object to specific oil operations. Nor were they invited to suggest solutions to or safeguards against the inequalities that the extractive industry both relies on and perpetuates. As Leonardo remarked, "They [the industry representatives] were more concerned about the macaw than about how they cause brothers to fight"—that is, than about how corporate activity has incited conflict and division among neighboring indigenous communities.

In point of fact, ARCO's oil operations in Ecuador were the model for establishing the EAP program. Based in Ecuador, OLADE was well acquainted with ARCO and its work. So was the World Bank. The World Bank prominently cited the multiple publications that champion the ingenuity of ARCO's technological operations in Ecuador on its website. The EAP program was modeled after ARCO's Technical Environmental Committee, which drew together corporate, state, and indigenous officials.

As is evident in Ecuador and will become more so with the World Bank–sponsored EAP, we embark on dangerous terrain when our understanding of the rain forest is overly determined by the precepts of conservation biology and geophysics. Without doubt, these sciences have enabled the oil industry to engage in immensely improved (and less polluting) practices. Yet an exclusive focus on biophysical phenomena can all-too-easily foreclose a simultaneous knowledge of the forest as a social landscape, a space whose complex and volatile lived history informs Leonardo's thinking and a broader indigenous politics. If we care about a more just, equitable, and sustainable forest, then we must be attuned not only to biophysical processes but also the contentious and disruptive social processes that shape the rain forest today.

FRANCIS E. PUTZ: Growing evidence that humans have had a hand in shaping many of what were formerly considered to be untouched forests makes it difficult for conservation biologists to disregard the social dimensions of these forests and the associated questions about rain forest images so important to the authors of this book. Surprisingly, not all conservation biologists agree that people need to be involved in any active way in rain forest production. Instead, some argue that we should lock up large tracts of pristine rain forest before it is too late. Rain forests are indeed going fast, taking with them staggering numbers of species. But who should be the keepers of the keys to these forests? And if, as growing evidence suggests, people shaped these forests, isn't their continued participation needed to reverse the current trends of destruction? I don't want to see conservationists demonized and want to keep strict protection in my portfolio of conservation options along with more active forms of management, but these and other similarly disturbing questions must be addressed.

As an ecologist and borderline misanthrope, I have also become increasingly concerned that even the scientific justification for rain forest protection has been weakened by scientific evidence. Despite the best efforts of "spin" experts in the science/conservation community, for example, field experiments have failed to demonstrate that all or even most species are needed for rain forests to function. Yes, ecosystem processes such as productivity and nutrient cycling do improve with increased numbers of species in experimentally created communities, but the benefits level off with just modest diversity. Species redundancy may provide ecosystems with resilience, but this is not the same as claiming that function derives from diversity. The idea of a "web of nature" is still compelling, but the fabric now seems a bit threadbare. And yes, rain forests harbor species that may provide cures for many human ailments, but designing drugs to suit particular purposes using molecular biology techniques appears to be at least as cost-effective as pharmaceutical exploration, especially if intellectual property right problems, such as those on which Alex Greene touches in his chapter, can thereby be avoided.

Today, mounting evidence suggests that many rain forest conservation programs fit comfortably under the now familiar rubric of "faith-based initiatives." Forests can provide sustainable livelihoods, recycle nutrients, and sequester carbon, but so can oil palm plantations. Certified timber might sell for higher prices than timber mined destructively, but natural forest management is likely to remain financially unattractive where soybeans could flourish for even a season or two. Timber production rates in natural forests can also be increased, but they will never attain plantation rates. But don't other species, cultures, languages, and ecosystems have fundamental rights to exist? I believe that they do, but this belief rests uncomfortably on faith, not science.

Integrating conservation and development in the tropics is a challenging goal made more difficult by the paucity of images of the future that faithfully reflect what we have learned. Many conservation biologists have been guilty of a narrowness of vision, but so have some social welfare advocates. Devolution of control over forests to indigenous groups and other rural communities seems fundamentally right, but will the forests provide opportunities for them to better their lives or just serve as poverty traps? And how much control are we willing to devolve? Seeing what happened to the forests in Indonesia after the fall of the Suharto regime and subsequent decentralization of the government does little to inspire confidence in local control. Similarly, the hoped-for conversion of communities of subsistence farmers into corporations of forest managers in Bolivia also now seems like a hard goal to reach. But what are the options for keeping some rain forests intact?

The emerging diversity of representations of rain forests is certain to cause some dismay among a number of conservation biologists especially those of us who chose to study nature so as to avoid the people we now need to consider in past, present, and future tenses. Dismay notwithstanding, these newly diverse images should ultimately make conservation more compelling to a wider audience, including the people who are most dependent on rain forests for survival. But before much progress can be made, we need to agree that distinctions between "natural" and "cultural" impacts on forests become irrelevant when the sweeps of time are large. From this broader perspective, active forms of management become more acceptable as conservation strategies to even the most ardent protectionists. I am convinced that a greater diversity of visions can inspire a greater range of solutions to the linked suite of social, economic, and biophysical problems that currently result in rain forest destruction.

ALEX GREENE: My account of the competing "voices" behind Ix Chel, the Maya goddess of healing, is part local intrigue, part literary critique. But is it of any use to those involved in the nuts and bolts of sustainable development or environmental activism? I would hope that some pragmatic lessons could be learned from Rosita Arvigo's trials in Belize since she is the very picture of both idealism and pragmatism, having seized all opportunities that came her way in pursuit of an alternative vision of progress. In the process, I think she succumbed to the global bioscript of the indigenous traditional healer, which in the end caused her to render many local healers mute, at least within her work—an occupational hazard for any well-intentioned international activist in a local setting.

Ethnobotany and bioprospecting are probably less important to development planning today than they were when I initiated my research in the 1990s. However, the notion that pharmaceuticals can be developed from local herbal remedies is still with us, and plans for attributing intellectual property rights to this or that community are still being proposed. I would hope that my chapter could serve as a cautionary tale about the dangers of such legal conceits.

I would also hope that any ethnobotanist or anthropologist reading the chapter might find new appreciation for the importance of theorizing culture and the idioms of culture's representation. As ethnobotany becomes increasingly focused on the molecular and genetic aspects of plants, even at the expense of traditional concerns such as taxonomy, considerations of representational power and local history are easily forgotten.

At the same time that the erasure of these considerations constitutes a potential failing, however, the beauty and strength of ethnobotany continues to reside in its detailed engagement with the material basis of a given culture. This engagement sometimes eludes anthropologists who are primarily concerned with popular representations. There were definitely points in my fieldwork when I longed for some botanical training in order to better understand a healer's subtle grasp of the nonhuman world. Better yet, to have been working with an ethnobotanist on a collaborative basis could have greatly improved my understanding of the ways nature is represented hand in hand with cultural identity on the most concrete and detailed level.

Ultimately, translating local reconstructions of nature into the language of professional scientists, and vice versa, can help to foster a more critical traffic in ideas. Because of the way it tacks back and forth between diverse bioscripts, ethnobotany offers the best potential for any critique of scientific process and thought to emanate from within the sciences. Such a critique could complement the discipline's more traditional mission, and place it on the cutting edge of anthropology and "science, technology, and society studies" as well.

Sandra Harding, the philosopher of science, writes in *Whose Science? Whose Knowledge?* that scientists of all stripes should adopt a position of "strong objectivity" toward their own work. This means, first of all, taking into consideration the social groups that one's work renders invisible or irrelevant, and placing one's own work in a larger social context. To me, this also suggests paying greater attention to how one represents oneself and the various others who comprise the other in any given study. At least for ethnobotanists and anthropologists, the best starting point is to pay more attention to the Romantic assumptions of "the folk" that underlie much ethnobotanical writing to date, and challenging those assumptions with more comprehensive models of how a "folk" can be both traditional and modern at the same time.

There is much to be said for any research, whether by academic ethnobotanists or shaman's apprentices like Arvigo, that illuminates the ways that different forest peoples lead their lives. This sort of work offers a crucial corrective to the nostalgia surrounding not just allegedly traditional peoples but also a supposedly ideal and, even today, still tenaciously unpeopled nature. If ethnobotany were to engage the anthropology of power and representation, and anthropologists of representation were to incorporate the environmental sciences, we would have a powerful argument against those who would defend a vision of wild purity at any price.

CLAUDIA ROMERO: Recently, several prominent preservationists backed by the huge budgets of international conservation organizations (ICOs) have begun to argue that the only tried-and-true way to conserve tropical rain forests is by locking them up in nature preserves. Although I acknowledge that purchasing land in developing countries may delay deforestation and forest degradation in a few select areas, the fallacy and danger of their representation of rain forest as essentially peopleless or selectively peopled has been underscored in different ways by all of the contributors to this volume. But sovereignty issues aside, what most concerns me is that the grossly asymmetrical negotiations between representatives of tropical countries and wealthy ICOs destroy the normal processes through which the fates of tropical forests should be decided. While perhaps expedient, the authoritarian preservationist approach of these ICOs results in the abortion of local social processes that favor conservation and sustainable development.

As a conservation biologist and a Colombian who has held various environmental posts in my own country, it is hard to find fault with the desire to preserve rain forests. However, the authoritarian preservationist approach rests on the flawed assumption that the tropics are for sale to the highest bidder. What happened to the view that tropical forests are an integral part of the cultural, economic, and natural patrimonies of the people of the tropics? And since when was there international acceptance that tropical forests can be considered global assets? Can the same be said for old-growth forests in Oregon and the Everglades of Florida? Would the citizens of these areas agree to international intervention in their management decisions? The perspective on rain forests that allows their purchase and protection by wealthy ICOs also sidesteps the question of why tropical countries have not, on their own initiative, declared those territories protected areas.

In the best of possible worlds, most of the remaining tropical forests will be protected from logging, will not be converted to legal or illegal crops, and will not be "preserved" only due to armed conflicts. But in the more complicated world of which this book offers varied glimpses, our best hopes for conservation are in community-based efforts that need work, but that have shown potential all over the world. In direct contrast, the preservationist approach precludes the possibility for real people to learn about conservation by engaging in conservation practices that, unfortunately, sometimes will fail. These failures should not be seen as final conclusions but as parts of a much longer, ongoing story about what rain forests are and could be.

Without tipping the power balance from an authoritarian outsiders' view to one exclusively dictated by local communities' desires, the complex challenge of tropical forest conservation calls for creative balances of complex solutions. First of all, rain forest–rich countries need substantial support for development, which necessarily includes conservation. This support needs to be provided in ways that respect the rights of local people to make choices, that are socially just, and that assign all stakeholders both rights and responsibilities. Given the dynamism of social, ecological, and political processes underway in the tropics, what is needed is a portfolio of approaches to conservation and development. Each approach needs to be implemented in a collaborative way as a long-term experiment in adaptive management. Reconciliation of the sometimes conflicting goals of tropical forest conservation and development—with due acknowledgment of concerns about intragenerational, intergenerational, and interspecific equity—demands rejection of the view that excludes people from tropical forests that can be purchased by anyone with sufficient funds. Making the contours of this view more apparent is a work in which conservation biologists, social scientists and humanists, environmentalists, and local peoples can and should all share.

NANCY LEE PELUSO: In thinking about the ways images influence policy and practice, it seems obvious that the violent images of people associated with particular rain forests should have a long-term impact on the ways we think about them. And yet, while this point is hardly surprising, it was brought home anew to me during a recent trip to Jakarta, where I was to attend a meeting assessing social forestry since the fall of Suharto. At the meeting were Indonesian foresters and academics, members of national and local NGOs, villagers from various forest areas of Indonesia, and a smattering of observers from international foundations and aid organizations. All the attendees had been involved in developing new, more local ways of managing Indonesia's forests over the previous four years of Indonesian political reform.

In my conversations in the capital city before the three-day meeting, I was struck by the frequency of stereotyped comments about Dayaks that people brought up when they learned of my experience in West Kalimantan. "Tell me, Nancy," someone would say, "are the Dayaks *really* as threatening as we have heard?" Or when speaking about Dayaks or the forested places in Kalimantan where they lived, the speaker would simply look one or another member of the audience in the eye, say, "You know, the place where . . . ," and draw a finger across his or her throat. At the meeting proper, I was surprised to find no Dayaks from West or Central Kalimantan participating, even though some of them had been key players in the civil society groups represented at this gathering, such as a national association of Indonesian indigenous peoples, a legal rights group, groups involved in community resource mapping, and so on.

It's conceivable, of course, that the conspicuous absence of Dayak participants from these recently violent Bornean provinces could be more the product of competing obligations than of their exclusion by the meeting's committee of NGO planners. Yet the apparently offhand, even flippant commentaries on the part of educated Indonesians in the nation's capital city suggest the intense effort that will have to be devoted to breaking down newly revived stereotypes of the Dayaks as violent, primitive people still at large in the modern world. These stereotypes—and their role in the ethnic violence in Kalimantan—are in part the fault of an uninformed or sensationalist media. However, the responsibility for them also lies with those Dayaks who have used them—explicitly or implicitly—to bolster their claims to the forest, land, and other resources. Moreover, this sudden return to antiquated images of the Dayaks as bloodthirsty headhunters underscores the difficulty of predicting the ways they will be included in planning futures for Borneo's rain forests.

The implications of the imagery used by local peoples to grasp at power brings us back to the politicized distinctions between jungles and rain forests and the importance of peopling these landscapes, of reclaiming rain forest landscapes as part of the realm of the social—both points that resurface in almost all of the essays in this book. If rain forests are, by definition, realms of nature that remain geographically separate from the people who happen to inhabit or use them, then these people are condemned to continue appearing as artifacts of a primitive jungle past. This timeless "primitivity"—whether in reference to people or the forests—contradicts the complex realities that we have tried to represent here.

PAUL GREENOUGH: The extinction of an indicator or keystone species such as the tiger—implying the collapse of its enfolded biodiversity and the erasure of its coded evolutionary history—would be a crisis for environmentalists in general and conservation biologists in particular. I see the problem of species extinctions and feel the anxiety. Yet in the year since completing my chapter on the Indian tiger reserves, in which, recall, I suggest that the Tiger in the Reserve is a new icon expressive of the weakness of tropical nature and its need for human help, I've had trouble explaining to science colleagues why I think that the system of reserves itself may be part of the problem and not, as they mostly believe, the only possible solution. Even suggesting that some reserves might profitably be thrown open to certain kinds of rural people causes them to go ballistic: "How dare you advocate the ruin of a Protected Area Network, so painstakingly assembled over decades, only to let in the Malthusian hordes who will plow up the trees and kill all the animals," they exclaim.

Frankly, their unanimity gives me pause. Like most environmental historians whose work has strayed into areas of technical expertise, I've had to work out for myself the science behind the PAN movement, and no doubt many biological/ecological subtleties go right over my head. I also readily admit that some of the people I most admire are wonderfully intelligent, physically active, and resourceful scientist-naturalists who have greater patience for the world of wet, smelly, snarly critters than I can muster. Thus, if arguments about the future of tiger conservation are going to be judged on the basis of who knows the most big-cat science and who really, truly most wants them to survive, then I have to yield. Nonetheless, I still have a couple of things to say.

For one, saving tigers and biodiversity is important to environmentalism, but this is not the only concern; there are other, competing absolute values besides preventing species extinctions. In my view, saving nature isn't a priori more important than saving Culture or serving Justice. A system of environmental protection that tramples on people's rights and dismisses their lifeways in order to institute a realm of pure nature isn't defensible, even though it may enjoy the backing of the state. Conservation needs to incorporate into its own genome an instinct for justice and a wider appreciation of human capacities or it will not flourish. For another thing, the process of building up national and international networks of reserves has never been *only* about science; it has always involved politics—and usually the politics involved have not been of the democratic sort. In every country where PANs are widespread, wily or powerful officials and socially prominent private persons as well as organizations have guided and cajoled politicians to appropriate public lands and forests (there may be exceptions, but I don't know them). This legacy of extrademocratic origins will continue to bedevil conservation as long as local people feel themselves aggrieved. Still another consideration looks to the sustainability of PANs, which even in wealthy countries, require constant funding, monitoring, and policing; these costly tasks are much more likely to be sustained where citizens value the animals and reserves, and therefore build a cherishing attitude toward them into daily routines and civic rituals. A final consideration is aimed directly at conservation biologists, whose own definition of the most protected of all the protected areas in the world, the category 1A nature reserves, is "a protected area managed mainly for science" (see World Conservation Monitoring Centre website). That is, the same people who have managed to exclude everyone else have kept the best for themselves. The most benign interpretation I can offer suggests an iconic rather than a purely rational excuse: biologists have been busy founding the new terrestrial paradise. While "a protected area managed mainly for science" may seem to be quite far from the *Oxford English Dictionary*'s early definition of paradise—"an Oriental park or pleasure-ground, esp. one enclosing wild beasts for the chase"—their cores are without doubt the lushest, most biodiverse spots on earth. It is in these magnificent and remote sanctuaries that biologists, and biologists alone, are said to be qualified to worship tigers, rhinos, and elephants on our behalf. Dare one ask, Are the keys to the kingdom in the right hands?

CANDACE SLATER: Reflecting on the various opinions expressed here, I'm struck by how much has happened not just since Eco 92 but in the scant two years since our last official meeting. These past two years have seen continuing wide-scale development in the Yucatán, new battles for indigenous peoples in the Amazon, giant fires in Colorado and Arizona, and frightening reports about new viruses linked to post–September 11 fears about terrorism. As I watch the television recaps of the UN Environmental Conference in Johannesburg, it's tempting to think about these recent developments as intensifications of the even larger changes that have taken place over the last decade.

I've already suggested that the glittering, biodiverse Rain Forest was the central icon of the Rio conference. It embodied its participants' awareness of the immensity of the Earth's riches. At the same time, it served as an effective expression of many people's anguished concerns about planetary preservation mingled with their intense optimism in the face of new ideas about sustainable development.

Rio Plus Ten, in contrast, offers a sobering retrospective in which no one image prevails. A sense of lost opportunities as well as new sorts of problems (including the U.S. government's refusal to commit on the highest levels to environmental preservation) have tempered the euphoric sense of possibility that permeated Eco 92. The conference's more somber air also reflects in part a growing appreciation of the profound intermingling of social, economic, and environmental factors that complicates easy solutions.

We began this book with the suggestion that the emerald green rain forest of the early 1990s has become considerably harder to define. No longer a single dazzling image, this increasingly fragmented forest makes multiple appearances in children's textbooks, different rain forest products, and an ongoing string of movies, documentaries, and newspaper articles.

The strength of approaches focused on environmental images and discourse is their ability to deal with exactly this sort of multiplicity and ongoing change. Good at making explicit unspoken assumptions, the essays in this book suggest some of the practical consequences of different images for particular groups of people. They leave no doubt about the impact of particular depictions of Borneo headhunters, Indian tigers, and Amazonian pipelines on local communities and individual lives.

A focus on representations is also effective in identifying patterns and divergences that invite comparisons over place and time. The reports on the huge fires the swept the U.S. West in summer 2002, for instance, take on a new dimension when seen beside accounts of the fires in the Amazon and Florida four years earlier. While the human role in the fires in Colorado and Arizona is far clearer in reports about these fires than it is in the Florida stories, the reporters' tendency to concentrate attention on individual perpetrators undercuts hard questions about new home construction bordering forests as well as the risks and benefits of controlled burns. The differences and similarities among these three sets of stories helps one to think more clearly about differing conceptions of wet and dry, tropical and temperate, so-called pristine and not-so-pristine forests. These same differences and similarities underscore the varied political and cultural backdrops for seemingly straightforward environmental processes such as fire.

And yet, if image-centered approaches have much to offer, they also have their limitations. Excellent at highlighting flaws and contradictions, they work best in hindsight. Extremely rich in implications for the future, they are less adept at furnishing specific directives. So then, one may ask, how exactly do we incorporate people into tiger reserves? What precise steps should pharmaceutical companies take to more constructively market native healing traditions? Which, in order of importance, are the most appropriate international measures to help prevent Amazonian fires? Although the essays have much to suggest about these sorts of practical questions, the answers they provide tend to be general in nature—the Maya forest demands redefinition as a series of gardens gone to seed; the trumpeting of "clean" technologies may serve to mute native opposition to development; and the violent images that local peoples seek to use to their own advantage can be projected back on them all too easily by a violent State. While these general principles could not be more important, it is the reader who must apply them to particular situations. For this reason, the value of approaches centered on representations depends very much on a series of next steps.

As an end in itself, environmental discourse analysis can do relatively little to help rain forests or rain forest peoples. Yet as a requisite beginning, and one absolutely crucial component of larger conversations involving rain forest peoples, academics from many different disciplines, NGOS, government agencies, and corporations eager to display their environmental concern, it is invaluable. The energetic integration of representations into ongoing debates about the present is particularly essential in the case of rain forests, where the long emphasis on nonhuman nature and the urgency of current issues have tended to conceal the practical (often negative) impact on local populations of recurring images such as golden treasure trove and hostile jungle.

The images we have examined in this book are both enduring and profoundly fluid. The newly fragmented rain forest at the beginning of the twenty-first century is clearly not the all-embracing, if endangered forest of Eco 92. And yet, if this forest has lost some of the thrill of exoticism, it has also acquired a growing familiarity and a wealth of new associations that make it a quintessential twenty-first-century ecology. Fractured and yet interconnected, simultaneously material and abstract, this shrinking and yet ever-more-diverse rain forest demands an active and respectful mixture of approaches that can do justice—in all senses—to its multiplicity.

CONTRIBUTORS

DAVID BARON, former science correspondent for National Public Radio, is a three-time recipient of the American Association for the Advancement of Science journalism award. His book *The Beast in the Garden: A Modern Parable of Man and Nature* (2003) tells the story of a fatal mountain lion attack near Boulder, Colorado, in 1991 through an investigation of the complex interactions between wildlife and humans in suburban America as some animals are displaced from their historic habitat and others are attracted to newly developed residential areas.

SCOTT FEDICK began his career as an archaeologist at about age nine, when he first dug up old trash pits in his neighbor's backyard in northern California. Since then, he completed graduate work at Arizona State University and is currently Professor of Anthropology at the University of California, Riverside. Fedick specializes in the archaeology of the Maya Lowlands with an emphasis on ancient agriculture and resource use, and is currently director of the multidisciplinary Yalahau Regional Human Ecology Project in Quintana Roo, Mexico.

ALEX GREENE grew up in rural Nebraska and first studied anthropology at New York University. A writer, independent scholar, and musician currently living in Memphis, he has published articles in *Culture and Agriculture, ETC.: A Review of General Semantics,* and *The Routledge Encyclopedia of Contemporary Latin American and Caribbean Cultures.* His doctoral dissertation for the University of California-Davis concerns the local repercussions of ethnobotanical research in Belize, where Greene conducted research from 1996–1999.

PAUL GREENOUGH is a historian of modern India with research interests in environmental and public health issues. He is the author of several dozen articles and studies, including *Prosperity and Misery in Modern Bengal* and *Imagination and Distress in Southern Environmental Projects* (coedited with Anna L. Tsing.) A Californian happily displaced in the Midwest, Greenough is a professor at the University of Iowa, where he directs that institution's Global Health Studies and Crossing Borders programs.

NANCY LEE PELUSO, Professor of Political Ecology in the College of Natural Resources at the University of California, Berkeley, has been conducting research in Indonesia for more than twenty years. Her works on environmental and social history in rural West Kalimantan began in 1990. In addition to a wide variety of journal articles, she has written or edited three books: *Rich Forests, Poor People; Borneo in Transition: People, Forests, Conservation, and Development* (with Christine Padoch); and *Violent Environments* (with Michael Watts). Peluso is currently writing a book with Peter Vandergeest that compares histories of forest politics in Indonesia, Malaysia, and Thailand.

FRANCIS E. PUTZ, Professor of Botany and Forestry at the University of Florida, conducts research on ways to manage forests for timber as a conservation strategy in the tropics. While primarily a silviculturalist, he has published on a wide range of issues ranging from forest economies to portrayals of tropical forests in art, literature, and film. His work in conservation biology has included topics such as the importance of suburban woodlots for children to cross-continental comparisons of the biodiversity impacts of tropical forest management. Putz is currently working on a defense of poor, Southern whites (commonly known as "crackers" or "rednecks") as ecosystem managers.

CLAUDIA ROMERO is a Colombian conservation biologist completing her Ph.D. at the University of Florida on the physiological ecology of tree bark. Before recommencing her formal studies, she directed a national park, supervised an international conservation project on protected areas and their zones of influence, and held various other environmental posts in governmental and nongovernmental organizations in Colombia. Her attention continues to be divided between evolutionary ecology and political ecology.

SUZANA SAWYER, the daughter of a former petroleum geologist, spent the majority of her early years in North Africa and South America, where her father searched for oil. Currently Assistant Professor of Anthropology at the University of California, Davis, her work has appeared in *Cultural Critique, Cultural Anthropology, Journal of Latin American Anthropology,* and *Latin American Perspectives.* She is presently concluding a book on oil and indigenous politics titled *Crude Chronicles.*

CANDACE SLATER, convener of the Rain Forest(s) Seminar and editor of the present volume, doesn't miss a movie with "rain forest" or "jungle" in the title. Marian E. Koshland Distinguished Professor in the Humanities, she teaches courses on Brazilian literature and culture, Latin American oral traditions, and environmental imagery at the University of California, Berkeley. She is also the director of the Townsend Center for the Humanities. The author of a half-dozen books and numerous articles, she has been doing research for the past fifteen years in Amazonia. The most recent of her books, *Entangled Edens: Visions of the Amazon* (2002), is a study of competing visions of the Amazon.

CHARLES ZERNER is Professor of Environmental Studies at Sarah Lawrence College, and the former director of the Natural Resources and Rights Program of the Rainforest Alliance, an international nongovernmental conservation organization. Trained as a lawyer, with a background in botanical art and lithography, he has conducted fieldwork on Indonesian fishermen's customary law of the sea and their conceptions of the marine environment in Sulawesi and the Maluku Islands. Zerner is contributing editor of *People, Plants, and Justice: The Politics of Nature Conservation* and *Culture and the Question of Rights: Forests, Coasts, and Seas in Southeast Asia.* He is now working on a book titled *Making Threats: Biofears and Environmental Anxieties* (with Elizabeth Hartmann and Banu Subramaniam).

INDEX

Adams, Richard, 149–50

Africa: Ebola virus and, 253–54; Robert Kaplan on, 256–57, 277nn35–36, 278n39; viral forest, 248, 257, 260–61, 266–69, 278n39, 278n44, 279n46, 279n48, 279n52

African viral forest: anthropomorphisms of, 283n71; outward trajectory of, 255, 257, 260–61, 266–68, 278n39, 278n44, 279n46, 279n48; prejudice against, 279nn51–52; sources of cures in, 280n53. *See also* Ebola virus; Maps and mapping; Rain forests

Agriculture: of Classic Maya period, 147; drought, 44, 63n8; home gardens, 154, 162n62; maize cultivation, 53, 146, 148, 153; Maya, 145–46, 146, 147, 149, 156, 164n71; overpopulation, 150–51; patterned ground, 148; soil quality in, 144–45, 146, 147, 294; swidden, 53, 145–46, 147, 149–51, 153; wetland cultivation, 148, 149, 150. *See also* Deforestation; Fires in the Amazon; Fires

in Florida; Human habitation; Maya; Maya forests; Maya Lowlands

AIDS, 250, 251, 255–56, 275n22

Anane, Mike, 278n44

ARCO: Block 10, 78m1, 83, 98n35; Buena Vecindad (Good Neighbor), 91; "buying consciences" strategy of, 89, 91; communities divided by, 85–87; Energy, Environment, and population (EAP), 292–93; environmental concerns of, 72, 76–77, 96n20; gifts from, 89, 91; land titles facilitated by, 84–85, 99n40; militarization of oil wells, 86–87, 99n44; Organizacíon de Pueblos Indígenas (OPIP), 84, 85–86, 89–90, 100n47; Quichua Indians and, 79, 85; Technical Environmental Committee, 90, 292–93. *See also* Oil industry; Villano Project

Army Medical Research Institute of Infectious Diseases (USAMIRID), 249, 261–62

Arnold, David, 273nn1–2

Arvigo, Rosita: and Michael Balick, 109, 125n5; funding for, 110–11; Ix Chel goddess appropriated by, 102, 109, 118, 122, 124; on Maya authenticity, 110, 115–18, 119, 120–22, 128n28, 130n37; media used by, 116–17; New Age sensibility of, 112–13; *One Hundred Healing Herbs of Belize*, 110; opposition to, 111–12, 120–22; Elijio Panti and, 101, 102, 108–9, 120–21; *Rainforest Remedies* (book), 110, 116, 117; Rainforest Remedies (business), 110, 122; tourism, 102, 112, 124n1; as traditional healer, 101–4, 108–9, 120–21, 125n5; Traditional Healers Foundation (THF), 112, 122–23. *See also* Ethnobotany; Traditional healers

Asociasción de Desarrollo Indígena Región Amazónica (ASODIRA), 90, 100n49

Asociasción Independiente Evangélica de Pastaza, Región Amazónica (AIEPRA), 90, 100n49

Balick, Michael, 109, 125n5

Baron, David, 286, 288–89

Belize: Garifuna or Garinagu, 106, 115; Maya authenticity in, 110, 115–18, 119, 120–22, 130n37; multiethnicity of, 105–6, 125n4; North American influence in, 110–11, 125n4; Elijio Panti and, 101, 102, 108–9, 115, 119–21; pharmaceutical companies in, 23, 113–14, 127n21; Terra Nova forest reserve, 109, 110, 111. *See also* Arvigo, Rosita

Belize Association of Traditional Healers (BATH), 109, 110, 111–12

Bio-irony: fires in Florida, 56; human expulsion for conservation, 170, 195n7

Bioscripts: indigenous concerns in, 87; interpretation of, 19–20; language of, 19; of multinational corporations, 11; rain forest as peopleless place, 71, 142, 228; Sacred Nature in, 19, 20; spectacles and, 8, 93; of the traditional healer, 115–16

Black Caribs. *See* Garifuna or Garinagu

Bock, Carl, 216–18, 234

Body and body parts: skin in epidemic narratives, 252–53; skulls, 213–14, 240n37; viral infection, images of, 259–60. *See also* Headhunting

Borders and boundaries: of African viral forest, 257, 260–61, 278n39, 278n44, 279n46, 279n48; alien invasions and, 269, 282n66; bio-invasion, 270, 283n71; Ebola virus, 271–72, 284nn72–73; militarization in West Kalimantan, 220–21; racism, 257–58, 269, 271–72, 275n22, 278n40, 282nn66–67, 283n71, 284nn73–74; in tiger reserves, 168, 192; in Turning Point Project advertisements, 266–67, 280nn55–56

Borneo. *See* Borneo Headhunter; Chinese population (West Kalimantan); Dayaks; West Kalimantan

Borneo Headhunter: cannibalism, 211, 226; and Chinese population, 223–24; colonialism and, 211–13, 218–19; in Dayak-Madurese conflict, 232–33; and decapitation, 233; Indonesian army's use of, 25, 207, 210, 223–26; orangutan, 213, 215, 216, 217, 241n45; Oriental savages as, 210; as tamable, 218–19; Victorian social theorists on, 211; violence as characteristic of, 233; Wild Man of Borneo and, 215–17

Brooke, James, 212, 215, 238n15

Cancún, 12, 14, 137, 158n3

Cattle ranchers, 44–45, 149

Centers for Disease Control (CDC), 249, 261

Chinese population (West Kalimantan): communists conflated with, 222; and Confrontation, 221; Dayaks and, 207, 222–27, 243n68; Demonstrasi Cina (Chinese demonstration), 225–26, 243n68; eviction from West Kalimantan, 225, 229–30, 243n69; as headhunters, 225–26; propaganda, 224, 226; return of, 229–30; wildness used against, 223–27

Coe, William, 144, 148

Colonialism: as authority over space, 79; in Belize, 125n4; Borneo, 212–13, 240n31; Dayaks' image as headhunters and, 218–19; on headhunting, 212–13, 240n31

Confrontation (Indonesia-Malaysia), 221

Conrad, Joseph, 255, 256, 276n29, 276n30, 276n32

Conservation: exclusion and enclosure model, 188–89; human expulsion for, 170, 195n7; the military, 287; programs for, 294, 297–98; protected area networks (PANs), 169, 195n6, 300–301; scientific knowledge and, 301–2

Conservation biologists, 5, 30n12

Cronon, William, 288

Davis, Wade, 125n5

Dayaks: agriculture, 229–31; blood revenge by, 223–24; Borneo Headhunter, 210; Chinese population and, 207, 222–27, 243n68; as exotics, 226, 234–35; headhunting and, 206, 211–12, 217–18, 231, 237n8, 241n51; as indigenous Indonesians, 227; as Indonesian army trackers, 222–23, 242n60; journalists on, 226–27, 243nn74–75; Madurese and, 204, 207, 229, 230–31, 238n11; nationalization of forests and, 227–28; and red bowl, 204, 206, 225, 230; *tariu* (ancestral spirits of war), 205, 215, 230, 237n2; victimization of, 207, 225–26, 238n11; Victorian social theorists on, 211–13, 215; violence, 241n51, 299–300; wildness of, 210–12, 227, 228, 231–33, 235

Deforestation: home construction and, 290–91; in Maya Lowlands, 148; migrations and, 229; nature preserves and, 297; and rescue of medicinal plants, 109; and road construction, 45, 82, 228–29, 290; slash pine plantations and, 46, 64n18; soil quality and, 134–35; timber industry and, 56, 61, 142, 179, 227–28

Demarest, Arthur, 150–51

Demonstrasi Cina (Chinese Demonstration), 225–26, 243n68

Dixon-Smith, Homer, 278n39

Ebola virus: as African predator, 254–55; AIDS and, 250, 251, 255–56, 275n22; as ancient, 254; biotechnology, 266–67, 280n55; Centers for Disease Control (CDC), 249, 261; clean up authority, 261–62; fearful images of, 254–55; Marburg virus, 257–58, 260; in the media, 261, 266–67, 274n13, 279n48, 280nn55–56; microscopic images of, 253; military response to, 249, 261–62; monkeys as carriers of, 249, 258–59, 263–64, 279n50; national security risks and, 258–59; outbreak in America suburbia, 258–59; trajectory of, 257, 260–61, 278n39, 278n44, 279n46, 279n48; in Turning Point Project advertisements, 266–67, 280nn55–56, 281nn60–61; in United States, 249, 258, 259, 261–62; xenotransplantation, 267–68

Ecuador. *See* ARCO; Villano Project

El Niño Southern oscillation, 44, 54, 57, 63n8

Energy, Environment, and population (EAP), 292, 293

Environmental scholarship: on environmental movements, 29n4; folk knowledge, 296–97; language of, 287; objectivity in, 296; rhetoric of, 287

Ethnobotany: Rosita Arvigo, 103–4, 125n5; culture and, 113, 116, 126n19, 128n28, 296; folk medicine and, 107, 109, 114, 176; the forest library and, 113–14, 127nn23–24; New Age audiences and, 113; and pharmaceutical companies, 23, 113–14, 127n21

Evangelical Protestantism, 104, 108

Fear and fear narratives: in American suburbia, 258–59; Ebola as icon of, 250, 256, 275n19, 278n44, 283n71; global transport

Fear and fear narratives (*cont.*)
  networks, 260–61, 278nn43–44; *Heart of Darkness* (Conrad) and, 255, 256, 276n29, 276n30, 276n32; national security, 269, 283n68; in Turning Point Project advertisements, 266–67, 280n55–56, 281nn60–61

Fires in the Amazon: Amazon as pristine, 56; as cataclysmic, 52–53, 65n36; farmers and, 21–22, 54, 57; frequency of, 46–47; government response to, 45, 58–60, 64n11; human intervention and, 54–55, 66n43; journalists' access to, 49; national security and, 59; newspaper coverage, 46, 47, 48–49; news report placement, 48; personification of, 51; prayer and, 60; Yanomami Indians and, 48–49, 52, 54–55, 57–58, 61, 65n36

Fires in Florida: closeup accounts, 49–52; as environmental fact, 52–53, 56–57; and Floridians' self-sufficiency, 58; frequency of, 46–47, 64n19; government response to, 47, 59; human intervention and, 53, 55, 56; loss and recovery in, 53; news report placement, 48; personifications of, 51; prayer and, 60–61; residential developments and, 47, 55, 64n19; slash pine plantations and, 46, 64n18

Folk medicine, 107, 109, 114, 176

Friedman, Thomas, 278n43

Galhano Alves, João Pedro, 186, 187, 188, 193

Garifuna or Garinagu, 106, 115

Garrett, Laurie, 261

Gender images: Ebola as she-devil, 276n32; in Maya culture, 118, 119; nature as woman, 56–57

Globalization of Ebola virus, 249; fears of, 251; virus transmission, 255–57, 260–61, 264–68, 278n39, 278n44, 279n46, 279n48, 283n71; xenophobia, 267–68, 282n67, 283n71

Gómez-Pompa, Arturo, 153

Greenough, Paul, 275n23

Hall, Stuart, 284n74

Haraway, Donna, 33n42

Harding, Sandra, 296

Headhunting: colonial rule on, 212–13; Dayaks and, 206, 210–12, 217–18, 231, 237n8, 241n51; defined, 206; journalists' descriptions of, 205–6; suppression of, 212–13, 240n31; trade dampened by, 212–13; war, 205, 225, 235, 237n5, 243n71; wildness, 209, 211–12, 238n15

Healing Forest Conservancy, 117–18

Heart of Darkness (Conrad), 255, 256, 276n29, 276n30, 276n32

Hose, Charles, 211–12, 218

The Hot Zone: AIDS and, 250, 251, 255–56, 275n22–23; alien invasions, 269, 282n66; Ebola as icon of fear in, 250, 256, 275n19, 278n44, 283n71; jungle images in, 252, 271; military intervention, 261–62, 270–71; monkeys as virus carriers, 249, 257, 279n50; negative images of Africa, 252, 271; plot summary, 249; rain forest as dangerous, 252, 271; viral forest, fear of, 264. *See also* Body and body parts; Ebola virus; Fear and fear narratives; *Outbreak* (movie); Virus transmission

HRI (Humanities Research Institute), ix

Human habitation: Amazon virgin forest and, 9, 51–52; Chinese eviction from West Kalimantan, 225, 229–30, 243n69; compensation for relocation, 196n17; conservation and, 170, 195n7; fires, 54–55, 56, 66n43; in Maya forest, 135, 151–54, 155, 156–157, 161n54; Maya temples ruins and, 133, 135, 143, 146, 147, 152; oil industry and, 71, 77–80, 82–83, 91, 93–94n4; and overpopulation, 150–51; protected area networks (PANS), 169, 195n6, 300–301; Quichua Indians, 79, 83–84, 85; in rain

Mathews, Jessica Tuchman, 277n35

Maya: archaeological investigations of, 145, 146, 148, 149–50; Arvigo opposed by, 111–12, 120–22; authenticity of, 110, 115–18, 119, 120–22, 128n28, 130n37; culture, 127n28; ethnic identity movement, 116, 128n28, 129n29; evangelical Christianity and, 104, 120; exploitation of, 14, 33n42; as extinct, 14, 33n41, 140; Garifuna culture and, 106, 115; gender relations among, 118, 119; healing practices, 23–24, 115–16; home gardens, 154–55, 156, 162n62; idealization of, 117–18, 128n28; Ix Chel, 102, 109, 118, 122, 123–24; San Antonio community, 120–21, 130n36; swidden agriculture, 145–46, 147, 149; temples of, 133, 135, 143, 146, 147, 152; traditional agricultural practices of, 291. *See also* Punta Laguna

Maya forests: changing public image of, 141–42, 159nn24–25, 160nn26–27; and ecotourism, 157; as garden gone to seed, 152, 161n54; home construction in, 290–91; and home gardens, 154–55, 156, 162n62; human habitation of, 135, 152–54, 155, 156–57, 161n54, 290–91; as jungle, 142, 145, 159n24, 160nn26–27; and maize cultivation, 53, 148, 153; road construction in, 290; sustainable homestead in, 291; and swidden agriculture, 153

Maya Lowlands: archaeological investigations in, 145, 146, 148, 149–50; ceremonial centers in, 146, 147–48; Classic Maya civilization in, 147–48; deforestation in, 148; and ecotourism, 142; farming in, 147–48; as jungle, 142, 145, 151–52; overpopulation in, 150–51; patterned ground, 148; rain forests and, 140, 141, 151, 159n21; soil quality in, 144–45, 146, 147; Tikal, 142–44; warfare in, 151

Maya Riviera, 136, 158n3

McDermott, Jeremy, 33n41

the media: Amazon fires in, 46, 47, 48–49;

Amazon images in, 289; Arvigo's ethnobotany in, 116–17; Dayaks in, 26, 299; Ebola virus in, 261, 266–67, 274n13, 279n48, 280nn55–56, 284n73; environmental rescue in, 21–22; farmer images in, 21–22; Maya agriculture in, 141–42, 146, 150–51, 159nn24–25, 160nn26–27; Turning Point Project advertisements in, 266–68, 280nn55–56, 281nn60–61; on Yanomami, 55

Meggers, Betty, 146–47

Migrations: of AIDS, 255–56; bio-invasion and, 283n71; Chinese eviction from West Kalimantan, 225, 229–30, 243n69; of illegal aliens, 269, 271–72, 282nn66–67, 284nn73–74; of Madurese, 229; Marburg virus, 257–58, 260; plants, 269, 283n68; spontaneous migrants (*transmigran spontan*) in West Kalimantan, 229; to United States, 258, 269, 282nn66–67; of viruses, 255–58, 260–61, 278n39, 278n44, 279n46, 279n48, 281nn60–61; xenophobia and, 267–68, 282n67, 283n71

the military: AIDS control by, 249, 261–62; conservation, 287; in Ebola virus interventions, 249, 261–62, 265–66, 270–71; immigration surveillance, 269; Indonesian, 25, 222–23, 226–28, 229, 232; oil production security and, 86–87, 99n44

Monkeys: African green monkeys, 257; as Ebola virus carriers, 249, 258–59, 279n50; in *Outbreak* (movie), 263–64; spider monkeys at Punta Laguna, 16, 17–19, 35n53, 35n57; violence and, 258–59; as virus transmittors, 263–64, 278n44

Morley, Sylvanus, 146

National Geographic: Maya forest in, 141–42, 159nn24–25, 160nn26–27; swidden agriculture of Maya, 146, 150–51; Tikal, 142–44

Naxalites, 177, 178, 198n31

New Age tradition, 112–13

Ngayau, 205, 225, 237n5, 243n71

Oil industry: biohazards of, 83, 92, 98n35; biophysical concerns of, 71, 87, 90, 93, 293; environmental concerns of, 72, 75–77, 96n20; indigenous development projects and, 89, 100n47; Latin American Energy Organization (OLADE), 292; rain forest as idyll, 22–23; site maps of, 77–78; social impact of, 71, 85–87, 87, 91, 93, 94n4, 293. *See also* Villano Project

Orangutan, 213, 215, 216, 217, 241n45

Organización de Pueblos Indígenas (OPIP), 84, 85–86, 89–90, 100n47; land titles, 99n40

Outbreak (movie): alien invasions, 269, 282n66; militarization of public health in, 262, 265–66, 270–71; monkeys as virus transmittors, 263–64; portrayals of Africans in, 279nn51–52; viral forest, fear of, 264

Palamau Tiger Reserve, 177–78

Pandanuque, 84–85, 99n40

Panti, Elijio: and Rosita Arvigo, 101, 102, 108; Black Carib as teacher of, 115; ethnicity of, 115, 119; on modernity, 109; *primicias* for, 121

Parry, Richard, 233–44

Perang. *See* War

Periwinkle Project (Rainforest Alliance), 280n53

Pharmaceutical companies: and ethnobotany, 113, 127n21; and intellectual property rights, 23–24, 118, 294, 295; Shaman Pharmaceuticals, 112, 117

Pollan, Michael, 287

Preston, Richard. See *The Hot Zone*

Pristine nature: Amazon as, 51–52, 56; and human intervention, 293; and jungles, 24–26, 36n69, 37n70, 142, 177; of Maya forest, 140, 152; of Maya tradition, 115–18, 119, 120–22, 128n28, 130n37; nostalgia for, 130n37; rain forests as, 51–52, 56, 74–75, 78,

95n11, 293; San Antonio (Maya community), 120–21, 130n37; and tigers, 190; tradition as, 120–21, 130n37

Project Tiger: popular support for, 191–92; reserve size, 171–73, 174, 179–80, 196n17; Van Gujars, 188. *See also* Tiger reserves; Tigers; Van Gujars

Propaganda: anti-Chinese, 223–27; Widodo (journalist), 226–27, 243nn74, 75; wildness in, 227

Protected area networks (PANS), 169, 195n6, 300–301

Punta Laguna: archaeological research at, 34n48; financial support for, 17, 34n52; outsiders' speech and, 19; as rain forest, 16, 34n49; spider monkeys and, 16, 17–19, 35n53, 35n57; Xcaret compared with, 17, 35n56

Pure nature. *See* Pristine nature

Putz, Francis, 286

Quichua Indians, 79, 83–84, 85

Racism: of AIDS, 275n22; alien invasions and, 269, 271–72, 282nn66–67, 284nn73–74; bio-invasion and, 283n71; environmental, 257–58, 278n40; mixed-blood Amazonians, 61–62; portrayals of Africans in *Outbreak*, 279nn51–52

Rainforest Café, 31n21, 137–38, 158n10

Rain forests: in Africa, 252, 255, 271; as biophysical realm, 71, 87, 90, 93, 293; in Borneo, 25–26, 36n68, 219; Cancún as iconic shadow of, 14; as communal property, 18; cultural patrimony of, 297–98; definitions of, 4, 31n15; destruction and disease outbreaks, 281n61; economic intervention in, 95n11; fragility of, 21–24, 247; government development programs in Borneo, 228–29; home gardens in, 154–55, 156, 162n62, 163n63; human intervention in, 22, 36n66, 71, 77–80, 97n28, 142, 228,

Rain forests (*cont.*)
275n18; iconic simplifications of, 7–9, 31n18; images of Ebola as, 253; as library, 113–14, 127nn23–24; Maya Lowlands and, 140, 141, 151, 159n21; people excluded from, 142–44; as pristine, 51–52, 56, 74–75, 78, 95n11, 293; Punta Laguna as, 16, 34n49; social character of, 71, 77–78, 87; as spectacles, 8; tropicality of, 4, 30n7, 273nn1–2; violence in, 208, 211

Ranthambhore Foundation, 182, 200n44
Reynolds, Percy, 107
Río Azúl, 148–49
Rio Plus Ten, 285, 302
Ritchie, Mark, 289n56
Road construction: in Borneo rain forest, 228–29; deforestation, 82; fire conduits and, 45; home construction, 290; Indonesian government, 228–29; oil industry, 82, 83

The Road to El Dorado: caricatures of native peoples in, 43, 63n7; Maya images in, 41–42, 136–37; nature in, 42–43, 62

Romero, Claudia, 286

Sanchez, Linda, 282n66
Sarawak People's Guerilla Force (PGRS), 221, 229
Sariska reserve: criminals discouraged from, 188; human traffic through, 183, 200n46; tiger population in, 183, 186, 201n53; Van Gujars in, 183–85

Sastun: My Apprenticeship with a Maya Healer (Arvigo), 102, 104–5, 125n5
Schaller, George, 177
Schama, Simon, 287–88
Semi-Terra (Landless Rural Workers') Movement, 55, 66n48
September 11, 2001, 284n72, 286
Sian Ka'an Biosphere Reserve, 127n21, 167, 291
Sontag, Susan, 275nn22–23
Spectacles, 8, 10–11, 93

Spider monkeys, 16, 17–19, 35n53, 35n57
St. John, Spencer, 211
Subramanian, Banu, 282n67
Suharto, 220–21, 227, 228, 239n20
Sukarno, 220
Swidden agriculture, 53, 145–46, 147, 149–51, 153

Tariu (ancestral spirits of war), 205, 215, 230, 237n2
Terra Nova forest reserve (Belize), 109, 110, 111
Tiger reserves: criminal incursions into, 172n1, 173–78; denotification, 178; drug trafficking in, 169, 195n4; government surveillance of, 174; human habitation in, 179–82, 188–89, 191, 196n17, 199n40, 200n42; island biogeograhy and, 172–73; models for, 168, 171–72, 192; poaching in, 173–76, 188, 197n18; population increase in, 179; prey species in, 179, 202n61; problems with, 300; Sariska reserve, 182–85, 188, 200n46; in South Asia, 170–71, 196nn11–12

Tigers: attacks by, 180–81, 199n40, 200n42; as charismatic megafauna, 171; divine status of, 188, 202n59; extinction of, 171, 190, 196n10; as indicator species, 170–71, 190, 203n69; as pure nature, 190; reproduction, 179; rules for tiger encounters, 193; Van Gujars and, 183–88, 193, 202n59

Tikal, 142–44
Timber industry: and Dayaks, 228; and deforestation in Florida, 56; Indonesian military and, 227–28; in rain forests, 61, 142; and tiger reserves, 179

Tourism: authenticity and, 122, 140; in Cancún, 12, 14, 136, 137, 158n3; ecotourism, 142, 143, 157; Maya, 12–15, 19–20, 128n28, 133, 140; and medicine trail (Belize), 102, 112, 122, 124n1; viral, 275nn22–23. *See also* Punta Laguna; Xcaret park

Traditional healers: intellectual property rights of, 118; Maya authenticity as, 110,

115–17, 119, 120–22, 128n28, 130n37; multi-ethnic identity of, 110, 115, 116, 130n37; Elijio Panti and, 101, 102, 108–9, 115, 119; rain forest image of, 103; San Antonio (Maya community), 120–21; shamans as, 205, 230, 237n2

Traditional Healers Foundation (THF), 112, 122–23

Tsing, Anna, 30n10, 118, 210

Turning Point Project advertisements: advocacy groups and, 266, 280nn55–56; AIDS in, 266, 267; on biotechnology, 266, 267; Ebola imagery in, 266, 280n56; fear in, 266–67; on viral transmissions, 266–67, 281n61; on xenotransplantation, 266, 267, 280n56, 281n58, 281n60

United States: alien invasions and, 269, 282n66; border controls for, 251; conservation of park land in, 189; fires in western, 303; and immigration control, 269, 282nn66–67; military response to Ebola outbreak, 249, 261–62, 265–66, 270–71

Van Gujars: lifestyle of, 185–86; protocol for meeting tigers, 186, 187–88, 193; public services denied to, 190; as residents of Sariska reserve, 188–89; and tigers as deities, 188, 202n59

Villano Project: Central Processing Facility, 81–82, 98n35; deforestation, 82; forest environment of, 73; invisible pipeline of, 80–83; security fence, 80

The Villano Project: Preserving the Effort in Words and Pictures, 69–71, 73, 87–89

Violence: the Borneo Headhunter and, 233; as conscious choice, 208–9; Dayaks and, 241n51, 299–300; images of lethality, 252; Indonesian military, 25, 222–23, 226; monkeys as, 258–59; in protected area networks (PANS), 169, 195n6, 300–301; in tiger reserves, 172n1, 173–78, 198n31; tropical forests as violent, 252; and Wild Man of Borneo, 215. See also War; Wildness

Virus transmission: of AIDS, 255–56, 279n46; and jet travel, 249, 260–61, 264–65, 279n46, 279n48; by monkeys, 263–64, 278n44; in Turning Point Project advertisements, 266–67, 281n61; written accounts of, 274n16

Wallace, Alfred Russel, 216, 219, 241n45

War: and environmental destruction, 25–26, 36n68; and headhunting (*ngayau*), 205, 225, 235, 237n5, 243n71; journalists' descriptions of, 205–6; and Maya Lowlands destruction, 151; and perang, 205, 225, 237n5; and red bowl, 204, 206, 225, 230; *tariu* (ancestral spirits of war), 205, 230, 237n2; violence in, 209–10

West Kalimantan: ethnic war in, 205, 221, 237n3; Indonesian control of, 220; migration to, 229; militarization and, 220–22; and resettlement (*transmigrasi*), 229; and road construction, 228–29; and spontaneous migrants (*transmigran spontan*), 229. See also Chinese population (West Kalimantan); Dayaks; Indonesian government; Indonesian military

Widodo (journalist), 226–27, 234, 243nn74–75

Wildlife Protection Act (India), 173

Wild Man of Borneo, 215–17

Wildness: of Chinese population, 223–27; of Dayaks, 211–12, 228, 231–33, 235; of Ebola virus, 253–54; and headhunting, 209, 211–12, 238n15; Indonesian military and, 222, 226–27; landscape as, 228; and orangutan, 213, 215, 216, 217, 241n45; skulls as symbols of, 213–14, 240n37; Victorian social theorists on, 211–13; Wild Man legend, 215, 216

Wilk, Richard, 128n28

Wilson, E. O., 30n12, 32n25, 172

World Bank, 292, 293

Library of Congress Cataloging-in-Publication Data
In search of the rain forest / edited by Candace Slater.
p. cm. — (New ecologies for the twenty-first century)
ISBN 0-8223-3205-1 (cloth : alk. paper)
ISBN 0-8223-3218-3 (pbk. : alk. paper)
1. Rain forests—Psychological aspects. 2. Rain forest
ecology. 3. Human ecology. I. Slater, Candace. II. Series.
QH86.15 2003   333.95—dc22   2003016614